Reactivity and Structure
Concepts in Organic Chemistry

Volume 1

Editors:

Klaus Hafner   Jean-Marie Lehn
Charles W. Rees   P. von R. Schleyer
Barry M. Trost   Rudolf Zahradník

Jiro Tsuji

Organic Synthesis
by Means of Transition Metal Complexes

A Systematic Approach

Springer-Verlag
Berlin Heidelberg New York 1975

Jiro Tsuji

Professor at the Faculty of Engineering, Department of Chemical
Engineering, Tokyo Institute of Technology

ISBN 3-540-07227-6 Springer-Verlag Berlin Heidelberg New York
ISBN 0-387-07227-6 Springer-Verlag New York Heidelberg Berlin

Library of Congress Cataloging in Publication Data. Tsuji, Jiro, 1927—. Organic synthesis by means of
transition metal complexes. (Reactivity and structure; v. 1). Bibliography: p. Includes index. 1. Chemistry,
Organic-Synthesis. 2. Transition metal compounds. I. Title. II. Series. QD262.T77. 547'.2. 75-14259.

Printed in Germany.

Monophoto typesetting and offset printing: Zechnersche Buchdruckerei, Speyer. Bookbindung:
Konrad Triltsch, Würzburg.

Preface

In 1970 I wrote a review article for "Topics in Current Chemistry", surveying the general reaction patterns of transition-metal complexes from the standpoint of organic synthesis. The article seems to have evoked wide interest, and I therefore thought it appropriate to treat the subject more comprehensively in the form of a monograph. Organometallic chemistry is a rapidly growing field of intensive research. In this field, the application of organometallic compounds to organic synthesis is an important subject and many unique synthetic methods have been discovered, especially in the last decade. Syntheses using transition-metal compounds are now in the foreground of organic chemistry.

The purpose of this monograph is to give a bird's-eye view of this field to both organic and inorganic chemists through a mechanistic approach. A systematic unification of the voluminous data accumulated in this field is now urgently required. This subject is discussed by classifying various reactions into general patterns and by illustrating them with a limited number of pertinent examples. Confining this kind of monograph to a certain size demands that several problems be considered. The subject requires considerable arbitrariness on the part of the author as to selection of topics, organization, and emphasis. The literature on transition-metal complexes applied to organic synthesis is now so voluminous that in many cases I have made no attempt to present detailed historical development or to treat each topic thoroughly. There are numerous synthetic reactions using transition-metal complexes and it is impossible to survey them all within the limits of this book. Therefore only typical general patterns are discussed. Many synthetic reactions involving transition-metal complexes can be analyzed and explained as the succession or combination of simple, fundamental reactions. These simple reactions are surveyed and supplemented with relevant examples. Reference citation is by no means exhaustive. It is not the purpose of this book to give a complete literature survey. Many other important and useful reactions are not included here because they cannot be explained by the general patterns. These reactions are found particularly in compounds of copper, titanium, tungsten, and oxygen complexes. I also had to omit many reactions of general interest, again simply because of the size of the book.

There are a few reactions that are not only unique but also useful as synthetic methods. First priority was given to these reactions. At the same time, plenty of reactions involving transition-metal compounds, though unique or interesting in themselves, have little or no practical value *per se* from the standpoint of organic synthesis, due to such defects as poor yields, difficulty in obtaining or preparing complexes, and technical and economic reasons. Most such reactions have been reported by inorganic chemists who have little interest in applying them to organic synthesis. Many papers are intended primarily for inorganic chemists and would not normally be read by organic chemists. However, some of these reactions may have a high potential value, or the principle involved could be applied to organic synthesis. A number of the reactions belonging to this category, are treated here in the hope of bridging the gap between organic chemistry and coordination chemistry.

I have to admit that the reader might consider my treatment of some subjects to be over-simplified, especially the discussions of mechanism. The mechanism proposed may be shown to be partially or totally incorrect as additional evidence accumulates. One of the purposes of the book is to convey to synthetic organic chemists a broad understanding of this field through a unified mechanistic approach. The chemistry of this field is still in its infancy and many data are accumulating. I therefore provide general patterns of reactions and suggest possibilities for future use, with the hope that this book will stimulate research and plans for further application.

I wish to take this opportunity to express my gratitude to Prof. R. Noyori, who read the entire manuscript and offered many useful comments and suggestions.

Kamakura, Japan   Jiro Tsuji
March, 1974

Table of Contents

Abbreviations

The following abbreviations have generally been used in the book, unless stated otherwise.

acac	acetylacetonate	DMSO	dimethylsulfoxide
Ar	aromatic group	HMPA	hexamethylphosphoramide
CDT	1,5,9-cyclododecatriene	L	undetermined ligand
COD	1,5-cyclooctadiene	Ph	phenyl
bipy	bipyridyl	R	alkyl
DMF	dimethylformamide	THF	tetrahydrofuran

I. Introduction

Metallic compounds have provided many useful synthetic tools. Historically Grignard reagent has been by far the most useful in organic synthesis. Compared to magnesium, the use of other metallic compounds in organic synthesis has for a long time been rather limited.

Historically speaking, transition metal complexes have exhibited their usefulness in industry rather than in chemical laboratories. The hydroformylation reaction (oxo reaction), a process for producing an aldehyde from an olefin, carbon monoxide, and hydrogen, established its industrial status in the 1930's and its catalyst was found to be cobalt carbonyl:

$$CH_2\!\!=\!\!CH_2 + CO + H_2 \xrightarrow{Co_2(CO)_8} CH_3CH_2CHO$$

The famous Reppe reactions were developed in the period between 1930 and 1940 and have shown useful reactivities of iron and nickel carbonyls as carbonylation catalysts. For example, the formation of acrylic acid from acetylene and carbon monoxide in water is catalyzed by nickel carbonyl; the carbonylation of propylene to form butanol is catalyzed by iron carbonyl. These reactions have been operated as large scale industrial processes. The remarkable cyclization reaction of acetylene to form cyclooctatetraene is catalyzed by certain nickel complexes:

$$CH\!\!\equiv\!\!CH + CO + H_2O \xrightarrow{Ni(CO)_4} CH_2\!\!=\!\!CHCOOH$$

$$CH_3CH\!\!=\!\!CH_2 + 3CO + 2H_2O \xrightarrow{Fe(CO)_5} CH_3CH_2CH_2CH_2OH + 2CO_2$$

$$4\,CH\!\!\equiv\!\!CH \xrightarrow{\text{Ni catalyst}} $$

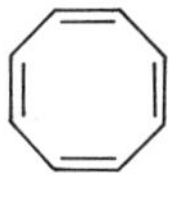

The discovery of these reactions constitutes the first stage of the history of organic synthesis by means of transition metal complexes. They have been extensively investigated, mainly as commercial processes in industrial laboratories and plants, though few academic studies have been carried out on them. Their mechanisms have still not been elucidated.

The second stage was initiated after World War II by the discovery of the famous Ziegler-Natta catalysts. A truly original reaction of olefin polymerization to form polyethylene, polypropylene, and later polybutadiene was discovered by using titanium compounds combined with alkylaluminums as catalysts.

$$CH_2{=}CH_2 \longrightarrow \text{polyethylene}$$

Because of the novelty of the reaction, there was an explosion of studies on the reaction of various olefins in the presence of similar catalysts in both industrial and academic laboratories. Cyclodimerization and trimerization of butadiene to form 1,5-cyclooctadiene (COD) and 1,5,9-cyclododecatriene (CDT) catalyzed by zerovalent nickel complexes may be considered to be an outgrowth of the Ziegler catalyst.

$$2\ CH_2{=}CHCH{=}CH_2 \longrightarrow$$

$$3\ CH_2{=}CHCH{=}CH_2 \longrightarrow$$

An interesting metathesis reaction of olefins was recently discovered and this can also be traced back to a Ziegler-type catalyst by critical scrutiny.

Another important contribution to the second stage in the development of organic synthesis using transition metal compounds was the discovery of the Wacker process; this made it possible to produce carbonyl compounds from olefins by using palladium chloride combined with cupric chloride as the catalyst. For example, acetaldehyde is produced by the oxidation of ethylene, and acetone from the oxidation of propylene.

$$CH_2{=}CH_2 + H_2O + PdCl_2 \longrightarrow CH_3CHO + Pd + 2\,HCl$$

The great industrial success of the Wacker process began a new era in organic synthesis for noble metal compounds, especially those of palladium. Again, the development of the use of transition metal compounds in the second stage occurred in the field of industrial chemistry. Development in the second stage was backed by the petrochemical industry and in turn contributed to the rapid development of the industry after World War II.

It is thus apparent that the early development of organic synthesis by means of transition metal compounds was achieved primarily in industry and received most attention from industrial laboratories. The situation, however, changed after the second stage. Studies on transition metal chemistry have been carried out in research laboratories in the last two decades, giving rise to a number of interesting synthetic reactions which would have been impossible by conventional organic methods. However, the intimate historical relationship between transition metal complexes and industry, and certain disadvantages such as toxicity and difficulty of handling due to instability, have made some organic chemists rather reluctant to use them in the laboratory. One mistaken supposition is that the reactions of transition metal compounds are quite different from usual organic reactions.

Transition metal compounds, unlike compounds of nontransition metals such as magnesium, lithium, or zinc, have several characteristic properties which contribute to their usefulness in organic synthesis. They have a strong affinity for simple substrates such as carbon monoxide, olefins, acetylenes, hydrogen, and other unsaturated compounds, and can activate them by forming complexes or coordination compounds. This process is indispensable for bringing these molecules and other organic substrates into reactions. In other words, a transition metal offers a site of synthetic reaction for many compounds which otherwise react only with difficulty; by forming intermediate complexes with many substrates, transition metals facilitate reactions that would be prohibitively endothermic in their absence. The transition metals also have the ability to stabilize a wide variety of species through coordination as σ- or π-bonded ligands; one of their functions is to assemble the reactants on them, thereby causing the reactants to react. This property, called the template effect, orients and activates the coordinated reactants for further reactions. Moreover, on complex formation, transition metals become soluble in many organic solvents and can be induced to react with other organic compounds.

In addition, transition metals have the ability to donate additional electrons or to accept electrons from organic substrates, and can readily change their valences, oxidation numbers, or coordination numbers. This facile, sometimes reversible oxidation-reduction process plays an important role in organic synthesis, especially in catalytic reactions. Their ability to serve as catalysts in organic reactions is the most unique property of transition metals.

In a survey of some general patterns of organic synthetic reactions using transition metal complexes, I cite typical examples of how σ-bonds are formed from simple molecules and metal complexes, and how they are then transformed into the final organic compounds. Many synthetic reactions involving transition metal complexes can be analyzed and explained as a succession or combination of simple, fundamental reactions. These simple reactions are reviewed here with related examples.

It should be pointed out that, to a large extent, the use of transition metals in organic synthesis remains empirical. There are many factors controlling the reaction course. At present, we cannot explain clearly why a specific reaction is achieved by a specific metal complex. For example, the hydroformylation reaction is catalyzed most efficiently by cobalt or rhodium carbonyl and not by other metal carbonyls, and the metathesis reaction of olefins is possible mainly with molybdenum and tungsten complexes. In addition, the effect of ligands is also crucial; even with the same metal, a reaction proceeds smoothly only when the metal is coordinated by particular ligands. Of course, the significant ligand effect can be partially explained in terms of steric and electronic effects, but the explanation is far from complete. Under the present circumstances, it is impossible to predict the catalytic activity or reactivity of transition metal complexes for new and unknown reactions. It is even more difficult to design a catalyst for certain desired reactions. This, then, is the present state of the chemistry of the transition metal complexes used in organic synthesis.

II. Comparison of Synthetic Reactions by Transition Metal Complexes with those by Grignard Reagents

Mechanisms of the usual organic reactions are now clearly established by classifying and analyzing vast data and facts accumulated over nearly 100 years. The reactions are simply classified as ionic, radical, and molecular in nature, though more detailed classifications have also been made. Mechanisms of many reactions involving nontransition metal compounds are quite clear, whereas those of organic reactions involving transition metal complexes are still ambiguous. Without doubt, the reactions proceed through the formation of a carbon-metal σ-bond, but chemical properties of these metal-carbon bonds remain obscure.

In order to understand more clearly the reactions carried out by using transition metal complexes, therefore, it is initially worthwhile to analyze and compare the reactions of Grignard reagents, which are very familiar to organic chemists. As is well known, in the first step of Grignard reactions, metallic magnesium reacts with alkyl halides to form alkylmagnesium halides, the so-called Grignard reagent. In this reaction, zerovalent magnesium is oxidized to a bivalent magnesium compound with the cleavage of the covalent carbon-halogen bond, and hence this step can be considered to be an oxidative addition of alkyl halides to metallic magnesium. The Grignard reagent, thus formed, is a carbanion source, and reacts with various electrophilic reagents such as carbonyl or nitrile compounds. This step can be understood formally as an insertion reaction of an unsaturated bond of the carbonyl group into the carbon-magnesium bond. In the latter process no change in oxidation state of magnesium is involved.

$$Mg^0 + R-X \longrightarrow R-Mg^{2+}-X \quad \text{Oxidative addition}$$

$$\begin{array}{c} R' \\ \diagdown \\ C=O \\ \diagup \diagdown \\ R'' \quad \diagdown \\ \quad R-Mg^{2+}-X \end{array} \longrightarrow \begin{array}{c} R' \\ | \\ R''-C-O-Mg-X \\ | \\ R \end{array} \quad \text{Insertion}$$

From these considerations, Grignard reaction can be explained by the combination of the two elemental processes, i.e. the oxidative addition and insertion reactions.

In the synthetic reaction of transition metal complexes, the elemental reactions given above are also widely observed. This means that there are some similarities between the reactions of Grignard reagents and transition metal complexes. However, more differences than similarities are found by further and more critical examination of the reactivities of magnesium and transition metals. In Grignard reaction, the oxidative addition takes place only with organic halides. It is known that magnesium reacts with conjugated dienes to form Grignard reagent, but in only a special case. The insertion reaction into the carbon-magnesium bond takes place mostly with carbonyl and nitrile groups, although the insertion of terminal olefins or conjugated dienes is possible with allylic Grignard reagents as an exceptional case. [1—3] It can be said that Grignard reactions cover rather limited numbers of reaction species, although the usefulness of Grignard reaction in organic synthesis is, of course, enormous.

On the other hand, the oxidative additions involving transition metal complexes are possible with a variety of compounds having covalent bonds to give metal σ-bonds, which exhibit various reactivities depending on the identity of metals. In addition, numerous reaction species participate in the insertion reactions to form transition metal σ-bonds. Olefins, conjugated dienes, acetylenes, carbon monoxide, carbonyl compounds, and other unsaturated compounds can insert. Thus apparently many organic reactions are possible through the combination of these two reactions. Also, the template effect, unique to transition metal complexes, plays an important role. It is understandable that so many kinds of synthetic reactions are possible using transition metal complexes. To summarize, the patterns commonly observed in synthetic reactions by means of transition metal complexes can be expressed by the following scheme:

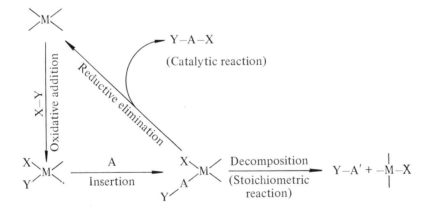

The transition metal complexes can be used in organic synthesis as stoichiometric reagents as well as catalysts. In stoichiometric reagents the original complex is not recovered at the end of the reaction as in Grignard reactions. Transition metal complexes are often recovered in a valence state higher than that of the original complexes, indicating stoichiometric consumption of the complexes. One of the most characteristic properties of transition metal complexes is the ability to carry out the reactions catalytically. In the catalytic processes, the complexes undergo the oxidative addition and insertion reactions. In the final step, an organic compound is liberated as a product from the metal complex. This last step, called the reductive elimination, at the same time regenerates the transition metal complex with reduction of the valence state to the original one and completes the catalytic cycle. Of course the catalytic reaction, if carried out smoothly, is synthetically the most desirable one, because only a small amount of each complex is necessary.

III. Formation of σ-Bonds Involving Transition Metals

1. Introduction

The first essential step in the synthetic reactions of transition metal complexes is the formation of σ-bonds, either directly or indirectly, with hydrogen, carbon, oxygen, nitrogen, halogens, and other elements. There are several ways in which σ-bonds are formed by the reaction of metal complexes with simple molecules. The most widely observed and important reaction is the oxidative addition reaction, which is discussed first. The σ-bond, once formed, is converted into a different σ-bond by insertion or other reactions which are discussed later.

2. Oxidative Addition Reactions

Oxidative addition reactions are a well-established class of reactions involving transition metal complexes, and several review articles have been published on this subject. [4—10] The reaction is widely observed with low-spin transition metal complexes, having d^8-d^{10} configurations and is most important for the direct formation of σ-bonds from metal complexes and simple molecules. This step, essential to a wide variety of catalytic reactions as well as to stoichiometric reactions, involves addition of a covalent molecule to a metal complex with the cleavage of the covalent bond of the addendum, and can be considered to be two electron oxidations of the metals.

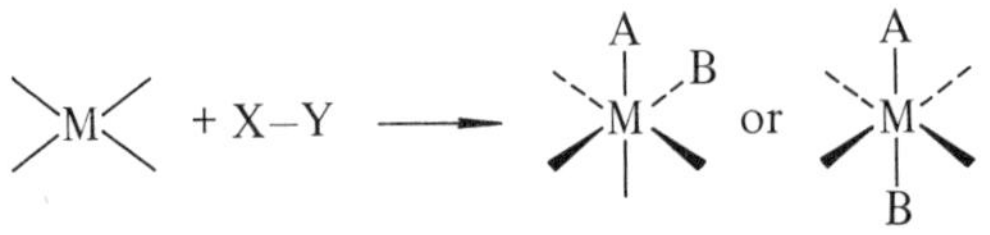

$$>\!M\!< \;+\; X\!-\!Y \;\longrightarrow\; >\!\overset{\overset{A}{|}}{M}\!\!<^{B} \quad \text{or} \quad >\!\overset{\overset{A}{|}}{\underset{\underset{B}{|}}{M}}\!\!<$$

Representative types of the oxidative addition of d^8 and d^{10} metals are the following.

$$d^8 \longrightarrow d^6$$

$$ML_4 + X{-}Y \longrightarrow X{-}ML_4{-}Y$$

example: $ML_4 = Ir(CO)Cl(PPh_3)_2$

$$d^{10} \longrightarrow d^8 \longrightarrow d^6$$

$$ML_4 \longrightarrow ML_2 + L_2$$

$$ML_2 + X{-}Y \longrightarrow X{-}ML_2{-}Y$$

$$X{-}ML_2{-}Y + X{-}Y \longrightarrow X_2{-}ML_2{-}Y_2$$

example: $ML_4 = Pt(PPh_3)_4$.

In the first reaction shown above, complexes of metals with a d^8 electronic configuration are converted to octahedral complexes with a d^6 configuration through addition of a covalent molecule. The oxidative addition is accompanied by an increase in the preferred coordination numbers of the metal atoms and hence by the incorporation of additional ligands into their coordination sphere. Since the cleavage of the covalent bond takes place by the reaction, simple and stable molecules are thereby activated, and are ready for further reactions.

3. What Conditions are Necessary for the Oxidative Addition?

There are several necessary conditions or rules for oxidative addition reactions to occur. One of them is that the transition metals should be at a low valent state. The metals at the low valent state behave as bases, or reductants, and the oxidative addition can be regarded as a process of removal of electrons from the electron-rich metal center. [10] Unlike Grignard reaction, with a few exceptions, it is usually not possible to use the transition metals in metallic state directly as the low valent state for the reaction. Metallic nickel or iron is in general not active enough to cause satisfactory reaction with organic compounds. In order to bring the transition metals into and keep them in the active low valent state, some ligands should coordinate with the metals to form certain complexes or coordination compounds. Neutral ligands such as carbon monoxide, phosphines, or tertiary amines are commonly used for complex formation.

Further consideration should then be given to coordination numbers. There is a rule for the formation of coordination compounds, called effective atomic number rule, which states that coordination compounds are formed by filling orbitals to attain a rare gas structure. The closed-shell configuration corresponding to 18 valence electrons is particularly stable and wide-spread. It is often said that "coordinative unsaturation" is essential for the oxidative addition reaction to occur. A metal complex with an inert gas configuration of 18 valence electrons on the metal, regardless of coordination number, is coordinatively saturated, and therefore does not undergo the oxidative addition reactions.

With the transition metal complexes, a saturated coordination number is determined by the configuration of the metal d electrons. Thus for d^6 configuration metals such as Mo^0, Rh^{3+}, and Ru^{2+}, six-coordination is regarded as saturated as shown in Table I, e.g. $Mo(CO)_6$ is a saturated and stable complex. Similarly, five-coordination for d^8 metals such as Fe^0, Ru^0, Co^{1+}, and Pd^{2+} is saturated. A typical example of the saturated d^8 complex is $Fe(CO)_5$. For d^{10} metals such as Ni^0, Pd^0, or Pt^0, four-coordination is general. Thus it is understandable that nickel forms $Ni(CO)_4$, but never $Ni(CO)_5$. The coordinative unsaturation means·that metals have vacant sites on it and can accept additional ligands for further coordination.

Table I

electronic configuration	saturated coordination number	examples
d^6	6	Rh^{3+}, Ir^{3+}, Co^{3+}, Pt^{4+}, Ru^{2+}, Mo^0, Cr^0
d^8	5	Fe^0, Ru^0, Os^0, Co^{1+}, Rh^{1+}, Ir^{1+}, Pd^{2+}, Pt^{2+}
d^{10}	4	Pd^0, Pt^0, Ni^0

The coordinative unsaturation includes potential, or solvent occupied, vacant sites. For example, in the four-coordinate complex of $RhCl(PPh_3)_3$ (1), which is one of the most active and versatile catalysts known so far and called the Wilkinson complex, the univalent rhodium has the d^8 configuration and the complex is apparently unsaturated. Furthermore, one mole of the triphenylphosphine dissociates in a solution [11], with an equilibrium constant of $(1.4 \pm 0.4) \times 10^{-4}$ M in benzene solution at 25°. [12] The dissociation is not especially extensive, but certainly sufficient to be catalytically important. In the presence of oxygen, or a coordinating solvent and molecule, the dissociation becomes more extensive. Thus the complex becomes highly unsaturated, making oxidative addition feasible.

$$RhCl(PPh_3)_3 \longrightarrow RhCl(PPh_3)_2 + PPh_3$$

1

This property partly explains the usefulness of this complex as a catalyst for various reactions such as hydrogenation and carbonylation.

On the other hand, with ordinary saturated complexes, a vacant site must be created in order for the oxidative addition reaction to occur. $Fe(CO)_5$, a saturated complex, may be made coordinatively unsaturated by expulsion of carbon monoxide initiated by heat or light. [13] The behavior of saturated metal carbonyls such as $Ni(CO)_4$ and $Mo(CO)_6$ as a catalyst can be similarly understood. [14] The induction period frequently observed in the reactions of these metal carbonyls is responsible for this process.

$$M(CO)_n \longrightarrow M(CO)_{n-m} + mCO$$
$$M=Fe, Mo, Ni, Cr, W$$

Reactivity of $Fe_2(CO)_9$ may be understood by the following dissociation to give an unsaturated species.

$$Fe_2(CO)_9 \longrightarrow Fe(CO)_4 + Fe(CO)_5$$

Some saturated complexes are easily converted into unsaturated species without an extra energy source. For example, four-coordinate platinum and palladium phosphine complexes, $M(PPh_3)_4$ ($M=Pt, Pd$), have d^{10} configuration and are regarded as saturated. However, the coordinated triphenylphosphines dissociate in solution, forming two-coordinate complexes which readily undergo oxidative addition reactions. [15, 16] Hence these potentially unsaturated complexes are active catalysts in several reactions.

$$Pd(PPh_3)_4 \longrightarrow Pd(PPh_3)_2 + 2PPh_3$$

Ligands can stabilize the low valent state due to the fact that donor atoms of the ligands have vacant orbitals in addition to lone pairs. σ-Donor property arises from lone-pair donation. The ligands which have ability to accept electron density of the metal into low-lying empty π-orbitals are called π-acceptor ligands. [17, 18] Triphenylphosphine has a strong σ-donor property, and the electron density of the palladium in the above complex becomes very high by the coordination of four moles of triphenylphosphine, and hence the triphenylphosphines tend to dissociate. Triphenylphosphine and carbon monoxide as ligands have somewhat different properties. The latter is a ligand with a strong π-acceptor property, rather than σ-donor. The following reaction can be explained by the different properties of the two important ligands. When $Ni(CO)_4$ is treated with triphenylphosphine, two moles of the coordinated carbon monoxide are displaced easily with two moles of triphenylphosphine. Bis(triphenylphosphine)nickeldicarbonyl *(2)* thus formed is a very stable, saturated complex

$$Ni(CO)_4 + 2PPh_3 \longrightarrow Ni(CO)_2(PPh_3)_2 + 2CO$$

2

and does not dissociate easily. A balancing effect of carbon monoxide and triphenylphosphine would account for the fact that this saturated complex is not very active as a catalyst. Rather drastic conditions are necessary to induce its reaction. Conversion of bicyclo[2,2,1]heptane-2,3-dicarboxylic acid anhydride to bicyclo[2,2,1]heptene (53% yield) with the complex (2) took place by heating up to 200°. [19]

The following coordinatively saturated ruthenium hydride complex *(3)* becomes unsaturated by partial dissociation of the coordinated triphenylphosphine. [20] Very interesting selective removal of the liberated triphenylphos-phine from the equilibrium mixture was achieved by means of reverse osmosis. [21] A reverse osmosis cell, fitted with a membrane of polyimide prepared from p,p'-diaminodiphenyl ether and pyromellitic anhydride, was used for the removal of triphenylphosphine dissociated from the ruthenium complex (3) in THF. The unsaturated complex thus formed could not be isolated, but converted into the isolable nitrogen complex (4) under nitrogen pressure.

$$RuH_2(PPh_3)_4 \longrightarrow RuH_2(PPh_3)_3 + PPh_3$$

$$3$$

$$\Big\downarrow N_2$$

$$RuH_2(N_2)(PPh_3)_3$$

$$4$$

Usefulness of the low valent transition metal complexes for many organic syntheses is a consequence of their coordinative unsaturation and the ease with which they undergo oxidative addition. The latter effect is determined partly by the nature of both metals and their ligands. Ligands which increase electron density at the central metal atom or favor higher oxidation states enhance the tendency of the metal to undergo the oxidative addition reactions. One of the triphenylphosphines in the Wilkinson complex (1) can be displaced very easily with carbon monoxide. This tendency is so great that the complex (1) can extract a carbonyl group from organic carbonyl compounds such as aldehydes. The resultant complex (5), though it is unsaturated, is stable and much less active for oxidative addition. Unlike the complex (1), neither dissociation of the ligands nor oxidative addition reactions of the complex (5) takes place under moderate conditions. The difference in the electronic nature of carbon monoxide and triphenylphosphine contributes to this behavior, although a steric effect is another factor to be considered.

$$RhCl(PPh_3)_3 + CO \longrightarrow RhCl(CO)(PPh_3)_2 + PPh_3$$

$$1 \qquad\qquad\qquad\qquad 5$$

Relative reactivities of different d^8 complexes toward oxidative addition decrease in going from left to right across the periodic table and with increasing initial oxidation number, and increase in going down a given group of the periodic table. Thus roughly the following order of reactivity is expected, assuming a constant ligand environment.

$$Os^0 > Ru^0 > Fe^0, \quad Ir^{1+} > Rh^{1+} > Co^{1+}, \quad Pt^{2+} > Pd^{2+} > Ni^{2+}$$

From the standpoint of organic synthesis, the stability of the product of the oxidative addition is another important factor for enabling further reaction to occur. If the product of the oxidative addition is stable enough to allow easy isolation, then such a complex is usually not useful for organic synthesis, at least as an efficient catalyst. For example, a coordinatively unsaturated iridium complex $Ir(CO)Cl(PPh_3)_2$ *(6)*, the so-called Vaska complex, undergoes a variety of oxidative addition reactions to give six-coordinate saturated complexes, which can be isolated as stable complexes in many cases. [22]

$$Ph_3P{\diagdown}{\diagup}CO \atop \underset{Cl \diagup \diagdown PPh_3}{Ir} \quad + X-Y \longrightarrow \quad Ph_3P-\underset{\overset{|}{Cl} \diagdown CO}{\overset{X \diagdown \diagup Y}{Ir}}-PPh_3$$

6

This property of the Vaska complex has made a great contribution to the development and understanding of the chemistry of oxidative addition reactions. Because of the stability of the products and no possibility of further addition of other ligands, the application of the complex to organic synthesis is rather limited.

On the other hand, the unsaturated $RhCl(PPh_3)_3$ *(1)* undergoes oxidative addition reactions with the elimination of one mole of triphenylphosphine; the products of the addition are five-coordinate d^6 complexes, which are still unsaturated. Therefore, they have the ability to undergo further transformation to form six-coordinate d^6 complexes by accepting another ligand.

$$RhCl(PPh_3)_3 \longrightarrow PPh_3 + RhCl(PPh_3)_2 \xrightarrow{X-Y} \underset{Y \diagup}{\overset{X \diagdown}{RhCl(PPh_3)_2}}$$

1

In general, the tendency for a transition metal having a d^8 configuration to become five-coordinate increases as one ascends a triad and as one passes from right to left in group VIII in the periodic table. This generalization is illustrated by pronounced tendency of a zerovalent iron complex to become five-coordinate, whereas bivalent platinum complexes are usually four-coordinate. The relative ease with which $Os(CO)_5$ loses carbon monoxide compared with $Fe(CO)_5$ is another example of this generalization.

Typical covalent compounds which add to metal complexes oxidatively are hydrogen, hydrogen halide, acyl halides, alkyl, aryl, and vinyl halides, halogens, silanes, hydrogen cyanide, nitriles, sulfonyl halides, and oxygen. There are some less common species which take part in the oxidative addition.

It should be noted here that, although a few mechanistic studies have been carried out, the mechanism of the oxidative addition reaction seems to be far from clear. One reason is that a variety of the addenda take part in the oxidative addition reactions. The nature of the covalent bonds of these adding molecules is quite different, and consequently the mechanism of the reaction should be divergent. The natures of the covalent bonds of hydrogen-hydrogen, hydrogen-halogen, halogen-sp^3 carbon, and halogen-sp^2 carbon are apparently widely different and it is difficult to imagine that the oxidative addition of these molecules even to the same complex can be explained by a single mechanism. Some reactions are explained by a S_N2 mechanism; addition of methyl iodide to the iridium complex is an example. The iridium behaves as a nucleophile, and the effect of the ligands can be explained on this basis. Other reactions can be understood by a free radical mechanism, which will be surveyed in III-8. Oxidative addition of hydrogen is explained by concerted three center-addition. There are at present only a few reactions which can be explained unequivocally by these mechanisms. Different mechanisms should be considered depending on the nature of the substrates. More experimental data should be accumulated before a clear classification of the oxidative addition is made from a mechanistic viewpoint.

4. Metal-Hydrogen Bond Formation by Oxidative Addition Reactions

a) Addition of Hydrogen; Homogeneous Hydrogenation

The first example of the oxidative addition is activation of hydrogen attached to various atoms by formation of metal-hydrogen bonds or metal hydrides. A variety of molecules of a general formula H—Z can react with metal complexes to form metal-hydrogen bonds. [23, 24] The most simple one is direct oxidative addition of hydrogen molecule to give dihydride complexes. Reaction of hydrogen with the Vaska complex *(6)* gives the dihydride *(7)*. [25]

$$Ir(CO)Cl(PPh_3)_2 + H_2 \longrightarrow HIrH(CO)Cl(PPh_3)_2$$

6	*7*

It seems likely that the term "oxidative addition" is somewhat confusing at first glance. It might be difficult for organic chemists who use molecular hydrogen for reduction, to understand that a transition metal is "oxidized" by the addition of hydrogen. The oxidative addition designates a wide-spread class of reaction, in which oxidation i.e., an increase in the oxidation number of the metal, is accompanied by an increase in the coordination number. In the reaction of the Vaska complex shown above, oxidation number of the iridium increased from 1 to 3.

The oxidative addition of molecular hydrogen is widely operative in the transition metal complex catalyzed hydrogenation of unsaturated bonds, although the intermediate hydride complex cannot be isolated in most cases. It should be pointed out that there is a striking similarity between oxidative addition of hydrogen to transition metal complexes and chemisorption on transition metal surfaces.

Homogeneous hydrogenation of various organic compounds catalyzed by transition metal complexes is of special interest in connection with the oxidative addition of hydrogen. Extensive studies on its practical application to organic synthesis and the reaction mechanism have been carried out involving various transition metal complexes. [26—29] Phosphine complexes of rhodium, ruthenium, platinum, and iridium are active catalysts for hydrogenation. Also Ziegler catalyst type complexes formed from nickel and other transition metal compounds with alkylaluminum compounds are effective. [30] It has been known for a long time that some metal carbonyls such as $Fe(CO)_5$ and $Co_2(CO)_8$ have a strong hydrogenating ability. For example, in the oxo reaction catalyzed by $Co_2(CO)_8$, alcohols are obtained as a byproduct via the hydrogenation of aldehydes.

The most extensively studied complex, among others used for homogeneous hydrogenation, is the Wilkinson complex *(1)*, [11] which readily reacts with hydrogen. [31, 32] With this complex, various olefins and acetylenes are hydrogenated at room temperature and under atmospheric pressure of hydrogen without concomitant attack on other functional groups such as carbonyl, nitrile, and nitro groups. The introduction of molecular hydrogen into unsaturated bonds with the Wilkinson catalyst can be understood by consideration of the following steps:

1. Dissociation of one of the triphenylphosphine ligands.
2. Hydrogen activation by coordination (oxidative addition of hydrogen).
3. Substrate activation by coordination.
4. Hydrogen transfer to the substrate
5. Liberation of the hydrogenated product and catalyst regeneration.

These steps in homogeneous hydrogenation of olefins can be expressed by the following scheme, where stereochemistry of the ligands is neglected: The di-

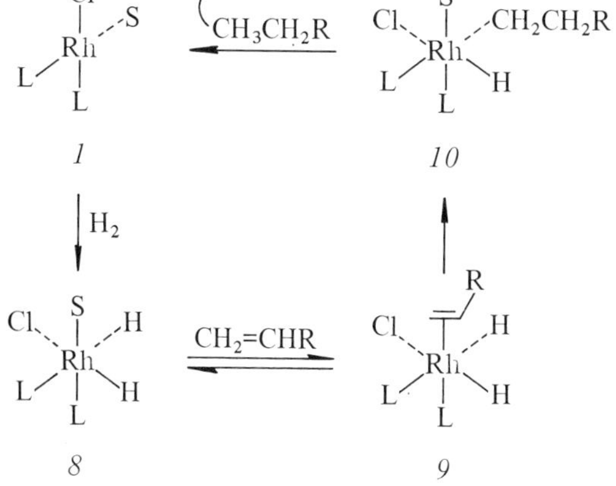

S: solvent, L: PPh_3

hydride complex *(8)* formation by oxidative addition is followed by olefin coordination *(9)*. Insertion of the coordinated olefin to the rhodium-hydrogen bond
gives the alkylrhodium complex (10). Finally, alkanes are liberated by the coupling
of the alkyl group and hydrogen attached to the same rhodium with regeneration
of the complex *(1)*.

Homogeneous hydrogenation with other complexes is also assumed to
proceed through oxidative addition of hydrogen followed by olefin insertion to
form alkyl complexes. Proof of this mechanism was provided by isolation of the
intermediate complex *(11)*, containing both hydrogen and π-bonded olefin on
the same metallic center. [33]

$IrHCl_2(COD)_2$

11

A model experiment of the hydrogenation of olefins with metal complexes
has been carried out. [34] At first, the thermally stable *cis*-dihydride complex
of molybdenum *(12)* was isolated and treated with olefins. Olefin insertion took
place to give *cis*-hydrido-σ-alkyl complex *(13)* which has been regarded as a
key intermediate of hydrogenation, indicating that the hydrogenation proceeds in a stepwise manner.

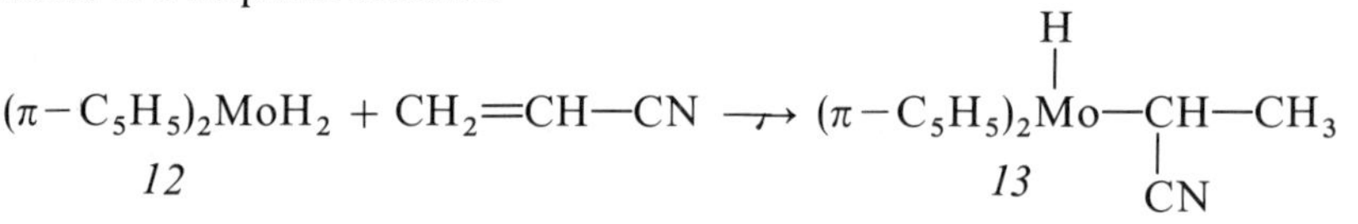

$$(\pi-C_5H_5)_2MoH_2 + CH_2{=}CH{-}CN \longrightarrow (\pi-C_5H_5)_2\overset{\overset{\textstyle H}{\textstyle |}}{Mo}{-}\underset{\underset{\textstyle CN}{\textstyle |}}{CH}{-}CH_3$$

12   *13*

The dihydride complex added to an acetylenic bond to give the following σ-bonded
olefin complex *(14)*, which was converted into an olefin π-complex *(15)* on
heating.

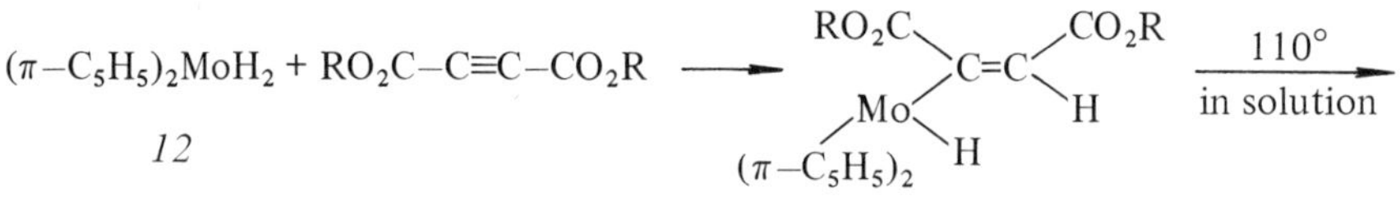

$$(\pi-C_5H_5)_2MoH_2 + RO_2C{-}C{\equiv}C{-}CO_2R \longrightarrow$$

12

14

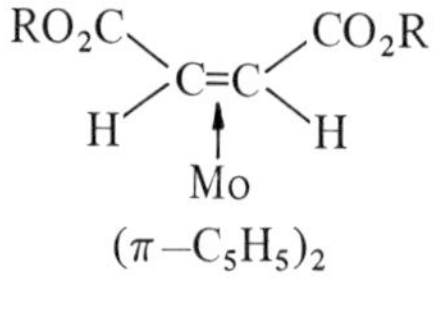

15

Similarly, diphenylacetylene was converted into a σ-bonded iridium-olefin complex *(16)*, which gave *cis*-stilbene on acidification. [35]

$$Ph-C\equiv C-Ph + HIrCl_2(DMSO)_3 \longrightarrow$$

(structure 16: a C=C double bond bearing Ph and Ph on one carbon, H and IrCl₂(DMSO)₃ on the other)

$$\xrightarrow{HX}$$

(structure: a C=C double bond bearing Ph and Ph on one side, H and H on the other)

Effects of various factors especially those of the ligands on the hydrogenation with the Wilkinson complex have been studied. [11] In a series of sterically similar triarylphosphines the order of activity was found to be as follows. [36]

$$(p\text{-CH}_3\text{OC}_6\text{H}_4)_3\text{P} > (\text{C}_6\text{H}_5)_3\text{P} > (p\text{-FC}_6\text{H}_4)_3\text{P}$$
$$\quad\;0.34 \qquad\qquad 0.25 \qquad\qquad 0.02$$

This is also the order expected for diminishing ease of the oxidative addition to give the dihydride complexes. The higher the electron density on the metal, the easier is the oxidative addition, and hence an electron-donating group on the ligands may accelerate the hydrogenation reaction. Alkylphosphines are considered to give higher electron density to the metal than arylphosphines by coordination, and hence a higher rate of hydrogenation is expected. Unexpectedly, however, the rhodium complexes coordinated by alkylphosphines are less efficient than arylphosphine coordinated ones for hydrogenation. [36] In the hydrogenation of cyclohexene, the triphenylphosphine complex absorbed 17.9 ml/min of hydrogen, while corresponding absorption by the triethylphosphine complex was only 0.5 ml/min. No satisfactory explanation for this result has been given.

The failure of a square planar phosphite complex, RhCl[P(OPh)₃]₃ to activate hydrogen at 1 atm may be attributed in part to the higher π-accepting property of triphenyl phosphite which lowers the non-bonding electron density on the metal as compared to triphenylphosphine [11]. However, the main difference seems to be the fact that the phosphite complex does not dissociate in solution, thereby failing to provide a vacant site for the coordination of the substrate.

For homogeneous hydrogenation using the rhodium complex, the solvent effect is also conspicuous, and rate of the hydrogenation changes depending on the solvents. Usually benzene in which the rhodium complex is easily soluble is used as the solvent. Addition of polar solvents such as alcohols and ketones accelerates the hydrogenation; a 3:1 mixture of benzene and ethanol is the one most widely used. The rate of hydrogenation of cyclohexene in benzene solution is increased by the addition of phenol, alcohols, acetone, or nitrobenzene, but decreased when solvents such as nitromethane, malonate, or DMF are added. The reaction almost stops in the coordinating solvents such as DMSO, acetonitrile, chlorobenzene, chloroform, acetic acid, and pyridine. [38] The following rates of the hydrogenation of 1-methylcyclohexene catalyzed by the rhodium complex have been reported. (K, min⁻1) [37, 38]

benzene-ethanol (3:1), 0.49; nitrobenzene, 0.45;   cyclohexanone, 0.39;
benzene, 0.13;   dichloromethane, 0.075;   1,2-dichloroethane, 0.038;
chlorobenzene, 0.025;   chloroform and benzonitrile, 0.

One of the useful properties of the Wilkinson catalyst in homogeneous hydrogenation is its high selectivity. Since the hydrogenation proceeds through the coordination of unsaturated compounds, steric effects play a decisive role. [39] Thus, terminal olefins are reduced more rapidly than internal olefins and cyclic olefins. *cis*-Olefins are hydrogenated faster than the *trans* isomers. By virtue of this selective property, homogeneous hydrogenation is superior to usual heterogeneous hydrogenation with metal catalysts. For example, an isopropenyl group in some terpenoids can be hydrogenated without attacking a double bond in the rings. It is possible to reduce only the isopropenyl groups of carvone *(17)*, eremophilone *(18)*, and γ-gurjunene *(19)* by selecting suitable reaction conditions. [40, 41]

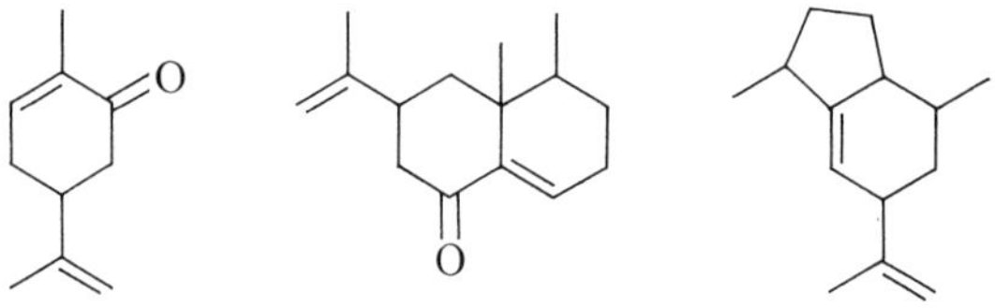

17   18   19

Since double bonds in various positions of steroids suffer different steric hindrance, selective hydrogenation of less hindered double bonds can be achieved by this method. [42—46] The double bonds of 1-, 2-, and 3-cholestene *(20)*, and Δ^1-3-ketosteroids were readily reduced, whereas more hindered 4-androstene, 14-ergostene, and $\Delta^{8(14)}$ ergostene *(21)* were not hydrogenated under the same conditions.

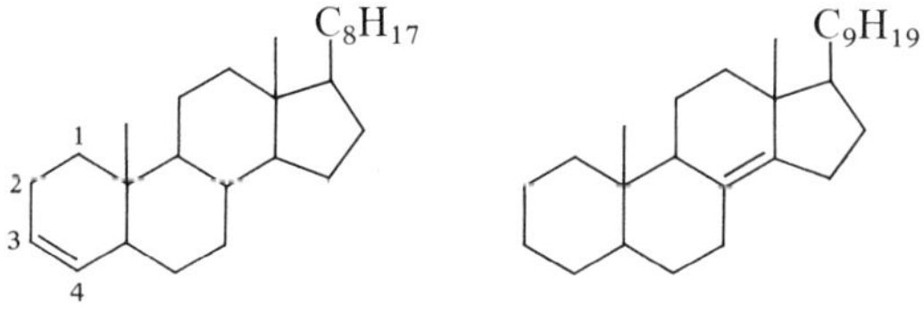

20   21

Thus both androsta-1,4-diene *(22)* and androsta-4,6-diene-3,17-dione *(23)* were hydrogenated to androst-4-ene-3,17-dione *(24)* in 75—85% selectivity. Cholesterol was not hydrogenated. These selective hydrogenations are not possible using common solid metal catalysts under heterogeneous conditions.

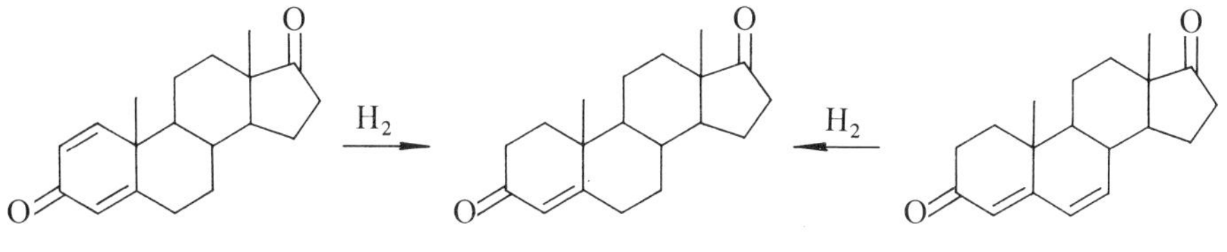

22   24   23

Unlike metallic palladium which afforded androst-4-ene-3,17-dione-1β-d$_1$ in only 15% by deuteration of androst-1,4-diene-3,17-dione *(22)*, the rhodium complex deuterated the same compound from the α side to give androst-4-ene-3,17-dione-1α,2α-d$_2$ *(25)* in 85% yield. [43]

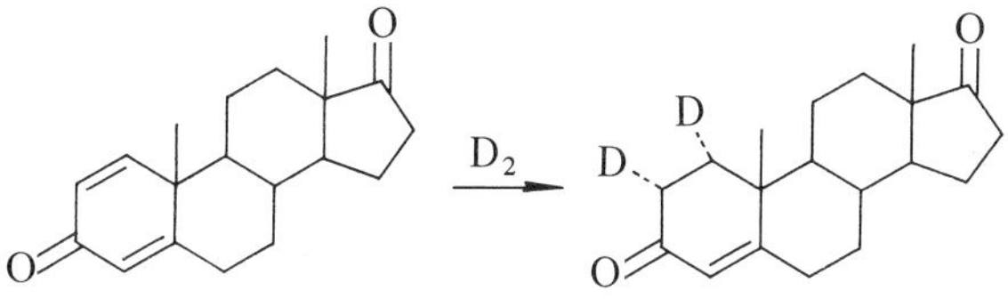

22   25

Also highly specific deuteration of 2-cholestene was effected in ethanol-benzene (1:1), benzene, and THF. But the specificity became lower in other solvents. [44] Santonin *(26)* was readily hydrogenated to give dihydrosantonin selectively. [47]

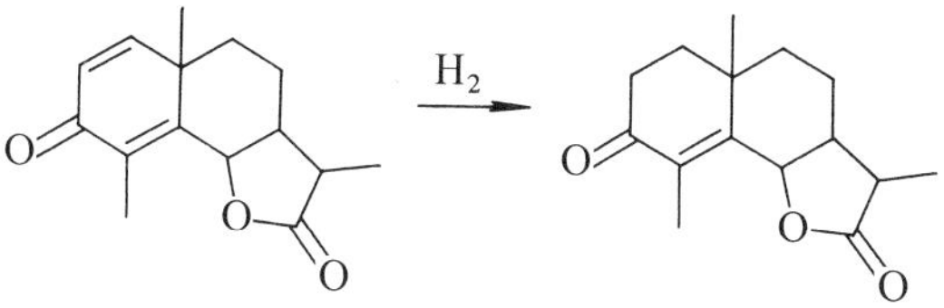

26

The conjugated diene system in α-phellandrene *(27)* was not hydrogenated, because the diene system behaves as a strong coordinating bidentate ligand and inhibits coordination of hydrogen to the catalyst. [41]

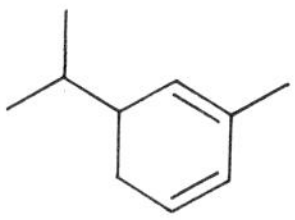

27

On the other hand, one of the double bonds of the conjugated diene system in thebaine *(28)* was hydrogenated to give dihydrothebaine *(29)* in 77% yield. [41] Again this reduction is different from metallic palladium catalyzed hydrogenation.

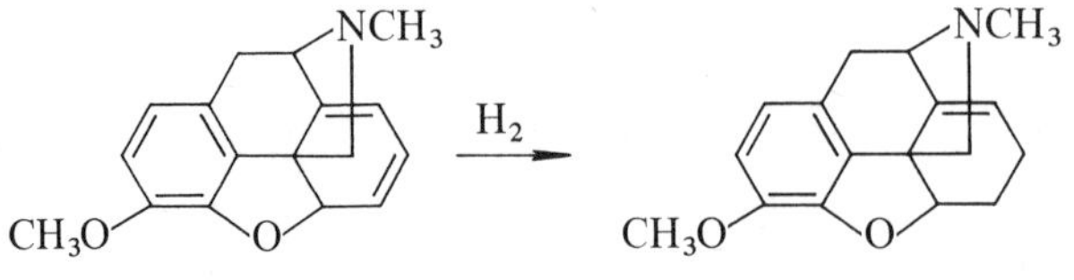

28   29

Various functional groups are tolerable under these catalytic conditions. [48] β-Nitrostyrene *(30)* was hydrogenated to phenylnitroethane *(31)*.

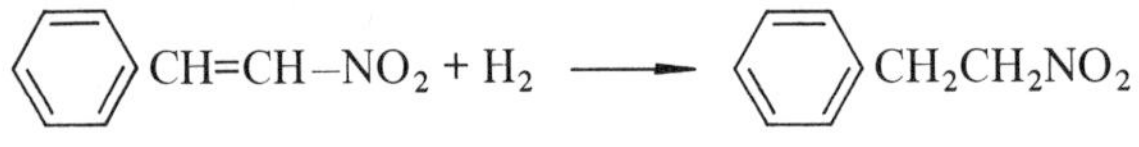

30   31

Similarly, double bonds of α,β-unsaturated acids, esters, nitriles, and ketones were easily reduced without attacking the functional groups.

$$RCH=CH-X + H_2 \longrightarrow RCH_2CH_2-X$$
$$X = COOH, COOR, CN, COR$$

Reduction of unsaturated aldehydes gives complex results due to the competing decarbonylation reaction of aldehydes, which is discussed later.

In the hydrogenation reaction using the rhodium complex, migration of double bonds takes place only to a minor extent. Thus the selective hydrogenation of 1,4-dihydrotetraline derivatives *(32)* was achieved. [47]

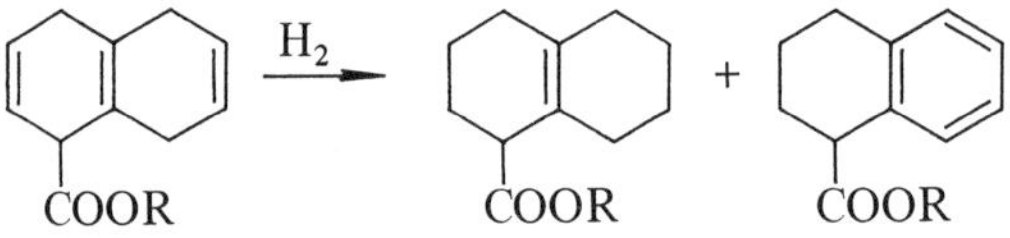

32   96%   4%

The thiophene ring remains intact and only a double bond of the side chain is reduced as shown by the following example. It should be pointed out that thiophene is not a catalytic poison for the rhodium complex. [49]

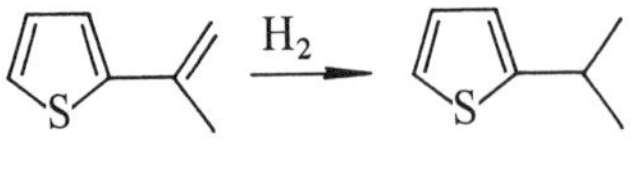

Reduction of allylic alcohols such as geraniol *(33)* and its *cis* isomer, nerol proceeded only to a small extent; instead $RhCl(CO)(PPh_3)_2$ was formed from $RhCl(PPh_3)_3$. [41] The reaction can be explained by assuming isomerization of the allylic alcohol part to an aldehyde followed by decarbonylation. [50, 51]

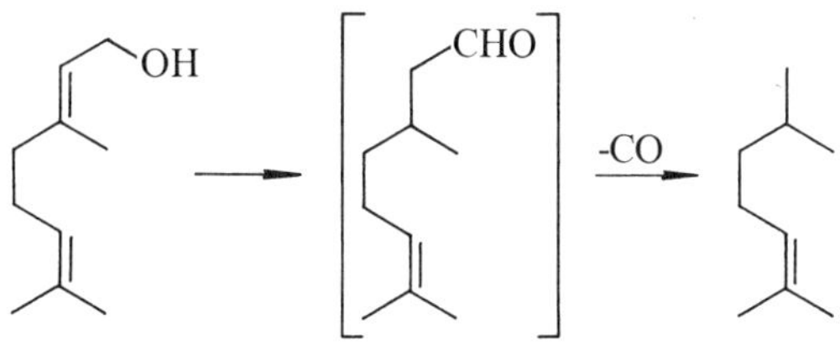

33

The terminal double bond of linalool *(34)* was saturated selectively.

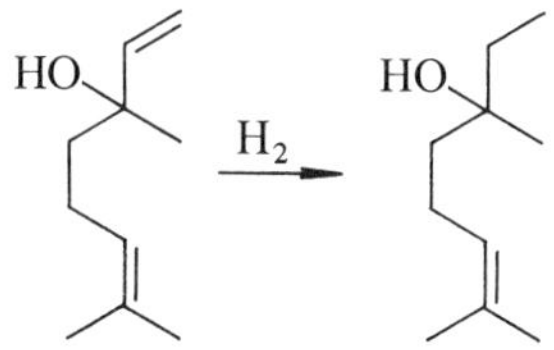

34

Acetylenic bonds are also reduced. The reduction proceeds *via* the formation of a double bond. [52] But the rate of the reduction of triple bonds seems to be somewhat lower than that of double bonds, and hence selective reduction of triple bonds to double bonds is not easy.

The most unique application of the homogeneous hydrogenation to organic synthesis is asymmetric hydrogenation of olefins using complexes coordinated by optically active ligands, giving optically active compounds. These chiral complexes have attracted much attention as a tool for stereochemical studies of olefins; their usefulness is based on the fact that they can chemically discriminate enantiomeric olefins and even two faces of prochiral olefins. This ability enables the resolution of enantiomeric olefins and asymmetric synthesis involving olefins, especially hydrogenation. The application of the former is the resolution of *trans*-cyclooctene *(35)* which has molecular asymmetry due to restriction of rotation. A mixture of diastereoisomers, *trans*-dichloro[(+)*trans*-cyclooctene] [(+)α-methylbenzylamine]platinum and *trans*-dichloro[(−)*trans*-cyclooctene] [(+)α-methylbenzylamine]platinum was obtained by the reaction

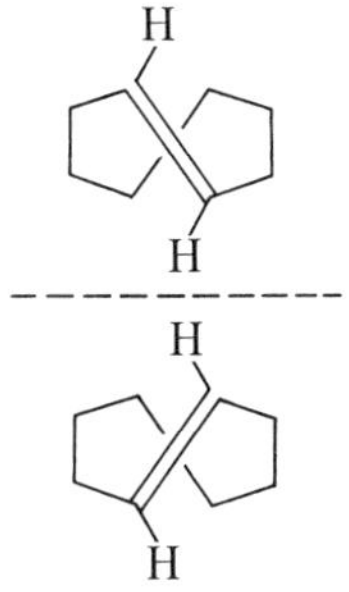

35

of *trans*-cyclooctene with *trans*-dichloro(ethylene)[(+)α-methylbenzylamine]-platinum, and separated simply by recrystallization. [53, 54, 55]

As a prerequisite to asymmetric hydrogenation, asymmetric induction is expected by the coordination of prochiral olefins to a complex coordinated by an optically active ligand. Actually this expectation has been confirmed. Asymmetric induction in π-coordination of prochiral olefins was observed when a platinum complex coordinated by an optically active amine was used. [56] An olefinic compound which has no asymmetric group and symmetry planes perpendicular to the plane of the double bond, posseses two enantiotopic faces. When the double bond, linked to two different substituents R and R', is π-complexed with transition metal compounds, each of the unsaturated carbon atoms becomes asymmetric. Paiaro and Panunzi predicted that a pair of enantiomorphs should be obtained if a prochiral olefin is π-bonded to a transition metal, and that if the olefin and optically active ligand such as an optically active amine, are coordinated to the transition metals, two different diastereoisomers are possible as shown by the structures *(36a)* and *(36b)*. [57, 58] Paiaro and

(S) or (R)-amine	(S) or (R)-amine	◯ Alkyl
36a	*36b*	◪ Hydrogen

coworkers actually prepared and resolved the diastereoisomeric pairs of *trans*-dichloro(olefin)(amine*)platinum bearing propylene, *trans*-2-butene, and styrene; (R) and (S)-α-methylbenzylamines were used as the optically active amine. The preferential crystallization of one diastereoisomer with extensive epimerization was achieved from suitable solvents.

trans-Pt(ethylene)(amine*)Cl_2 + olefin $\longrightarrow$

$trans$-Pt (olefin)(amine*)Cl_2 + ethylene

Furthermore, they discovered that asymmetric induction is possible when an olefin is complexed *cis* to an optically active amine in a platinum complex. [59, 60] In other words, for *cis*-coordinated complexes, the equilibrium constant of the following equation is different from unity.

$$\begin{array}{c}Cl\\|\\amine(S)\text{-}Pt\text{—}Cl\\|\\olefin(R)\end{array} \rightleftharpoons \begin{array}{c}Cl\\|\\amine(S)\text{-}Pt\text{—}Cl\\|\\olefin(S)\end{array}$$

For these types of *cis* complexes of *trans*-2-butene, *trans*-3-hexene, and *trans*-1,4-dichlorobutene, an excess of 25% or more of the (−)diastereoisomers was formed in acetone at room temperature. This observation indicates that asymmetric induction occurs in the coordination of olefins, under the influence of a chiral ligand. In homogeneous hydrogenation of olefins, the insertion of olefin into the metal-hydrogen bond takes place under the influence of the chiral ligand. At the moment of hydrogenation, the olefin, hydrogen, and the chiral ligands are all present on a single metal species.

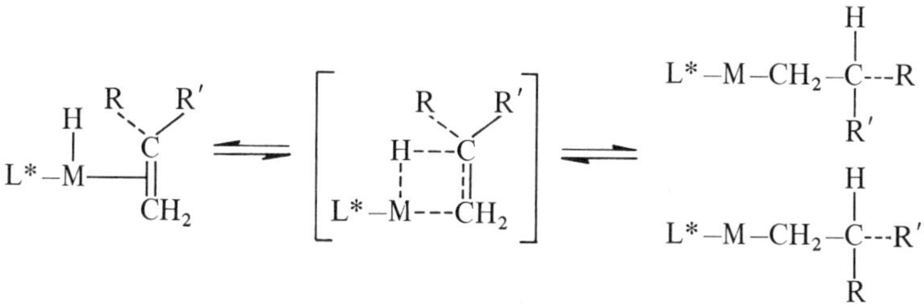

$$L^*-M-CH_2-\overset{\overset{\textstyle H}{|}}{\underset{\underset{\textstyle R'}{|}}{C}}\text{---}R$$

$$\begin{array}{c}H\\|\\L^*-M-CH_2-C\text{---}R'\\|\\R\end{array}$$

The problem thus becomes how to select and synthesize proper chiral ligands for the asymmetric hydrogenation catalyst. A synthetic method of optically active phosphines has been developed. [61] The phosphines prepared by this method have three different groups which make the phosphorus the chiral center. Based on this background, many attempts for asymmetric hydrogenation have been carried out. [62] Proper selection of both the chiral ligands and substrates is essential for obtaining a high degree of the asymmetric hydrogenation. Some examples are shown below.

Hydrogenation of α-phenylacrylic acid *(37)* with the Wilkinson catalyst coordinated with optically active (R)-(−)-methylphenyl-n-propylphosphine gave hydratropic acid *(38)* in an optical yield of 22%. [63]

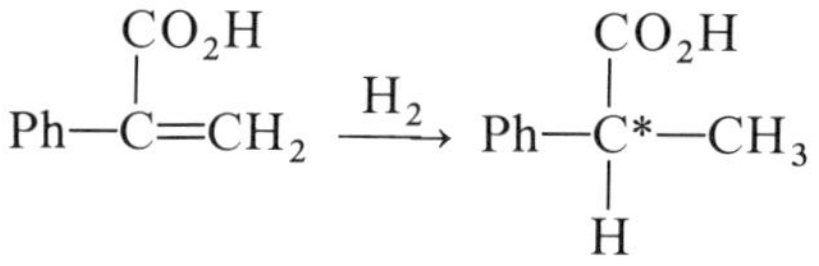

$$\underset{37}{\overset{\overset{\textstyle CO_2H}{|}}{Ph-C=CH_2}} \quad\xrightarrow{H_2}\quad \underset{38}{\overset{\overset{\textstyle CO_2H}{|}}{\underset{\underset{\textstyle H}{|}}{Ph-C^*-CH_3}}}$$

Optically active *o*-anisylcyclohexylmethylphosphine *(39)* and other related phosphines were found to be better. It was discovered that substituted α,β-unsaturated acids, even with a trisubstituted double bond, were hydrogenated rapidly with the rhodium complexes containing these chiral phosphines. This result shows that the selection of the substrates is very important. Steric interaction seems to have a crucial effect. The asymmetric hydrogenation was carried out successfully

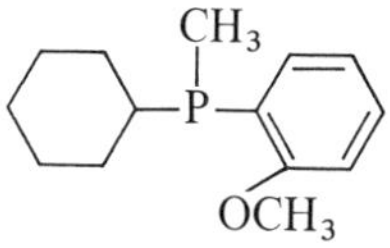

39

with a high optical yield using α-acylaminoacrylic acids *(40)* as the substrates. [64] This result opened a new avenue of approach to the asymmetric synthesis of optically active amino acids. As shown in Table II, 80—90 % optical yields were realized.

Table II

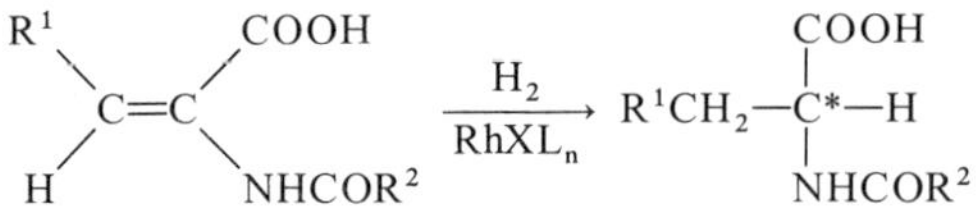

$$R^1\diagdown C=C \diagup COOH \qquad \xrightarrow[\text{RhXL}_n]{H_2} \qquad R^1CH_2-\underset{NHCOR^2}{\overset{COOH}{C^*-H}}$$

$$H\diagup \qquad \diagdown NHCOR^2$$

40   L=(R)(39)

R^1	R^2	optical purity %	resulting amino acid
3-MeO-4-OH—C$_6$H$_3$	Ph	90	L-Dopa
3-MeO-4-AcO—C$_6$H$_3$	Ch$_3$	88	L-Dopa
C$_6$H$_5$	CH$_2$	85	L-phenylalanine
C$_6$H$_5$	Ph	85	L-phenylalanine
p-Cl—C$_6$H$_4$	CH$_3$	77	p-Cl-L-phenylalanine
3-(1-Ac-indolyl)	CH$_3$	80	L-tryptophane
H	CH$_3$	60	L-alanine

The high optical yield obtained with the chiral phosphine *(39)* may be ascribed to the rigidity provided by the close interaction of oxygen of the methoxy group in the ligand with the amide group of the substrate. The optical yield was only 1 % when α-phenylacrylic acid *(37)* was hydrogenated under the same reaction conditions. The hydrogenation was carried out at 25° with 0.05% of the rhodium catalyst in alcoholic solvent.

In order to achieve a high degree of asymmetric hydrogenation, it is not always necessary to use a ligand that is asymmetric at the phosphorus atom.

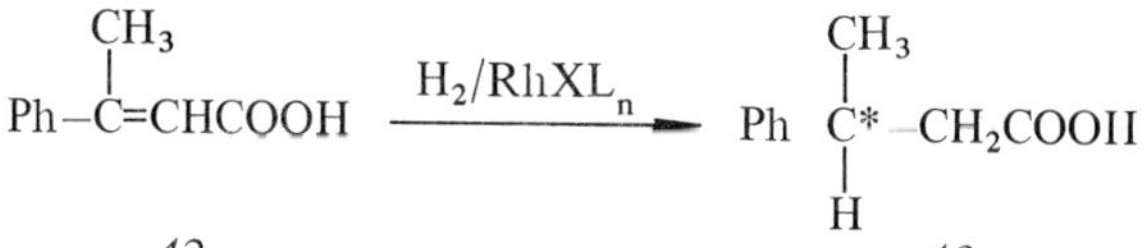

$$Ph-\underset{}{\overset{CH_3}{C}}=CHCOOH \quad \xrightarrow{H_2/RhXL_n} \quad Ph-\underset{H}{\overset{CH_3}{C^*}}-CH_2COOH$$

42   *43*

L=

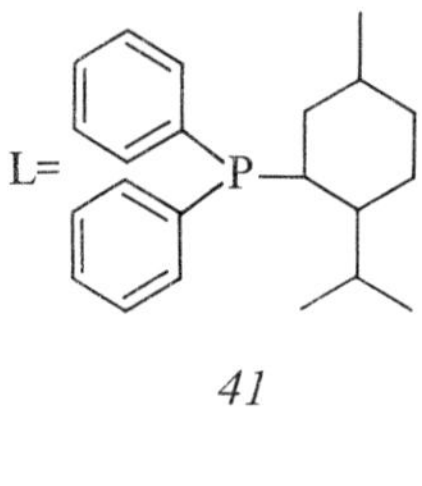

41

Another kind of optically active phosphines used has the chiral center at the carbon in the alkyl groups attached to phosphorus. The asymmetric alkyl groups used for this purpose were derived from optically active natural products. For example, diphenylneomenthylphosphine *(41)*-rhodium complex was used for the reduction of (E)-β-methylcinnamic acid *(42)* to give (S)-3-phenylbutyric acid *(43)* in an optical yield of 61%. [65]

Also a 1:1 complex of rhodium and optically active 2,3-O-isopropylidene-2,3-dihydroxy-1,4-bis(diphenylphosphino)butane *(44)* was used for a similar amino acid synthesis with an optical yield of 70—80%. [66] Optically active (+)-tartaric acid *(45)* was used for the synthesis of the phosphine *(44)*.

COOH
|
H–C–OH
|
HO–C–H
|
COOH

45

H   Ph
| /
H₃C   O–C–CH₂–P–Ph
   \C/ |
H₃C   O–C–CH₂–P–Ph
      | \
      H   Ph

44

(R)-N-acetylphenylalanine was obtained in an optical yield of 72%. Rigidity of this phosphine seems to be the contributing factor for the high optical yield. While the preparation of phosphine ligands which are asymmetric at phosphorus is tedious, many ligands that contain a chiral center remote from phosphorus are readily prepared without the tedious procedure of resolution from naturally occurring optically active compounds.

The above rhodium catalysts with chiral ligands for asymmetric hydrogenation are prepared by the reaction of rhodium cyclic olefin or diene complexes such as $[Rh(1,5\text{-hexadiene})Cl]_2$, or rhodium trichloride with two moles of the chiral ligands, followed by brief prehydrogenation. The active catalyst seems to be octahedral; two chiral phosphine ligands bound securely, and the remaining four coordination sites are in dynamic equilibrium with hydrogen, substrate, product, and solvent. Stereospecificity is achieved by the hindered rotation of the *cis* phosphines. The reaction is carried out in the absence of oxygen which may deteriorate the phosphine ligands. One serious problem involved in the homogeneous hydrogenation is the difficulty in recovering the catalyst dissolved in solution together with the products without lowering the activity.

The successful asymmetric hydrogenation will cause an immediate impact on the manufacture of amino acids in which a high optical yield is required. Clearly, whenever optically active compounds are needed, this approach may be considered a promising alternative to the biochemical route.

In addition to the rhodium complexes, there are a number of complexes used for homogeneous hydrogenation. Another very active hydrogenation catalyst is $HRuCl(PPh_3)_3$ formed by treatment of $RuCl_2(PPh_3)_3$ with hydrogen in ethanol in the presence of amine. The complex catalyzes hydrogenation of 1-olefins at 25° and 1 atm very rapidly and selectively. [67—69]

$$RuCl_2(PPh_3)_3 + H_2 + (C_2H_5)_3N \longrightarrow HRuCl(PPh_3)_3 + (C_2H_5)_3NHCl$$

Mechanistic studies indicate that one mole of triphenylphosphine dissociates to form a coordinatively unsaturated and highly active complex. Subsequent olefin insertion gives the alkyl intermediate, which reacts with hydrogen in the rate determining step to give hydrogenated product accompanied by the re-generated catalyst.

$$HRuCl(PPh_3)_3 \rightleftharpoons HRuCl(PPh_3)_2 + PPh_3$$

$$HRuCl(PPh_3)_2 + olefin \rightleftharpoons RuCl(alkyl)(PPh_3)_2$$

$$RuCl(alkyl)(PPh_3)_2 + H_2 \longrightarrow HRuCl(PPh_3)_2 + alkane$$

$RuCl_2(CO)(PPh_3)_2$ is a less active catalyst, but the complex can be used for selective hydrogenation of certain dienes. 1,5,9-CDT was reduced selectively to cyclododecene *(46)* with 98.5% yield at 125° and 7 atm of hydrogen. The catalyst can be conveniently prepared from ruthenium trichloride, triphenylphosphine, and carbon monoxide under the hydrogenation conditions of 1,5,9-CDT. [70,71]

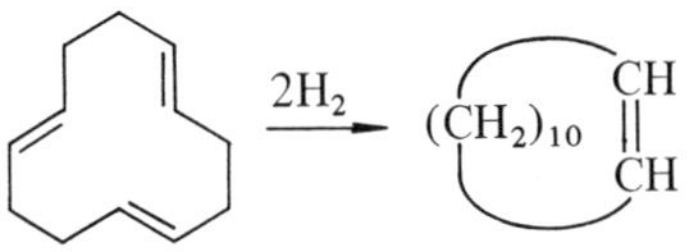

46

The molybdenum dihydride complex *(12)* is an active catalyst for the selective homogeneous hydrogenation of 1,3- and 1,4-dienes to give monoenes. [72] As a useful application, dicyclopentadiene *(47)* was hydrogenated at 180° under 160 atm of hydrogen in the presence of 0.2 mole% of the complex *(12)* to give cyclopentene selectively with 55% yield. Almost no cyclopentane was formed. The relatively high temperature, a condition of hydrogenation, causes retro-Diels-Alder reaction of dicyclopentadiene to give cyclopentadiene monomer, which in turn is hydrogenated to cyclopentene.

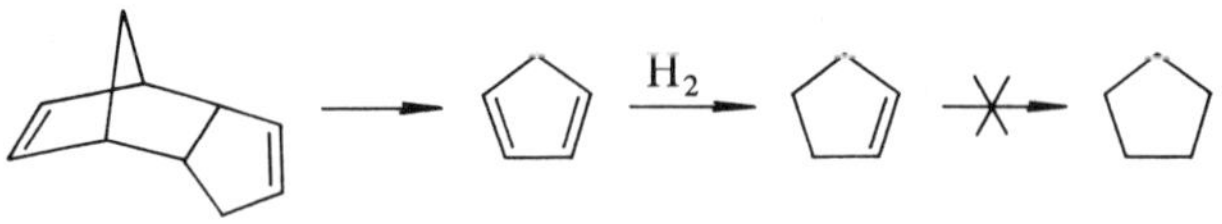

47

Similarly, 1,3- and 1,4-cyclohexadienes, 1,3-cycloheptadiene, norbornadiene, and 1,3-cyclooctadiene were hydrogenated at 140—180° to the corresponding cyclic monoenes. Also, double bonds of acrylate, crotonate, crotonaldehyde, and mesityl oxide were hydrogenated.

b) Oxidative Addition of other Hydrogen-Containing Compounds

In addition to the hydrogen molecule, other hydrogen-containing compounds form metal hydride complexes by oxidative addition reactions. Hydrogen abstraction from various activated carbon-hydrogen bond is an interesting reaction from the standpoint of organic synthesis. Indeed, the carbon-hydrogen bonds are ubiquitous in organic compounds, and unusual ability of transition metal catalysts to assist their breaking and forming is particularly useful.

Oxidative addition of aldehyde groups is expected to occur from the mechanism of their decarbonylation reaction catalyzed by rhodium and palladium catalysts as is discussed later. [73—75] The formation of the following diacyl complex *(48)* by the reaction of aldehydes with $Pt(PPh_3)_4$ was reported. [76]

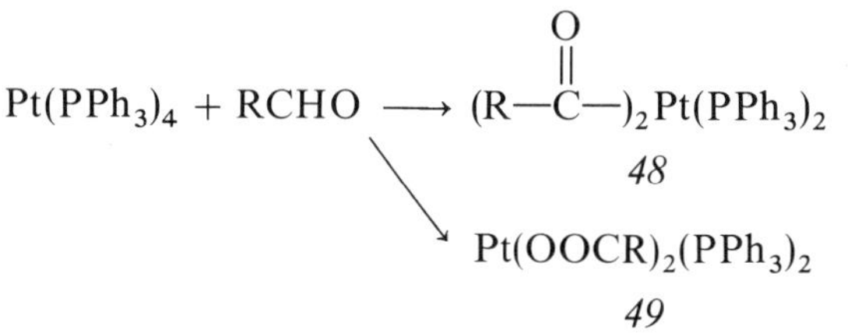

However, Tripathy and Roundhill proposed that the product of the reaction was not the diacyl complex *(48)*, but dicarboxylate *(49)*, and that reexamination of the reaction is required. [77]

A clear example of the carbon-hydrogen bond cleavage is observed by the reaction of acetylenes having a terminal hydrogen to give a hydride complex as shown below. The complex is probably an intermediate of polymerization of terminal acetylenes catalyzed by transition metal complexes. The following are typical examples. [78—80]

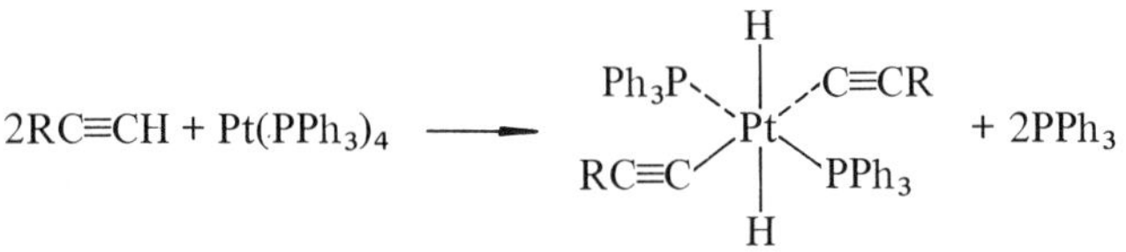

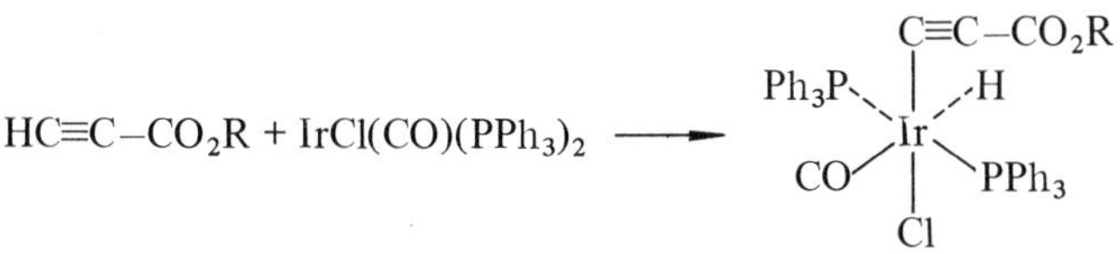

The following intramolecular isomerization provides an example of the cleavage of a saturated carbon-hydrogen bond of a methyl group. [81] The four-coordinate ruthenium complex *(50)* has the unsaturated d^8 configuration, and hence tends to form a saturated six-coordinated d^6 complex *(51)* by oxidative

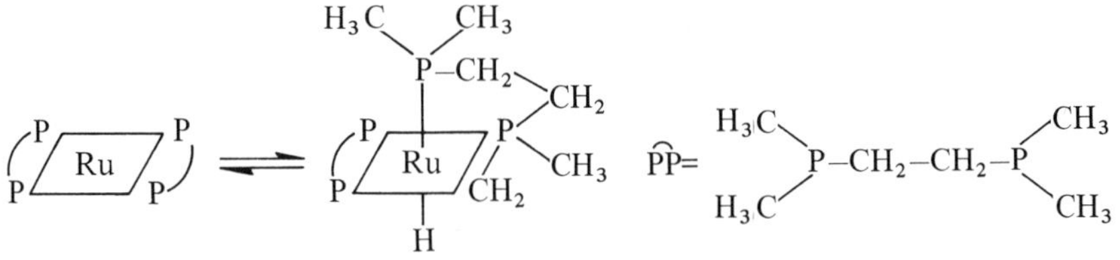

$$\left(\begin{smallmatrix} P & & P \\ & Ru & \\ P & & P \end{smallmatrix}\right) \;\rightleftharpoons\; \left(\begin{smallmatrix} P & & P \\ & Ru & \\ P & & P \end{smallmatrix}\right) \begin{smallmatrix} H_3C \; CH_3 \\ P-CH_2 \\ CH_2 \\ CH_3 \\ CH_2 \end{smallmatrix} \qquad \widehat{P}\widehat{P} = \begin{smallmatrix} H_3C & & CH_3 \\ P-CH_2-CH_2-P \\ H_3C & & CH_3 \end{smallmatrix}$$

50   *51*

addition. This tendency is the driving force of the cleavage of the stable carbon-hydrogen bond of the methyl group.

Carbon-hydrogen bonds in aromatic compounds, when they are suitably located, can be cleaved by the intramolecular oxidative addition reaction. [82] The four-membered ring *(52)* was formed by replacement of one of the *ortho*-phenyl hydrogens of the coordinated triphenylphosphines in the complex by the iridium through carbon-hydrogen bond cleavage. [83, 84] A similar reaction was observed with a zerovalent iron complex. [85]

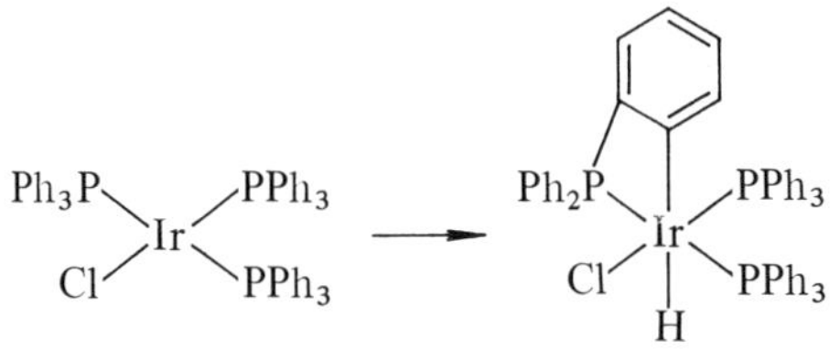

$$\begin{smallmatrix} Ph_3P & & PPh_3 \\ & Ir & \\ Cl & & PPh_3 \end{smallmatrix} \quad \longrightarrow \quad \begin{smallmatrix} Ph_2P & & PPh_3 \\ & Ir & \\ Cl & | & PPh_3 \\ & H & \end{smallmatrix}$$

52

Treatment of dichlorobis(triphenylphosphine)palladium *(53)* with lithium acetate caused *ortho*-substitution to give the four-membered ring *(54)*. The chlorine was introduced at the *ortho* position by treating the complex with carbon tetrachloride to give *(55)*. [86]

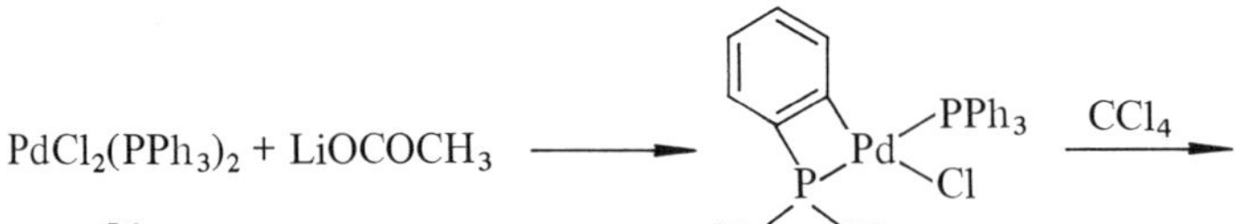

$$PdCl_2(PPh_3)_2 + LiOCOCH_3 \quad \longrightarrow \quad \begin{smallmatrix} & & PPh_3 \\ & Pd & \\ P & & Cl \\ Ph & Ph \end{smallmatrix} \quad CCl_4 \longrightarrow$$

53   *54*

$$PdCl_2(PPh_3)[PPh_2(o{-}Cl{-}C_6H_4)]$$

55

Similarly, a five-membered ring *(56)* was formed by cleavage of a carbon-hydrogen bond in the benzene ring of the triphenyl phosphite complex *(57)* with formation of a metal-carbon bond. [87]

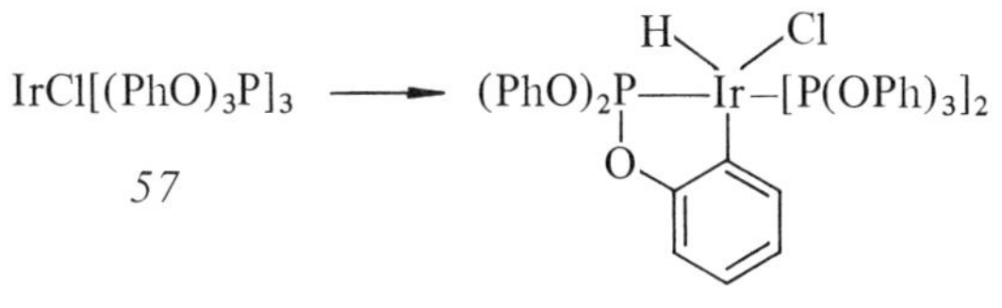

IrCl[(PhO)$_3$P]$_3$ $\longrightarrow$ (PhO)$_2$P—Ir—[P(OPh)$_3$]$_2$

57

56

With the hydride complex *(58)*, *ortho*-substitution took place with evolution of hydrogen to give the five-membered ring *(59)*. [88] The reaction is reversible, and the original hydride complex *(58)* was regenerated under hydrogen pressure.

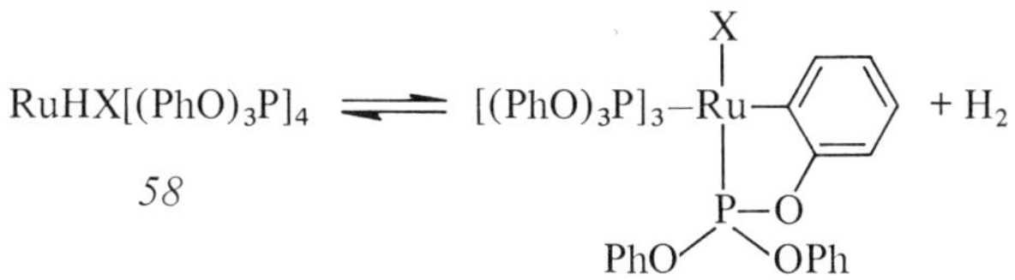

RuHX[(PhO)$_3$P]$_4$ $\rightleftharpoons$ [(PhO)$_3$P]$_3$—Ru + H$_2$

58

59

The complex *(60)* was reported to suffer from displacement of the methyl group giving methane and the complex *(61)*, which has the phenyl-metal σ-bond. [89]

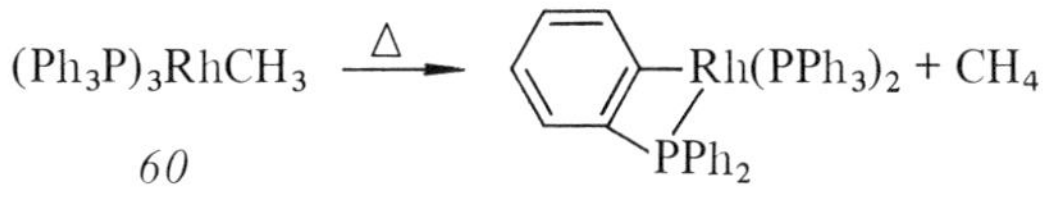

(Ph$_3$P)$_3$RhCH$_3$ $\xrightarrow{\Delta}$ —Rh(PPh$_3$)$_2$ + CH$_4$

60

61

When triphenylphosphine was treated with osmium carbonyl, the hydride complex *(62)* was formed. [90] The source of the hydrogen is triphenylphosphine undergoing the oxidative addition.

Os$_3$(CO)$_{12}$ + PPh$_3$ $\longrightarrow$ H—Os$_3$(CO)$_9$(PPh$_3$)(PPh$_2$C$_6$H$_4$)

62

Similar activation and cleavage at *ortho* positions of aromatic rings with transition metal complexes have been widely observed. [82]

These *ortho*-substitution reactions with transition metals can be regarded as a new aromatic substitution reaction, and are called *ortho*-metalation reactions. Here some other examples are cited, though some of them are not oxidative addition reactions from a mechanistic viewpoint. Azobenzene *(63)* or N,N-dimethylbenzylamine reacted with palladium chloride and other metal compounds to give the complex *(64)*. [91] Depending on the nature of the metals, the reaction can be explained as either an electrophilic or nucleophilic attack on the aromatic ring by the metal. [92]

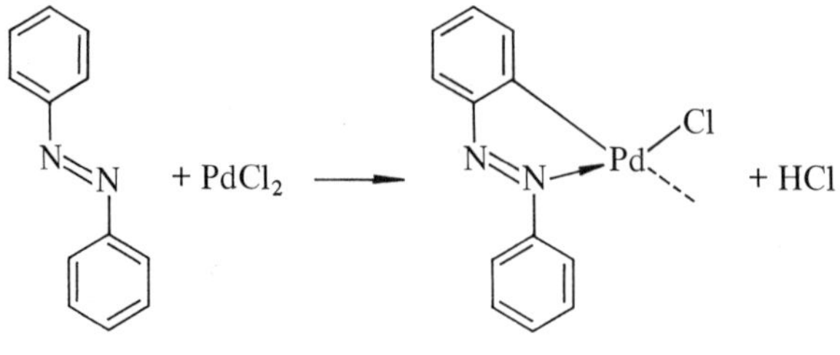

63   64

The *ortho*-substitution was involved in the reaction of thiobenzophenone *(65)* and $Fe_2(CO)_9$ to give the complex *(66)*. [93]

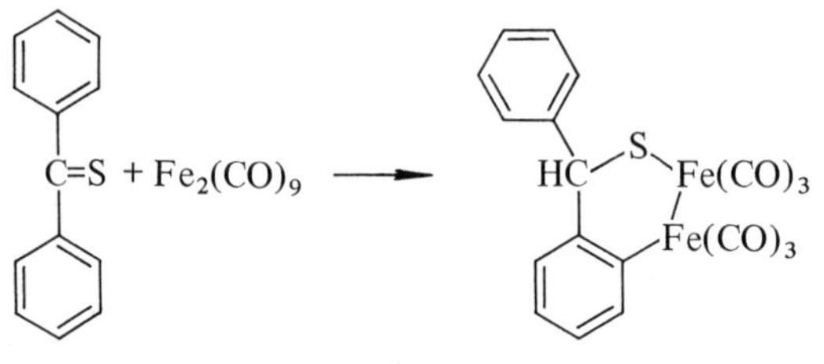

65   66

These carbon-hydrogen bond cleavage reactions suggest the possibility of using transition metal complexes for aromatic substitution reactions.

Related reactions, such as oligomerization and cooligomerization of various olefinic compounds to form unsaturated acyclic oligomers and cooligomers, involve a carbon-hydrogen bond cleavage. In the mechanism of these reactions, hydrogen transfer is an important step. This process proceeds through the carbon-hydrogen bond cleavage to give an intermediate with a metal-hydrogen bond. The hydrogen then migrates intermolecularly to another olefin molecule. For instance, the deuterated product *(67)* was obtained by dimerization of deuterated butadiene *(68)* catalyzed by a cobalt complex. The result clearly shows an inter-molecular hydrogen shift *via* metal-hydrogen bond formation. [94]

$$2 D_2C=CHCH=CD_2 \xrightarrow{CoC_2H_5(bipy)_2} D_2C=CHCHCD=CHCH=CD_2$$

68   67

with CD_3 group on the central carbon of 67.

The reaction can be explained by the following mechanism. Butadiene reacts with the cobalt-deuterium bond to form a π-allylic complex *(69)*. Insertion of butadiene into the π-allylic complex forms another π-allylic complex *(70)* con-taining eight carbon atoms. Liberation of the dimer *(67)* from the complex *(70)* regenerates the catalyst having a cobalt-deuterium bond. The coordinated deuterium is then intermolecularly transferred to a fresh butadiene.

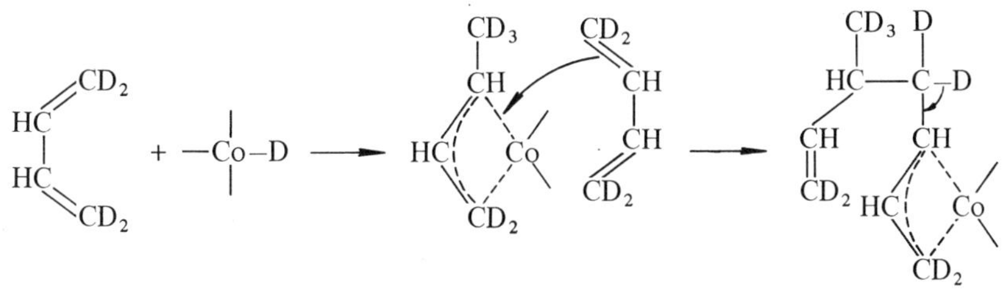

$$69 \qquad\qquad 70$$

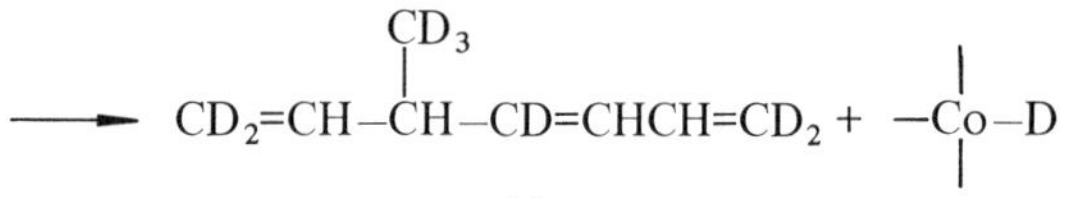

$$\longrightarrow \quad CD_2=CH-\underset{\underset{CD_3}{|}}{CH}-CD=CHCH=CD_2 + \ -\underset{|}{\overset{|}{Co}}-D$$

$$67$$

Hydrogen halides also serve as an active substrate for oxidative addition reactions. Anhydrous hydrogen chloride reacted with the Vaska complex to give $HIrCl_2(CO)(PPh_3)_2$. [95] The reaction with the iridium complex *(71)* is quantitative, and is reversed by treating with a base such as sodium methoxide or triethylamine to give the original iridium complex *(71)*. [96]

$$\textit{trans-}Ir(CO)Cl[P(CH_3)_2Ph]_2 + HCl \underset{\text{base}}{\rightleftharpoons} IrH(CO)Cl_2[P(CH_3)_2Ph]_2$$
$$71$$

Reaction of the Wilkinson complex with hydrogen chloride in the presence of an excess of triphenylphosphine gave the following addition product *(72)*. [97]

$$RhCl(PPh_3)_3 + HCl \longrightarrow HRhCl_2(PPh_3)_2 + PPh_3$$
$$72$$

The coordinatively unsaturated nickel complex *(73)* reacted with hydrogen chloride in toluene to give the nickel hydride complex *(74)*. [98]

$$Ni[P(cyclohexyl)_3]_2 + H-Y \longrightarrow HNiY[P(cyclohexyl)_3]_2$$
$$73 \qquad\qquad\qquad\qquad 74$$

$$Y=Cl, CH_3CO_2, PhO, C_5H_5, CN$$

In addition to hydrogen halide, some protic organic compounds add oxidatively to the above unsaturated complexes. Phenol, acetic acid, hydrogen cyanide and cyclopentadiene undergo the addition reactions with the above zerovalent nickel complex *(73)*. [98] Strongly electron-donating tricyclohexylphosphine as a ligand makes the oxidative addition feasible. In the oxidative addition of acids, it was found that the greater the pk_a value of the acid, the easier the oxidative addition occurs. It was also found that the tendency of the iridium complexes $Ir(CO)XL_2$ *(75)* to become protonated with benzoic acid increases with the ligands, X: $Cl < Br < I$, and L: $PPh_3 < AsPh_3 < P(CH_3)Ph_2 < P(C_2H_5)_2Ph < P(CH_3)_2Ph < As(CH_3)_2Ph < P(CH_3)_3$. [99]

$$PhCOOH + Ir(CO)XL_2 \longrightarrow HIr(CO)XL_2(PhCOO)$$
$$75$$

In some cases further reaction takes place after the oxidative addition of hydrogen halide. By undergoing oxidative addition, the proton of protic acids is incorporated into the complex as a hydride. Therefore, the hydrogen often acquires an anionic character, reacting with another proton to yield hydrogen gas and the metal with an increase of two charge units.

$$-\overset{|}{\underset{|}{M}}- + H^+ \longrightarrow -\overset{|}{\underset{|}{M}}{}^+\!-H \xrightarrow{H^+} -\overset{|}{\underset{|}{M}}{}^{2+}- + H_2$$

Addition of hydrogen halide to the zerovalent palladium phosphine complex causes the oxidation to form the bivalent palladium halide complex. [100]

$$Pd(PPh_3)_4 + 2HCl \longrightarrow PdCl_2(PPh_3)_2 + H_2 + 2PPh_3$$

Hydride complexes may be formed by oxidative addition of sulfur-hydrogen bonds to metal complexes. For instance, hydrogen sulfide and mercaptans are known to react with the Wilkinson complex. [101]

$$RhCl(PPh_3)_3 + RSH \longrightarrow HRhCl(SR)(PPh_3)_2 + PPh_3$$

Some nitrogen-hydrogen bonds are also capable of undergoing oxidative addition; amines and imides add to the platinum complex with the formation of the metal-hydrogen and metal-nitrogen bonds. [102, 103]

$$R_2NH + Pt(PPh_3)_4 \longrightarrow R_2N\!-\!Pt\!-\!H(PPh_3)_2 + 2PPh_3$$

Reactions of hydrosilanes and hydrogen have similar aspects. Oxidative addition of silicon-hydrogen bonds is an important step in transition metal catalyzed hydrosilylation of olefins. Actually, products of the oxidative addition reaction were isolated from the zerovalent platinum complex and trichlorosilane. [104] Hydrogen evolution was observed during the reactions. The addition to the Wilkinson complex is also known. [105]

$$2HSiCl_3 + Pt(PPh_3)_4 \longrightarrow Pt(SiCl_3)_2(PPh_3)_2 + H_2 + 2PPh_3$$
$$HSiR_3 + RhCl(PPh_3)_3 \longrightarrow R_3SiRh(H)Cl(PPh_3)_2 + PPh_3$$

The cyclohexyne phosphine complex of platinum *(76)* was protonated with remarkable ease by active hydrogen compounds such as acetone, nitromethane, and acetophenone to give the complexes having a metal-carbon σ-bond *(77)*. [106, 107] The reaction is assumed to proceed *via* the oxidative addition of these active hydrogen compounds with the acidic hydrogen-carbon bond cleavage.

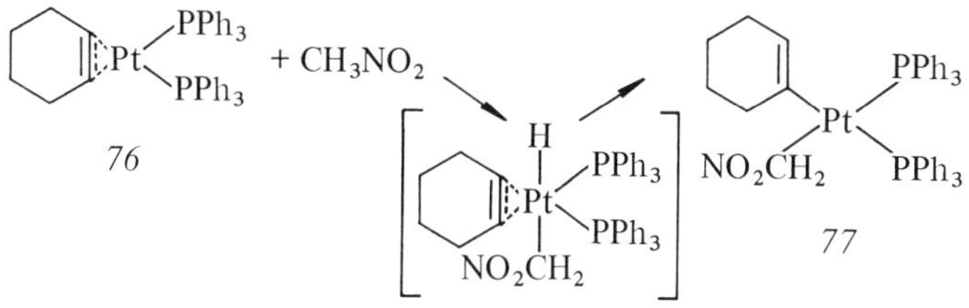

In the conversion of various olefins to substituted π-allylpalladium chloride, an allylic hydrogen is abstracted, probably through an addition-elimination mechanism. [108—114]

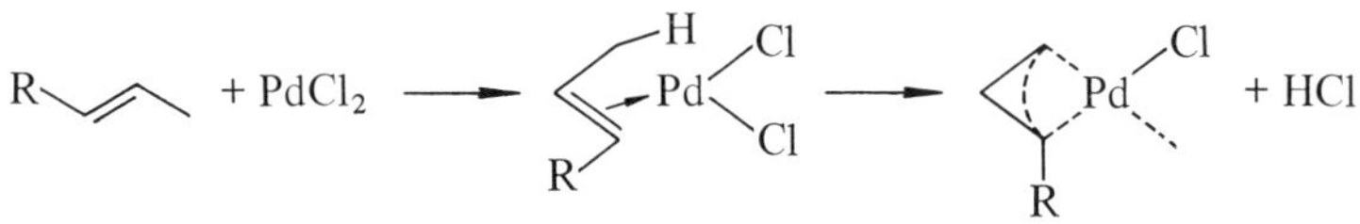

In addition to the above-mentioned active hydrogen compounds that are capable of undergoing oxidative addition reactions, some other active hydrogen compounds, such as alcohols and water, are expected to undergo the reactions, although little evidence for this is available. Some reactions involving these compounds can be explained by assuming the oxidative addition with cleavage of the hydrogen-oxygen or hydrogen-carbon bond.

5. *Metal-Carbon Bond Formation*

Formation of carbon-metal σ-bonds is the crucially important process, since most synthetic organic reactions involve the formation of certain types of carbon-metal σ-bonds. There are a variety of metal-carbon bond forming reactions. The reaction of active hydrogen compounds and many other examples are described earlier. In this section, the reaction is surveyed with additional examples.

Carbon-metal bond formation *via* carbon-carbon bond cleavage, if possible, would be the most interesting from a synthetic point of view. Unfortunately, however, only few examples of such a reaction are known so far, because of both thermodynamic and kinetic stability of the carbon-carbon bond. As an example, hexaphenylethane *(78)* undergoes carbon-carbon bond cleavage with a zero-valent nickel complex. [115, 116] This reaction is rather exceptional and is a consequence of the large steric hindrance released upon bond cleavage.

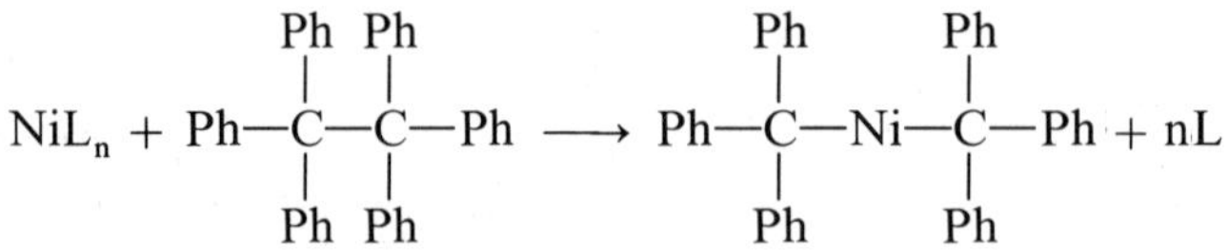

$$\text{NiL}_n + \text{Ph}-\underset{\underset{\text{Ph}}{|}}{\overset{\overset{\text{Ph}}{|}}{\text{C}}}-\underset{\underset{\text{Ph}}{|}}{\overset{\overset{\text{Ph}}{|}}{\text{C}}}-\text{Ph} \longrightarrow \text{Ph}-\underset{\underset{\text{Ph}}{|}}{\overset{\overset{\text{Ph}}{|}}{\text{C}}}-\text{Ni}-\underset{\underset{\text{Ph}}{|}}{\overset{\overset{\text{Ph}}{|}}{\text{C}}}-\text{Ph} + n\text{L}$$

78

The carbon-carbon bond cleavage could also be achieved by electronic activation. Somewhat surprisingly, carbon-carbon bond cleavage reactions were observed in the reaction of oxalate *(79)* and benzil *(80)* with the zerovalent triphenylphosphine complex of platinum. [117]

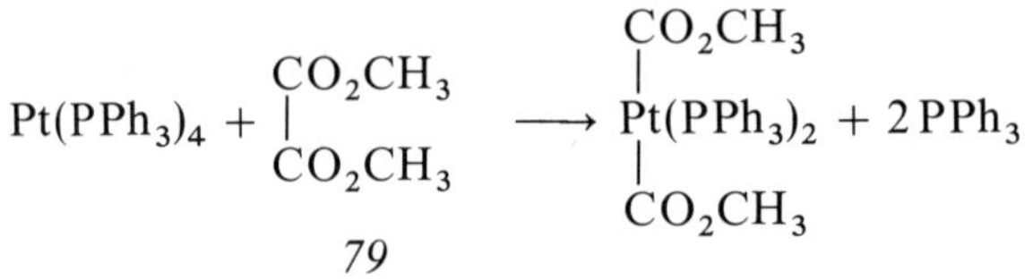

$$\text{Pt}(\text{PPh}_3)_4 + \underset{\underset{\text{CO}_2\text{CH}_3}{|}}{\overset{\overset{\text{CO}_2\text{CH}_3}{|}}{}} \longrightarrow \text{Pt}(\text{PPh}_3)_2 + 2\,\text{PPh}_3$$

with CO_2CH_3 above and CO_2CH_3 below Pt.

79

$$\text{Pt}(\text{PPh}_3)_4 + \underset{\underset{\text{Ph}-\text{C}=\text{O}}{|}}{\overset{\overset{\text{Ph}-\text{C}=\text{O}}{|}}{}} \longrightarrow \text{Pt}(\text{PPh}_3)_2 + 2\,\text{PPh}_3$$

with COPh above and COPh below Pt.

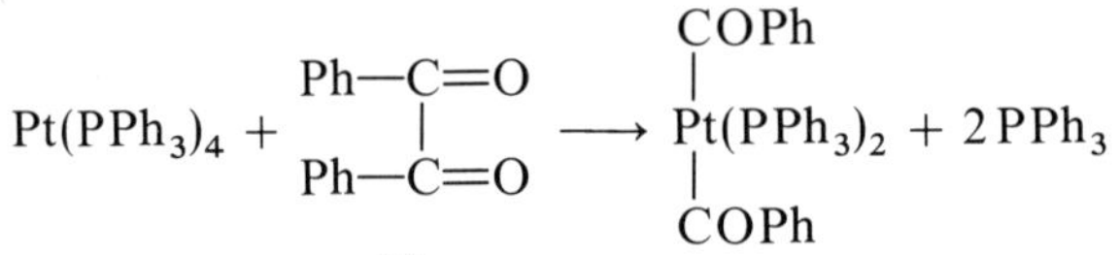

80

1,2-Benzocyclobutadienequinone *(81)* underwent the following carbon-carbon bond cleavage reaction with the platinum complex. [118]

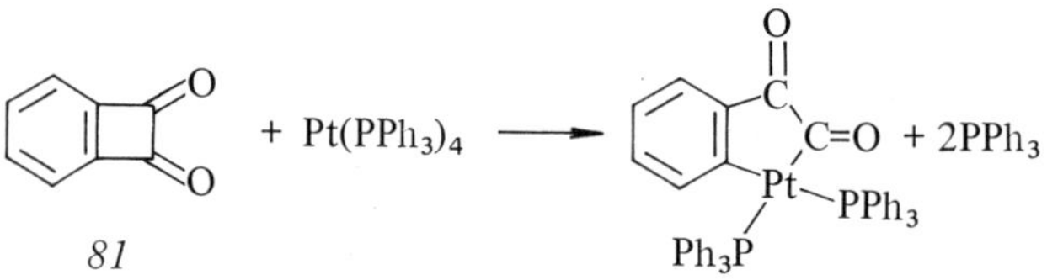

81 Ph_3P

Oxidative addition of cyanogen, a pseudo-halogen compound, is known to occur onto triphenylphosphine complexes of nickel, palladium, and platinum with the cleavage of the carbon-carbon bond. [119]

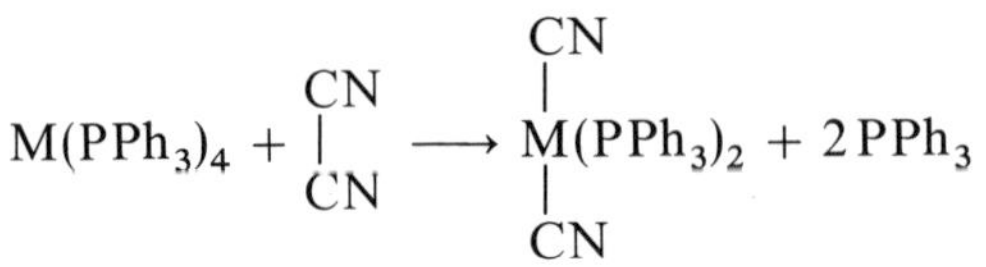

$$\text{M}(\text{PPh}_3)_4 + \underset{\underset{\text{CN}}{|}}{\overset{\overset{\text{CN}}{|}}{}} \longrightarrow \text{M}(\text{PPh}_3)_2 + 2\,\text{PPh}_3$$

with CN above and CN below M.

$$\text{M} = \text{Ni, Pd, Pt}$$

Another example of the pseudo-halogenic character exhibited by nitrile compounds is the oxidative addition of certain nitriles with the carbon-carbon bond cleavage. Benzonitrile reacted with zerovalent platinum or the corresponding nickel complexes. [120]

$$\text{Pt}[\text{P}(\text{C}_2\text{H}_5)_3]_3 + \text{Ph}-\text{CN} \longrightarrow \text{Ph}-\text{Pt}(\text{CN})[\text{P}(\text{C}_2\text{H}_5)_3]_2 + \text{P}(\text{C}_2\text{H}_5)_3$$

The effect of the ligand was crucial and no reaction took place with the corresponding triphenylphosphine complex. On the other hand, one of the carbon-cyano bonds of tricyanoethane and tetracyanomethane is cleaved by the triphenylphosphine complex. [121, 122]

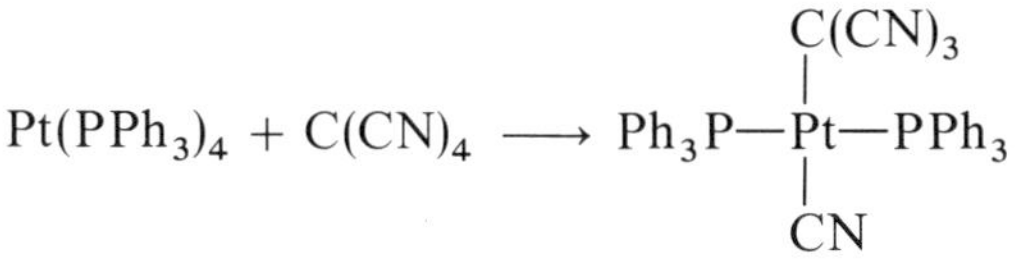

$$Pt(PPh_3)_4 + C(CN)_4 \longrightarrow \underset{\underset{CN}{|}}{\overset{\overset{C(CN)_3}{|}}{Ph_3P-Pt-PPh_3}}$$

Oxidative addition of π-coordinated tetracyanoethylene by UV irradiation has been reported. [123]

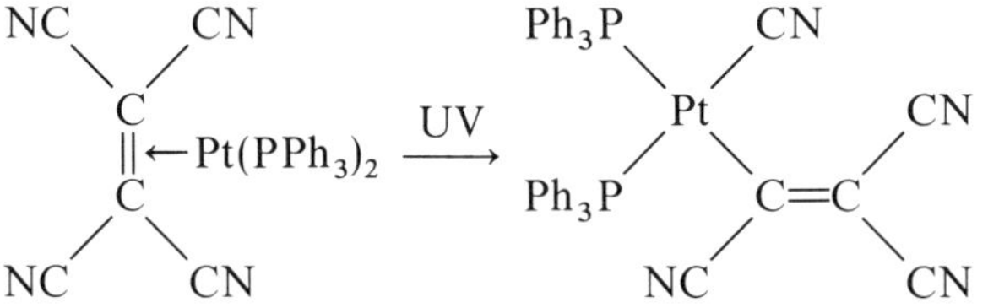

Unlike halogens, a simple nitrile group is not easily removed from organic compounds by the usual methods of organic chemistry. It can be done by using sodium in HMPA. [124] A method of facile removal of a nitrile group was found by using an iron complex formed from iron acetylacetonate and sodium. [125] In this case, the oxidative addition of the nitrile group takes place and then intramolecular hydrogenolysis of the nitrile group with the hydrogen supplied from the methyl group proceeds. The reaction proceeded better with aliphatic nitriles than with aromatic derivatives.

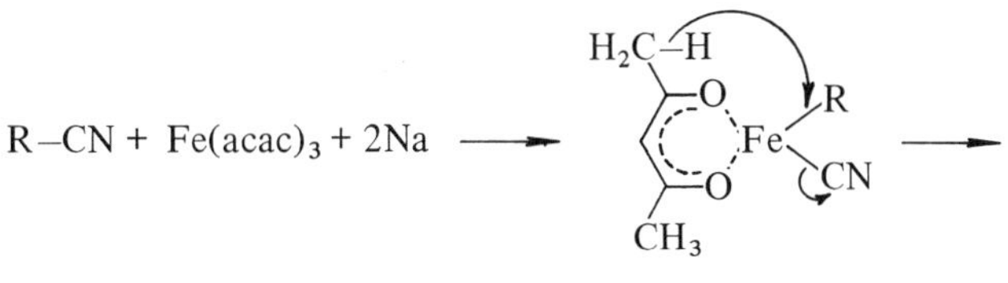

$$RH + CN^- + 2Na^+ + Fe(acac)_2C_5H_6O_2^-$$

Intramolecular rearrangement catalyzed by transition metal compounds or valence isomerization of highly strained molecules is a topic of active studies. The rearrangement is now believed to proceed through a metal-carbon bond formation. [8] Valence isomerization of cubane *(82)* to *syn*-tricyclooctadiene *(83)* catalyzed by univalent rhodium complexes of the type [Rh(diene)Cl]$_2$ was shown to involve oxidative addition of the cubane through a carbon-carbon bond cleavage. [126] The direct demonstration of this oxidative addition of cubane was presented by isolation of the acylrhodium adduct *(84)*, formed by the oxidative addition followed by carbon monoxide insertion.

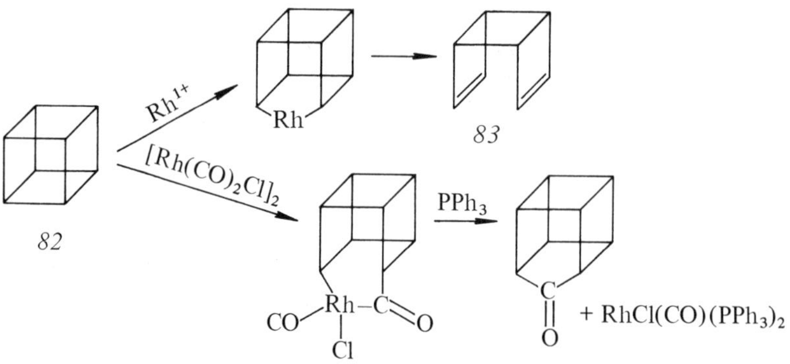

Another example is the rearrangement of bicyclo[2,1,0]pentane *(85)* catalyzed by several transition metal complexes. [127] The first step of the reaction is probably the oxidative addition across the central bond of the pentane. Abstraction of the *exo*-hydrogen would generate an allylrhodium hydride complex *(86)*, which decomposes to cyclopentene and a metal complex.

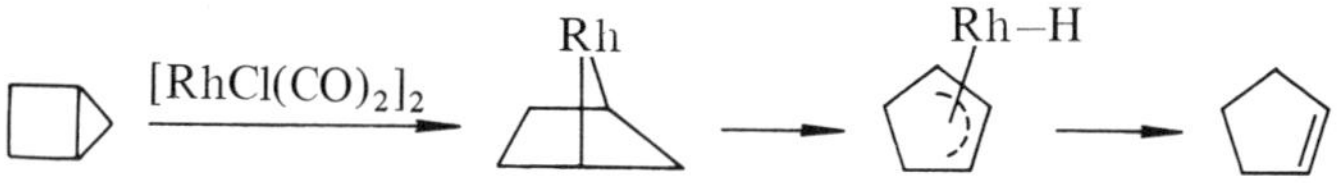

The most versatile method of carbon-metal bond formation is cleavage of carbon-halogen bonds. Many types of organic halides can be used as starting material for organic synthesis. Mechanistically oxidative addition is the first step of this reaction. The first example is the oxidative addition of acyl halides. The product of the reaction is called an acyl metal complex, which is sometimes isolated as a stable complex; RhCl(PPh$_3$)$_3$ reacts with benzoyl chloride or long chain aliphatic acyl chloride to give the five-coordinated acylrhodium complexes *(87)*. [128, 129] A zerovalent nickel complex also undergoes a similar reaction. [130]

$$\text{RhCl(PPh}_3)_3 + \text{RCOCl} \longrightarrow \text{RCORhCl}_2(\text{PPh}_3)_2 + \text{PPh}_3$$

87

$$(\text{t-BuNC})_4\text{Ni} + \text{PhCOCl} \longrightarrow \text{PhCONiCl(t-BuNC)}_3 + \text{t-BuNC}$$

The acyl complexes play a key role in carbonylation and decarbonylation reactions catalyzed by transition metal complexes. This topic is discussed later (p. 148). The acyl complexes formed sometimes spontaneously undergo a decarbonylation reaction to give the alkyl complexes; the product of the reaction of RhCl(PPh$_3$)$_3$ with acetyl chloride was a methylrhodium complex *(88)* rather than an acetylrhodium complex. [129]

$$RhCl(PPh_3)_3 + CH_3COCl \longrightarrow CH_3Rh(CO)Cl_2(PPh_3)_2 + PPh_3$$

$$88$$

Chloroformate and chloroformamide underwent oxidative addition with cleavage of their carbon-chlorine bonds. [131—133]

$$Pt(PPh_3)_4 + ClCOOCH_3 \longrightarrow PtCl(COOCH_3)(PPh_3)_2 + 2\,PPh_3$$

Alkoxycarbonyl complexes of palladium and nickel formed by the oxidative addition reaction of chloroformate underwent facile decarboxylation or decarbonylation. [134]

$$Ni(PPh_3)_2 + ClCOOCH_3 \longrightarrow \begin{array}{c} Ph_3P \quad\quad COOCH_3 \\ \diagdown \quad \diagup \\ Ni \\ \diagup \quad \diagdown \\ Cl \quad\quad PPh_3 \end{array} \xrightarrow[+PPh_3]{-CO} NiCl(PPh_3)_3$$

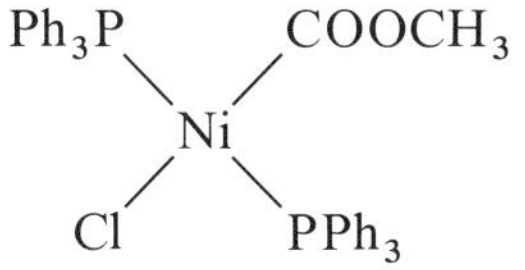

The dithioethoxycarbonyl complex *(89)* was obtained by the reaction of chlorothioformate with $RhCl(PPh_3)_3$. [135]

$$RhCl(PPh_3)_3 + ClC(S)SC_2H_5 \longrightarrow H_5C_2S(S)C-RhCl_2(PPh_3)_2 + PPh_3$$

$$89$$

Reduction of acyl halides to aldehydes is possible by using certain hydride-forming compounds; reaction of aromatic acid chlorides with triethylsilane in the presence of *cis*-$PtCl_2(PPh_3)_2$ gave aldehydes. [136]

$$PhCOCl + HSi(C_2H_5)_3 \longrightarrow PhCHO + ClSi(C_2H_5)_3$$

In this case the reaction of the silane with the platinum complex takes place to give the platinum hydride *(90)* first.

$$PtCl_2(PPh_3)_2 + HSi(C_2H_5)_3 \longrightarrow HPtCl_2[Si(C_2H_5)_3](PPh_3)_2$$

$$\longrightarrow HPtCl(PPh_3)_2 + ClSi(C_2H_5)_3$$

$$90$$

Subsequent oxidative addition of the acid chloride gives an acyl metal bond, which is reduced with the hydride to give aldehydes. When aromatic acid chlorides with an electron-withdrawing substituent were treated with triethylsilane in the presence of $RhCl[P(C_2H_5)Ph_2]_2(CO)$, the corresponding aldehydes were obtained. But the reaction of aromatic acid chlorides with an electron-donating group gave corresponding ketones. The formation of the ketones proceeds through the formation of an acylalkylrhodium complex by oxidative addition.

These Rosenmund-type reductions of acyl halides to aldehydes present an example which shows a similarity between the oxidative addition of transition metal complexes and chemisorption on the surface of transition metals. Reduction of acyl halides to aldehydes with hydrosilanes is also possible by using platinum on carbon as a catalyst. [137] The proposed mechanism of the Rosenmund reduction with palladium metal catalyst assumes the formation of an acyl-palladium complex. [138]

$$RCOCl + Pd \longrightarrow RCO-Pd-Cl \xrightarrow{H_2} RCO-\overset{\overset{\displaystyle H}{|}}{\underset{\underset{\displaystyle H}{|}}{Pd}}-Cl \longrightarrow RCHO + Pd + HCl$$

Oxidative addition reactions of alkyl, aryl, and vinyl halides are potentially important and simple methods for the synthesis of carbon-metal bonds. Alkyl halides, especially alkyl iodides, react with metal complexes to give stable alkyl complexes. This reaction has been studied extensively. The methylation of the Vaska complex is explained by a bimolecular displacement reaction at the carbon, in which the iridium atom acts as a nucleophile. In the reaction of methyl iodide with $Ir(CO)X(PPh_3)_2$ the rate at which methyl iodide adds to the complex was found to follow the order, $Cl > Br > I$. This is the reverse order observed for the oxidative addition of hydrogen to the same complexes. [139] The result suggests the oxidative addition of hydrogen is of a different nature to that of methyl iodide. Also the rate of the addition of methyl iodide changed with solvent; the reaction was much faster in DMF than in benzene.

Addition of alkyl halides to certain carbonyl complexes affords acyl complexes rather than alkyl complexes. An example is the reaction of $Pd(CO)(PPh_3)_3$ with various organic halides, such as vinyl, allyl, benzyl, methyl, and phenyl halides. [140]

$$Pd(CO)(PPh_3)_3 + R{-}X \longrightarrow PdX(COR)(PPh_3)_2 + PPh_3$$

Very high reactivity toward oxidative addition was observed with the unsaturated d^8 rhodium complex (91). The complex reacts with acetyl chloride, benzoyl chloride, methyl iodide, 6-bromohexene, cyclohexyl bromide, and even with neopentyl bromide to give trivalent rhodium complexes very readily. [141]

91

Vinyl halogens attached to sp² carbons are regarded as inert in the usual organic reactions, and direct displacement of vinyl halogen with a nucleophile does not occur easily. On the other hand, vinyl and aryl halides are more reactive than ordinary alkyl halides attached to sp³ carbons toward transition metal complexes. This is one of the characteristic properties of transition metal complexes. Thus vinyl chloride adds to $Pd(PPh_3)_4$ with cleavage of the carbon-chlorine bond. [133, 142]

$$H_2C=C\begin{smallmatrix}H\\\\Cl\end{smallmatrix} + Pd(PPh_3)_4 \longrightarrow H_2C=C\begin{smallmatrix}H\\\\Pd\end{smallmatrix}\begin{smallmatrix}PPh_3\\\\Cl\end{smallmatrix} + 2\,PPh_3$$

$$\underset{Ph_3P}{}$$

A variety of vinyl and aryl halides react with zerovalent complexes; $Ni(PR_3)_2(C_2H_4)$ forms a stable carbon-metal σ-bond. [143, 144] Tetrachloroethylene afforded *trans*-chloro(trichlorovinyl)bis(triethylphosphine)nickel *(92)*.

$$\underset{Cl}{\overset{Cl}{}}C=C\underset{Cl}{\overset{Cl}{}} + [(C_2H_5)_3P]_2Ni(CH_2=CH_2) \longrightarrow \underset{Cl}{\overset{Cl}{}}C=C\underset{Ni}{\overset{Cl}{}}P(C_2H_5)_3$$

$$\underset{(C_2H_5)_3P}{}\overset{}{\underset{92}{}}\underset{Cl}{} + CH_2=CH_2$$

The reaction which might involve the oxidative addition of vinyl halides is the coupling of vinyl halides in the presence of bis(1,5-COD)nickel. [145] Various vinyl halides were coupled by the reaction of the nickel-COD complex especially in the presence of a donor ligand such as triphenylphosphine at room temperature. The coupling was stereospecific as shown by the coupling of *cis-β-*bromoacrylate *(93)*.

$$\underset{Br}{\overset{H}{}}C=C\underset{CO_2CH_3}{\overset{H}{}} + \xrightarrow{Ni(COD)_2,\ PPh_3} H_3CO_2C\underset{C=C}{\overset{H\quad\quad H}{}}CO_2CH_3$$

93

Coupling of certain aromatic halides proceeded in high yields to give biaryl by using the same nickel complex. [146] The nickel complex is inert to functional groups such as carbonyl and nitrile. This method is superior to Ullmann reaction which uses copper as the coupling agent because of moderate reaction conditions. A mechanism suggested by Semmelhack involves the oxidative addition of two moles of aryl halides to the zerovalent nickel, followed by reductive elimination with coupling of the coordinated aryl groups. Selection of a proper solvent is crucial in this reaction; the reaction proceeds only in DMF.

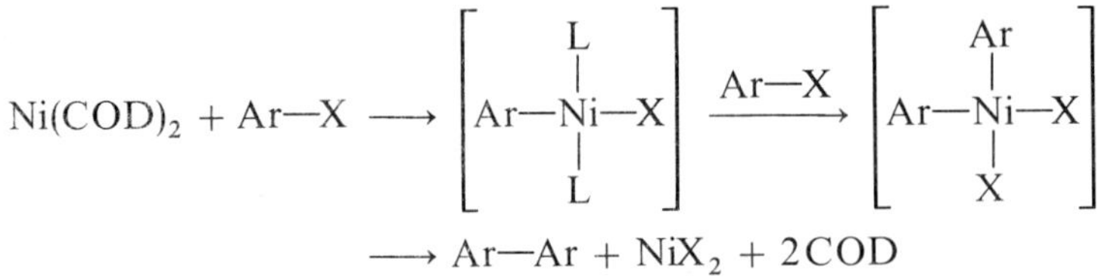

$$\text{Ni(COD)}_2 + \text{Ar-X} \longrightarrow \begin{bmatrix} & L & \\ \text{Ar-Ni-X} & \\ & L & \end{bmatrix} \xrightarrow{\text{Ar-X}} \begin{bmatrix} & Ar & \\ \text{Ar-Ni-X} & \\ & X & \end{bmatrix}$$

$$\longrightarrow \text{Ar-Ar} + \text{NiX}_2 + 2\,\text{COD}$$

Nucleophilic substitution reaction of aromatic halides can be carried out efficiently by using the zerovalent nickel complex. [147] Reaction of phenyl-nickel iodide complex *(94)* formed from iodobenzene and Ni(PPh$_3$)$_4$, with lithium salt of acetophenone at 25° gave 65% yield of benzyl phenyl ketone based on consumed iodobenzene and a yield of 250% based on the nickel catalyst.

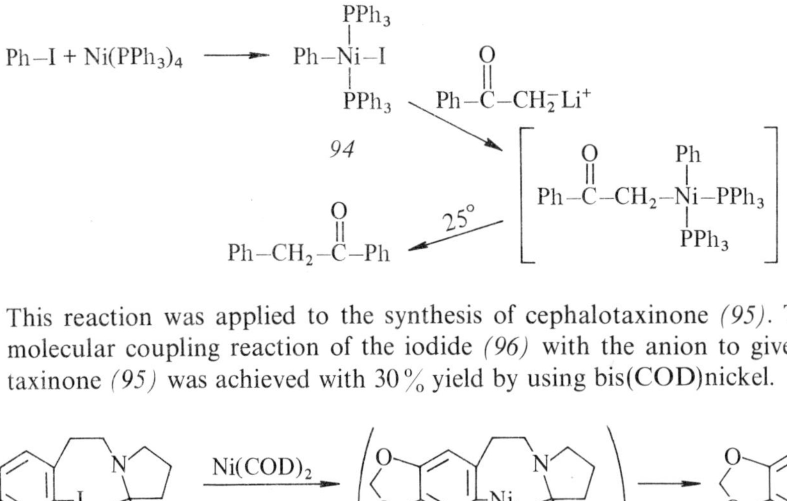

This reaction was applied to the synthesis of cephalotaxinone *(95)*. The intramolecular coupling reaction of the iodide *(96)* with the anion to give cephalotaxinone *(95)* was achieved with 30% yield by using bis(COD)nickel.

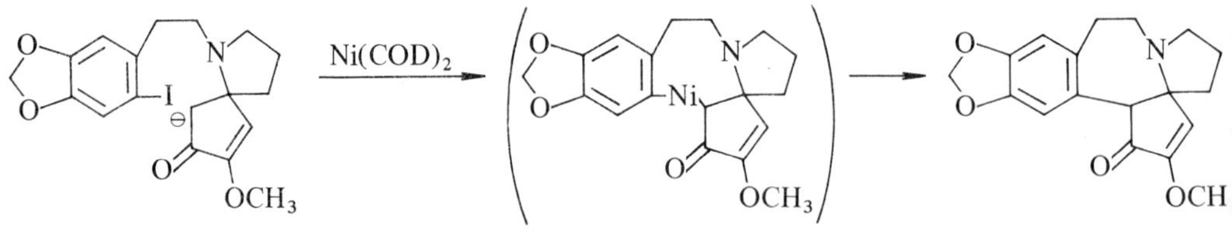

Reactions of halogen-containing compounds with metal carbonyls such as Ni(CO)$_4$, Fe(CO)$_5$, Fe$_2$(CO)$_9$, Fe$_3$(CO)$_{12}$ have been widely explored as useful synthetic methods. Compounds having relatively active halogens of the following types take part in the reactions. [148, 149]

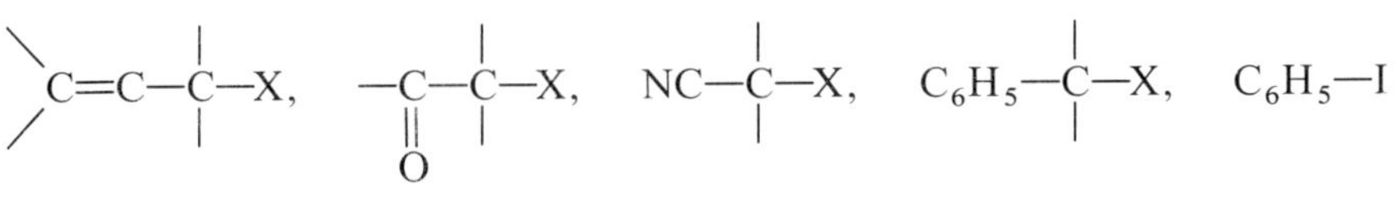

Simple alkyl halides, except methyl iodide, do not react with these metal carbonyls. Although there is no direct evidence, formation of the intermediate metal-carbon σ-bonded complexes by the oxidative addition reaction is assumed in these reactions. Sometimes, the oxidative addition of halides is followed by insertion of coordinated carbon monoxide to afford acyl complexes, rather than expected alkyl or aryl complexes. From these acyl complexes, many useful carbonyl compounds can be synthesized. In the following, these synthetic reactions are exemplified.

Reaction of allylic halides with transition metal complexes is quite easy due to the formation of the π-allylic complexes (97).

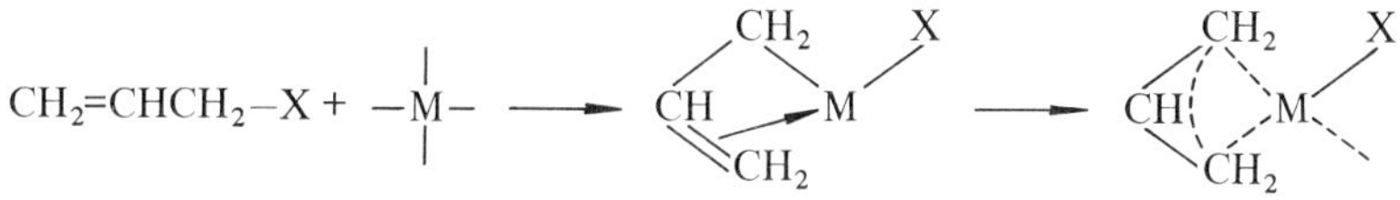

97

Allylic complexes, especially those of nickel obtained by the reaction of various allylic halides with Ni(CO)$_4$ are very useful compounds, although the very high toxicity of Ni(CO)$_4$ prevents its wide use as a synthetic reagent. [148—153]

An equilibrium exists between the π-allylnickel halide (98) and bis(π-allyl)nickel (99). [154]

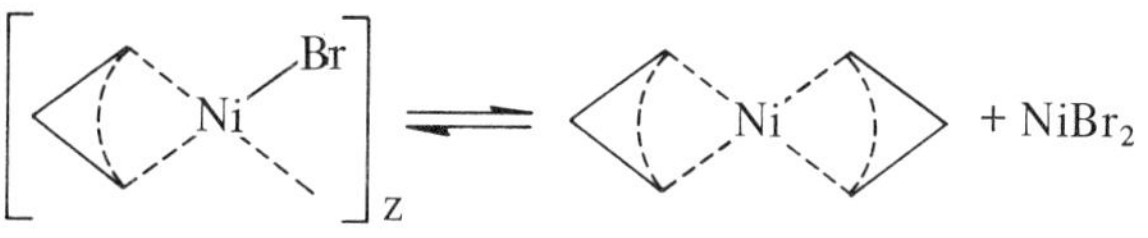

98   99

π-Allylnickel halides offer three types of useful synthetic reactions. They react with carbon monoxide, organic halides, and reactive carbonyl groups. Reaction of allyl halide with Ni(CO)$_4$ gave 3-butenoate (100) by carbon monoxide insertion at the allylic position of allylnickel halide. [155]

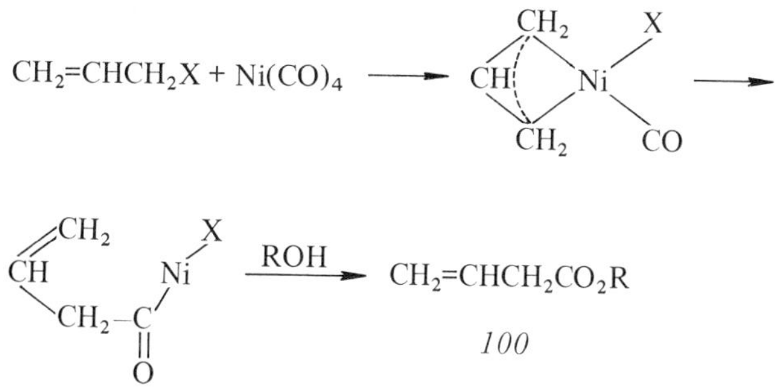

100

The same carbonylation reaction of allyl halides is possible using palladium chloride as a catalyst. [156, 157]

Carbon-carbon bond formation takes place by the reaction of π-allylnickel halides with allyl, alkyl, and aryl halides. Coupling with allylic halides is a useful synthetic method for 1,5-diene systems. Cyclic dienes can be synthesized by the intramolecular coupling of bis-allyl bromide by using $Ni(CO)_4$ in polar solvents such as DMF. No reaction takes place in non-polar solvents. The coupling might proceed via bis(π-allyl)nickel complex. Application of this coupling method to the preparation of twelve-, fourteen-, and eighteen-membered ring compounds has been reported. Yields of these ring compounds were in a range of 60—80% when high dilution technique was used to inhibit intermolecular coupling. [158]

$$
\begin{array}{l}
\text{CH}=\text{CHCH}_2\text{Br} \\
(\text{CH}_2)_n \qquad + \text{Ni(CO)}_4 \longrightarrow (\text{CH}_2)_n \\
\text{CH}=\text{CHCH}_2\text{Br}
\end{array}
\qquad
\begin{array}{l}
\text{CH}=\text{CHCH}_2 \\
| \\
\text{CH}=\text{CHCH}_2
\end{array}
$$

n = 6, 8, 12

Unsymmetric π-allyl systems attached to nickel have two positions for reaction and the cyclization does not always proceed at the terminal position. Usually a mixture of products is obtained, the distribution of which is determined by the ring size. This method of cyclization was also applied to the last step of the synthesis of humulene (101), a unique natural sesquiterpene. Reaction of the dibromotriene (102) with $Ni(CO)_4$ gave a mixture of products, from which humulene (101) was obtained in an overall yield of 10% after isomerization of the cis double bond by irradiation. [159]

102   101

Another example is the synthesis of dl-elemol. [160] Reaction of the dibromide (103) with an excess of $Ni(CO)_4$ in N-methylpyrrolidone gave 83% yield of a mixture of cyclized products which consisted of the ten-membered ring and the stereoisomers of the six-membered ring. Chromatographic separation gave the ester (104) in 32% yield, which was converted to dl-elemol (105) by treating with an excess of methylmagnesium bromide.

The intermolecular coupling of allylic groups is useful. Homo-coupling proceeds satisfactorily, but the cross-coupling of allylic groups does not always proceed selectively. Even in the reaction of allylnickel halide with allylic halide, a halogen-nickel exchange takes place and a mixture of all the possible coupling products is obtained. For example, the reaction of π-(2-methallyl)nickel bromide with allyl bromide produced all three possible coupling products in 95% yield and roughly the statistical distribution. [161]

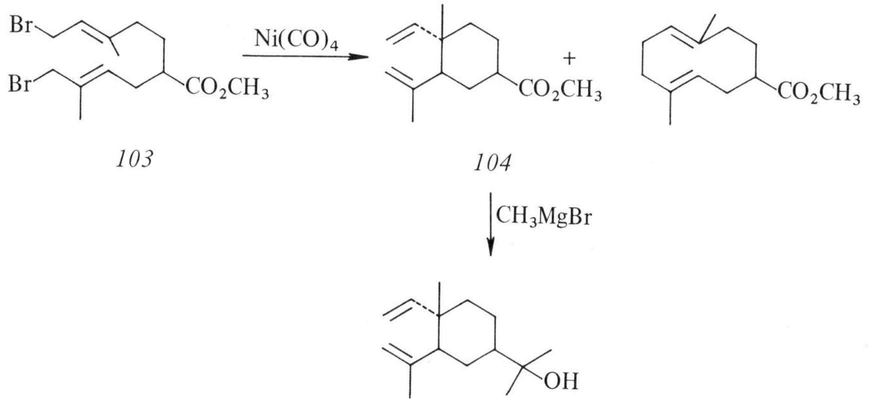

The structure of the allylic groups is crucial for selective cross-coupling. Chiusoli found that cross-coupling takes place selectively between the π-allylnickel halides having electron-withdrawing substituents, such as carboxy or cyano group and usual allylic halides; an ester of farnesoic acid *(106)* was synthesized by the following reaction. [162]

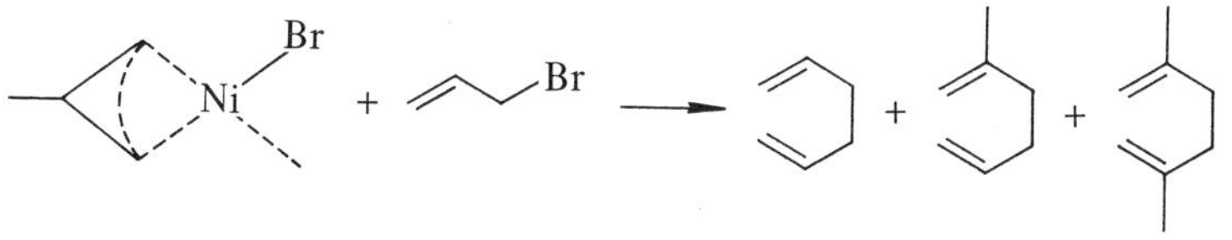

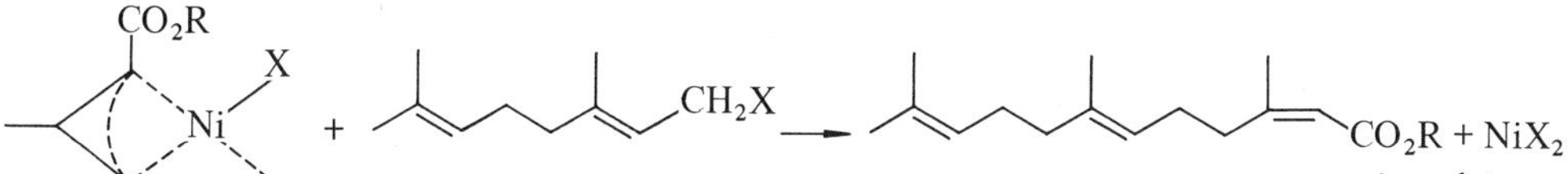

106

cis and *trans*

π-(1,1-Dimethallyl)nickel bromide *(108)* prepared from prenyl bromide *(107)* does not undergo self-coupling and is useful for the synthesis of isoprenoids. The reaction with 4-acetoxy-1-bromo-2-methyl-2-butene *(109)* gave geranyl acetate *(110)* in 60% yield. [163]

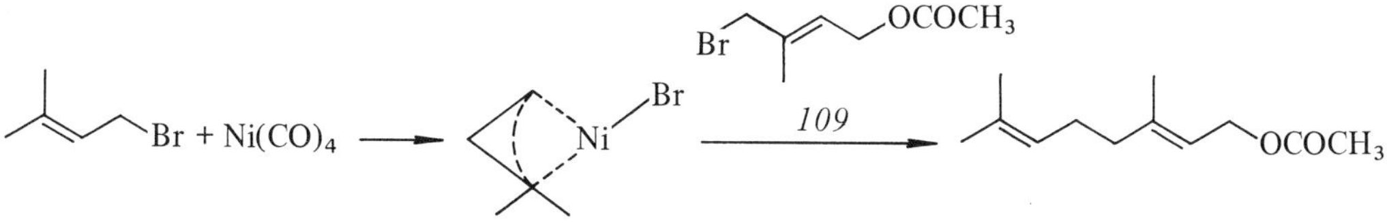

107   *108*   *109*   *110*

Isoprene unit can be further lengthened to farnesyl acetate *(112)* by the reaction of 4-acetoxy-1-bromo-2-methyl-2-butene *(109)* with the π-allylic complex formed from Ni(CO)$_4$ and geranyl bromide *(111)*.

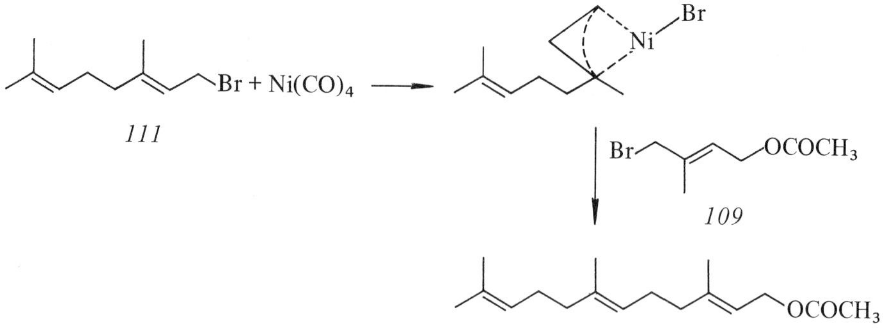

111

109

112

The reaction of the π-allylic nickel complexes with various alkyl, aryl, and vinyl halides proceeds in polar, coordinating solvents such as DMF, HMPA, and N-methylpyrrolidone. [164]

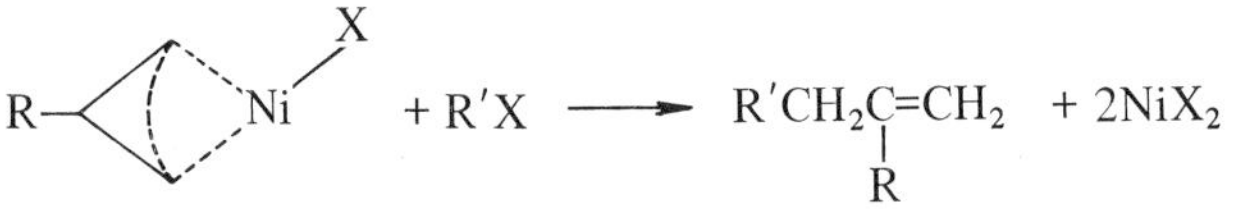

$$R-\underset{Ni}{\diamondsuit}^{X} + R'X \longrightarrow R'CH_2\underset{R}{C}=CH_2 + 2NiX_2$$

α-Santalene *(113)* was synthesized by the reaction of π-(1,1-dimethallyl)nickel bromide *(108)* and the tricyclic iodide *(114)*. [164]

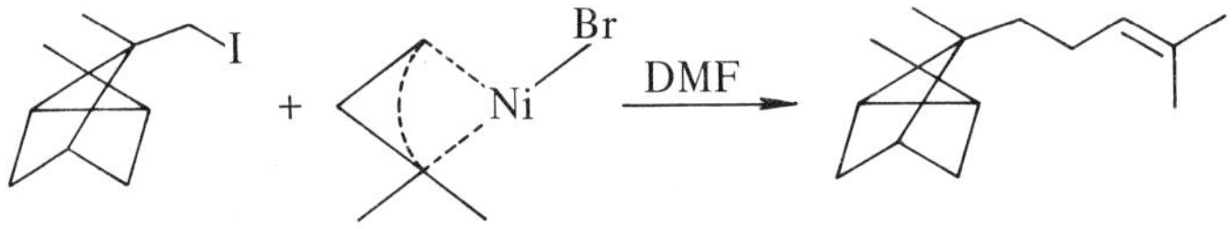

114   *108*   *113*

Similarly, β-santalene *(115)* was synthesized by the following sequences. [165]

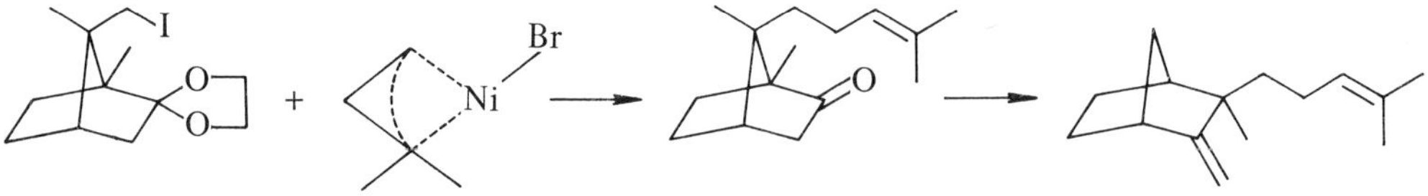

108   *115*

The coupling of aryl halides with π-allylnickel halides can be applicable to the synthesis of coenzyme Q. The reaction of π-(1,1-dimethallyl)nickel bromide *(108)* with the bromobenzene derivative *(116)* in HMPA at 60° gave 2,3-dimethoxy-5-methyl-6-prenyl-1,4-diacetoxybenzene *(117)* with 60% yield. Deacetoxylation and oxidation with ferric chloride gave the quinone coenzyme Q_1 *(118)* with 66% yield. [166]

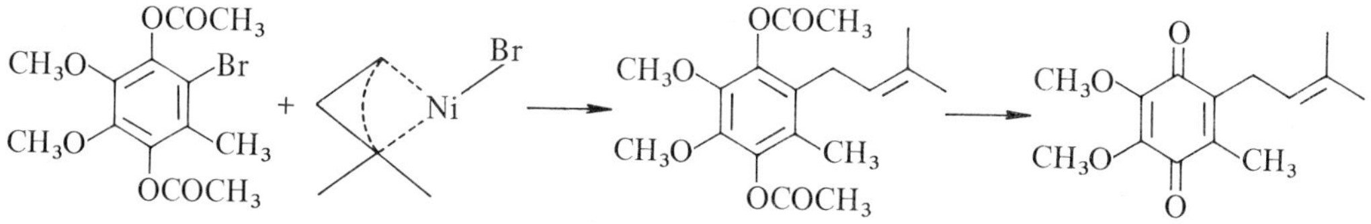

116   108   117   118

Similarly, the isoprene units can be lengthened by using the π-allylic nickel complex *(119)* to give coenzyme Q_{10} *(120)*.

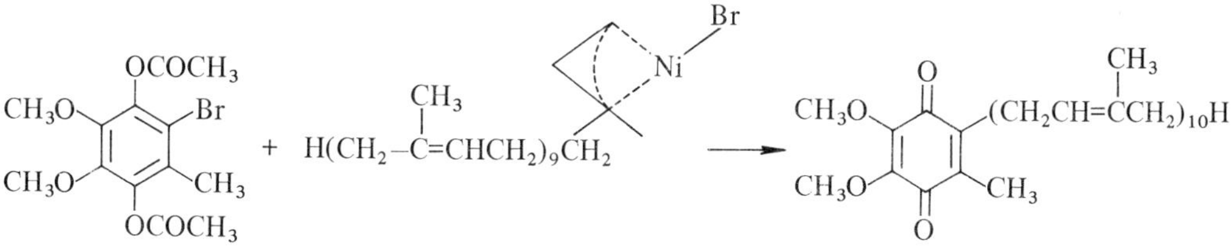

116   119   120

Synthesis of vitamin K *(123)* was carried out in the same way. The coupling of the π-allylic nickel complex *(121)* with 3-bromo-2-methyl-1,4-diacetoxy-naphthalene *(122)* in HMPA, followed by hydrolysis of the protecting group and oxidation gave vitamin K *(123)*. [166]

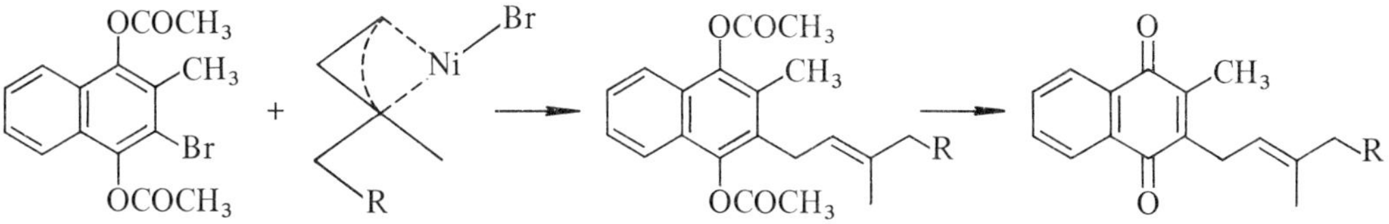

122   121   R=H, $C_{15}H_{31}$ 123

The nucleophilic character of the π-allylic moiety of nickel complexes is apparent by the Michael-type addition reaction to acrylonitrile. [167]

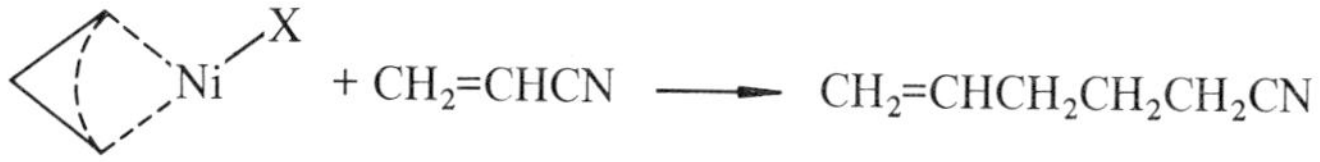

As a related reaction, the addition of π-allylnickel bromide to *p*-benzoquinone at room temperature to give the allylated hydroquinone *(124)* in a high yield has been reported. [168]

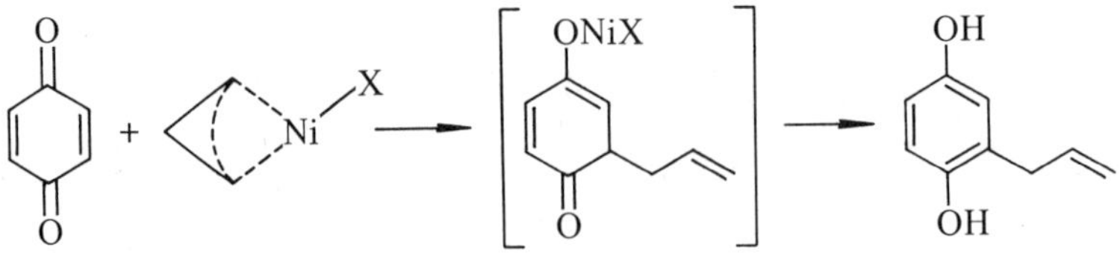

124

Methylbenzoquinone and 1,4-naphthoquinone underwent the same allylation reaction. The products were the quinones *(125)* and *(126)* rather than the hydroquinone.

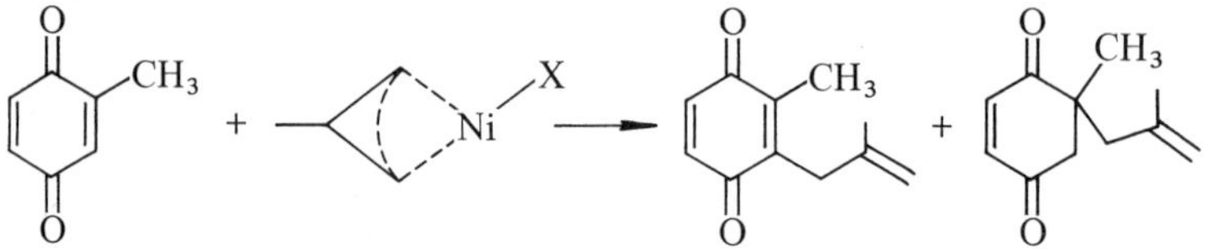

125   *126*

This reaction offers another synthetic route to vitamin K and coenzyme Q *(118)*. Usually the conventional syntheses utilizing acid catalyzed addition of phytol or polyprenyl alcohols to quinones, give rise to a mixture of products. The π-allylnickel method gives better selectivity than the conventional method for introducing allylic groups.

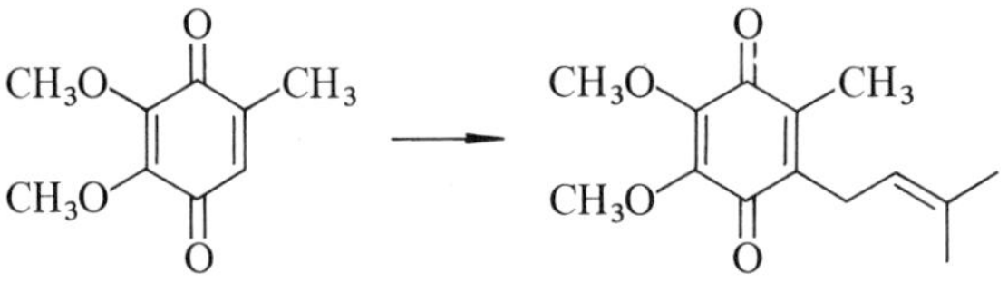

118

The π-allylic nickel complexes shown above are prepared by the oxidative addition of allylic halides to $Ni(CO)_4$. In general, the oxidative addition re-actions are usually possible with complexes coordinated with certain ligands, mostly carbon monoxide or phosphine. Zerovalent, metallic transition metals without ligands do not undergo smooth oxidative addition. This is one of the major differences between magnesium and transition metals. However, there are some exceptions. For example, metallic palladium reacts with allyl bromide to give π-allylpalladium bromide. [169] Also some aromatic aldehydes were subjected to attack by allyl and alkyl complexes formed *in situ* by the re-action of the corresponding alkyl or allyl halides with activated cobalt or nickel metal. [170] These metals were prepared in alcoholic solution, even in the presence of a small amount of water, by reduction of the salts with manganese-iron alloy. Certain weakly bonded ligands are necessary in order for the reaction to occur. Various types of compounds such as amides or nitriles are effective ligands. The overall reaction of benzaldehyde and allyl chloride can be shown by the following equation.

$$CH_2=CHCH_2Cl + PhCHO + M \xrightarrow{H_2O} PhCHOHCH_2CH=CH_2 + M(X)(OH)$$

This reaction is analogous to Grignard reaction. The product obtained from the reaction of crotyl bromide and benzaldehyde is the same branched isomer as can be obtained by Grignard reaction. The fact that the reaction takes place in a protic solvent indicates that a metal-carbon bond of the intermediate formed is quite stable to hydrolysis. This is markedly different from Grignard reagent. This reaction suggests an interesting property of transition metals. However, use of transition metal complexes in organic synthesis sometimes presents some difficulties. Some complexes are unstable and difficult to prepare. $Ni(CO)_4$, although a versatile reagent, is highly toxic. These disadvantages can be overcome if transition metals can be used in their metallic state as shown above. But so far the reaction of transition metals in the metallic state has given a rather poor yield of products. Therefore some efficient ways of activation are needed. This is one of the important problems to be solved in the future.

Iodobenzene reacted with $Ni(CO)_4$ in benzene to form benzoylnickel carbonyl iodide *(127)* as an intermediate complex. [171] Its thermal decomposition gave benzil *(128)* and stilbenediol diester *(129)*. An ester of benzoic acid was obtained by alcoholysis of the complex *(127)*.

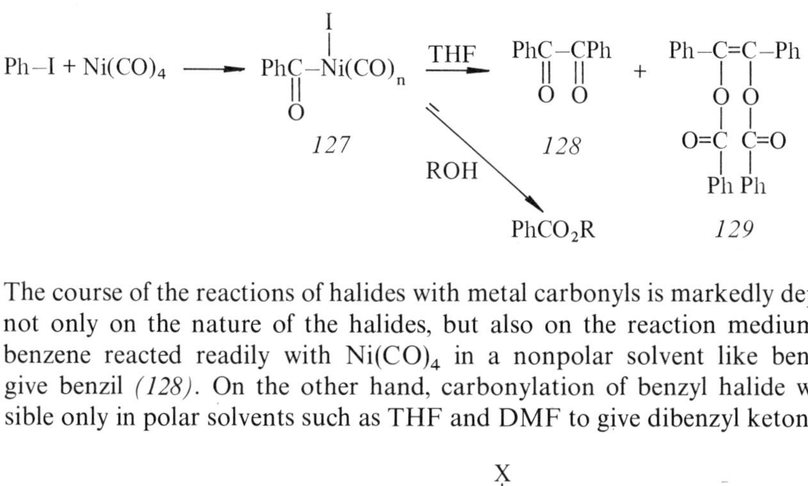

The course of the reactions of halides with metal carbonyls is markedly dependent not only on the nature of the halides, but also on the reaction medium. Iodobenzene reacted readily with $Ni(CO)_4$ in a nonpolar solvent like benzene to give benzil *(128)*. On the other hand, carbonylation of benzyl halide was possible only in polar solvents such as THF and DMF to give dibenzyl ketone *(130)*.

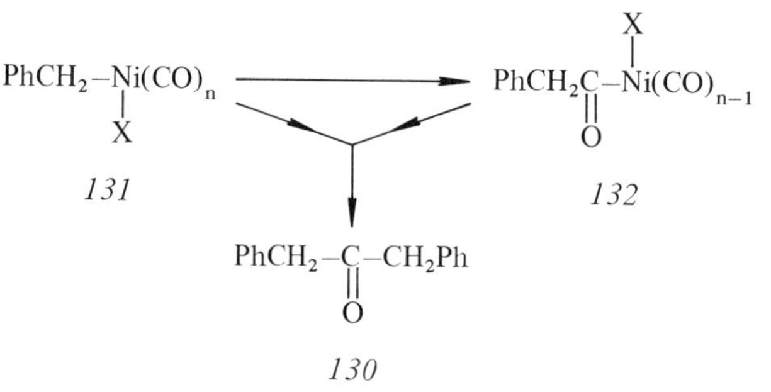

In benzene, only Friedel-Crafts reaction took place. Ketone formation in polar solvents can be explained by the coupling of the benzylnickel complex *(131)* and phenylacetylnickel complex *(132)*.

$Fe_3(CO)_{12}$ also reacted with iodobenzene in toluene to produce benzophenone. [172] Dibenzyl ketone *(130)* was formed from benzyl halide.

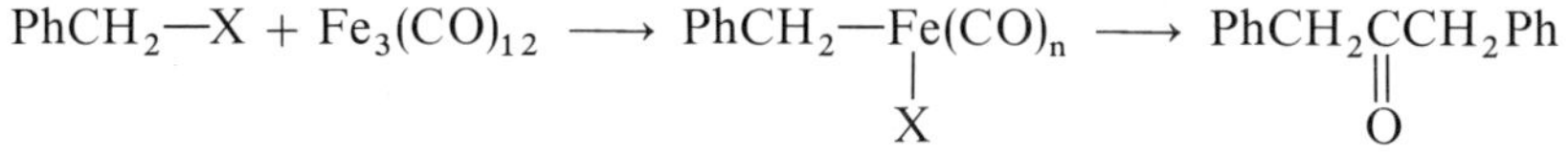

$$PhCH_2-X + Fe_3(CO)_{12} \longrightarrow PhCH_2-\underset{\underset{X}{|}}{Fe}(CO)_n \longrightarrow PhCH_2\underset{\underset{O}{||}}{C}CH_2Ph$$

130

α-Halo ketones are also reactive halides and undergo various interesting reactions with metal carbonyls. Furan formation took place by the reaction of α-bromo ketone with $Ni(CO)_4$. [173, 174] The products of the reaction were different depending on the solvents used. The reaction of phenacyl bromide *(133)* in DMF gave 2,4-diphenylfuran *(134)*. On the other hand, 1,4-diphenyl-1,4-diones *(135)* was obtained in THF. Cyclization of the latter gave 2,5-diphenylfuran *(136)*.

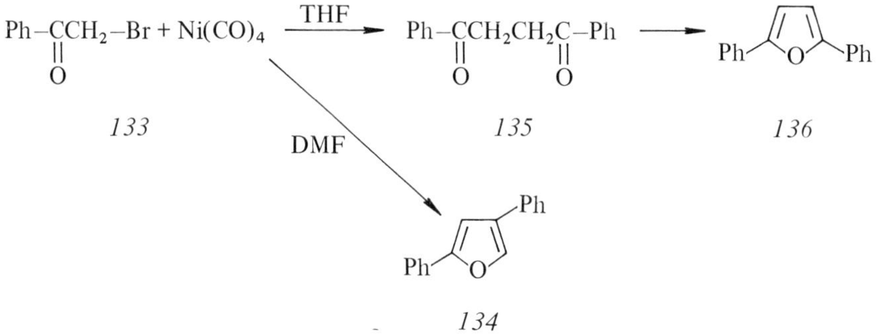

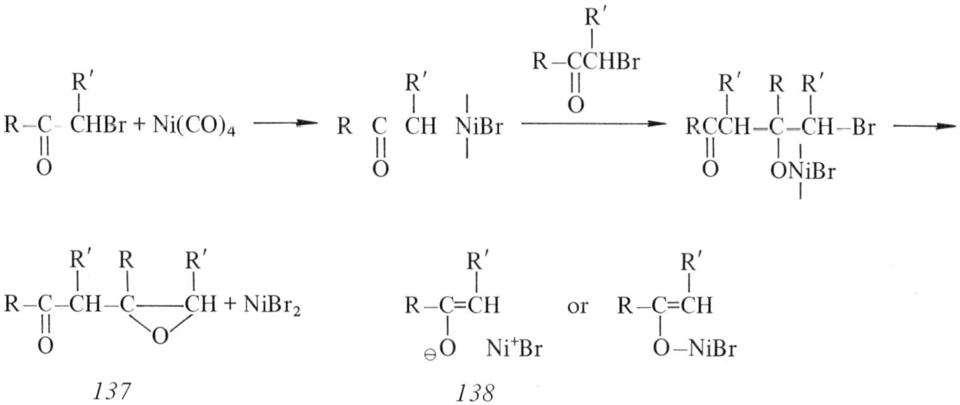

As an intermediate of the furan formation, the epoxy compound *(137)* was isolated. The reaction can be explained by the oxidative addition of the carbon-bromine bond to form a nickel-carbon bond followed by an intermolecular nucleophilic attack of the complex on the carbonyl of another molecule of the bromo ketone to give the epoxy ketone *(137)*. The fact that the reaction proceeds better in polar solvents suggests the polar nature of the intermediate complex formed by the oxidative addition of the carbon-bromine bond. Alternatively, the intermediate complex can be written as the enolate form *(138)*.

Actually, the reaction with bromomethyl t-butyl ketone in DMF gave 2-t-butyl-5,5-dimethyl-1,2-epoxyhexan-4-one *(139)* in 61% yield. Conversion of the epoxy ketone to the furan proceeds without help of the metal carbonyl. Unsymmetrically substituted furans are prepared by this method.

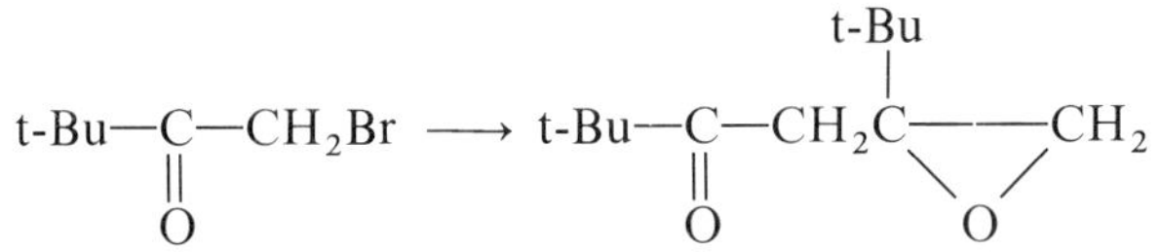

$$\text{t-Bu—C—CH}_2\text{Br} \longrightarrow \text{t-Bu—C—CH}_2\overset{\overset{\displaystyle \text{t-Bu}}{|}}{\text{C}}\text{———CH}_2$$

139

It should be pointed out that similar unsymmetrical furans were prepared with other products by Grignard and Reformatsky type reactions of α-bromo ketones using magnesium [175, 176] and zinc. [177] These reactions show similarity in reactivity to zerovalent nickel, magnesium, and zinc.

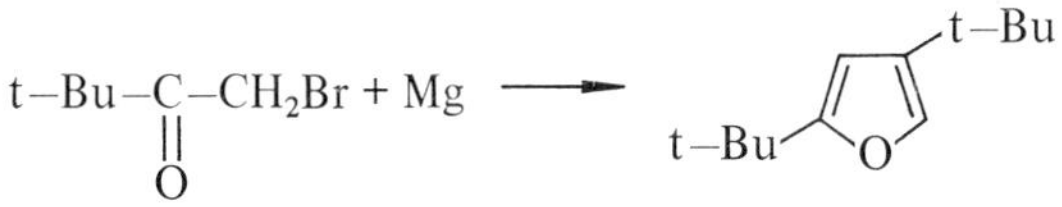

$$\text{t–Bu–C–CH}_2\text{Br + Mg} \longrightarrow$$

The reaction of α-halo ketones with $Fe(CO)_5$ in DMF gave the 1,4-diketones *(140)*, monoketones *(141)* and epoxy ketones *(142)*. [178] In this reaction, $Fe(CO)_5$ and $Ni(CO)_4$ show similar behavior.

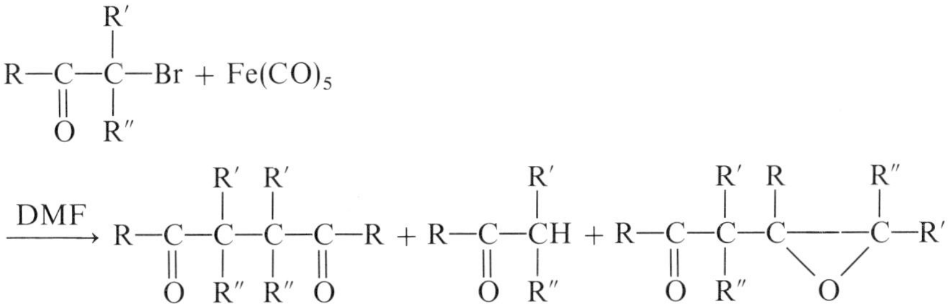

$$\overset{\overset{\displaystyle \text{R}'}{|}}{\text{R—C—C—Br + Fe(CO)}_5}$$

$$\xrightarrow{\text{DMF}} \text{R—C—C—C—C—R} + \text{R—C—CH} + \text{R—C—C—C———C—R}'$$

140   *141*   *142*

The ratio of these products changes with varying the structure of the halo ketones and reaction conditions. The following mechanism was proposed, in which the first step is the oxidative addition of the bromide to $Fe(CO)_5$ to form the σ-bonded complex *(143)*.

The reaction of α,α'-dibromo ketones and unsaturated molecules with the aid of $Fe_2(CO)_9$ provides a versatile method for the construction of various

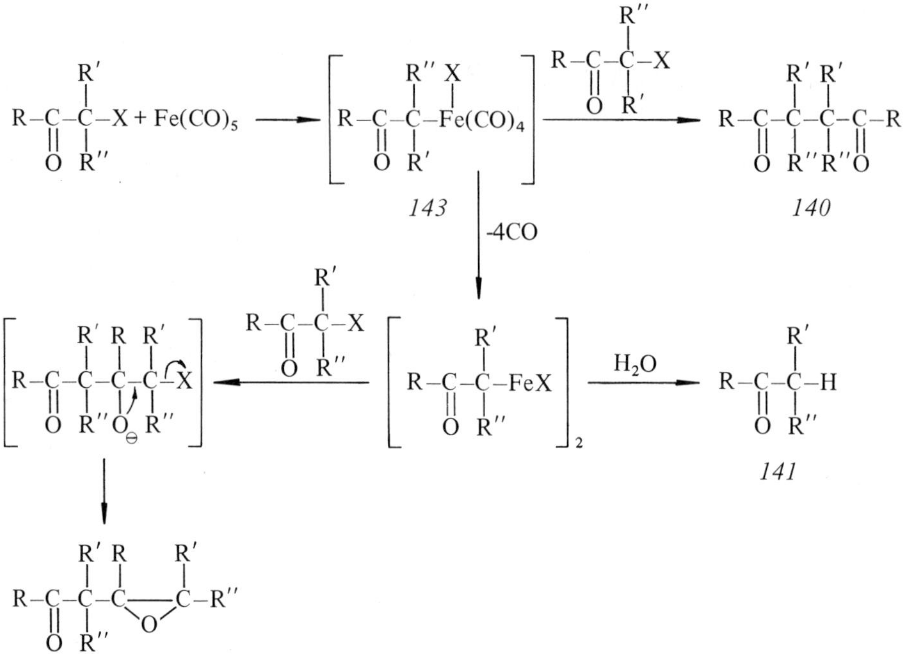

$$R-\underset{\underset{O}{\|}}{C}-\underset{\underset{R''}{|}}{C}-X + Fe(CO)_5 \longrightarrow \left[R-\underset{\underset{O}{\|}}{C}-\underset{\underset{R'}{|}}{C}-Fe(CO)_4 \right] \xrightarrow{\;R-\overset{\overset{R''}{|}}{\underset{\underset{O}{\|}}{C}}-\underset{\underset{R'}{|}}{C}-X\;} R-\underset{\underset{O}{\|}}{C}-\underset{\underset{R''}{|}}{\overset{\overset{R'}{|}}{C}}-\underset{\underset{R''}{|}}{\overset{\overset{R'}{|}}{C}}-\underset{\underset{O}{\|}}{C}-R$$

143 *140*

143 → (−4CO) →

$$\left[R-\underset{\underset{O}{\|}}{C}-\underset{\underset{R''}{|}}{\overset{\overset{R'}{|}}{C}}-Fe-X \right]_2 \xrightarrow{H_2O} R-\underset{\underset{O}{\|}}{C}-\underset{\underset{R''}{|}}{\overset{\overset{R'}{|}}{C}}-H$$

141

$$\left[R-\underset{\underset{O}{\|}}{C}-\underset{\underset{R''}{|}}{\overset{\overset{R'}{|}}{C}}-\underset{\underset{O}{|}}{\overset{\overset{R}{|}}{C}}-\overset{\overset{R'}{|}}{C}\overset{\curvearrowright}{-}X \right] \longleftarrow$$

$$R-\underset{\underset{O}{\|}}{C}-\underset{\underset{R''}{|}}{\overset{\overset{R'}{|}}{C}}-\overset{\overset{R}{|}}{C}\underset{O}{\diagdown}\overset{\diagup}{C}-R''$$

142

cyclic frameworks. The method is particularly useful for the synthesis of seven- and five-membered ketones. The reaction with conjugated dienes leads to 4-cycloheptenones *(144)* in moderate to high yields. [179] The cycloheptenones produced by this method can be converted into tropones *(145)* and tropolones *(146)* by bromination and dehydrobromination.

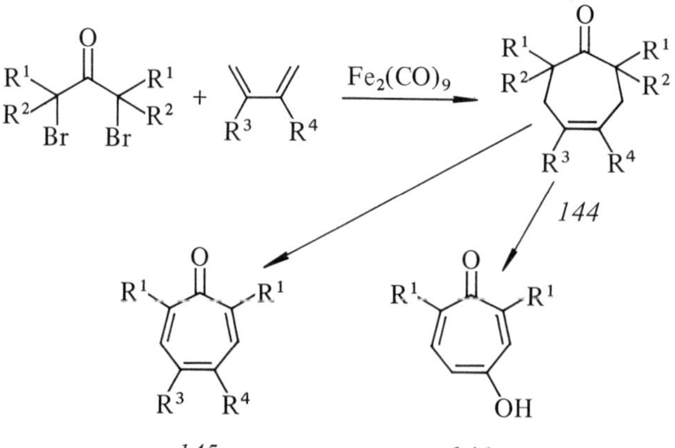

$$\underset{Br\;\;\;\;Br}{R^2-\underset{}{\overset{\overset{O}{\|}}{C}}-R^2} + \underset{R^3\;\;\;R^4}{\diagup\!\!\diagdown} \xrightarrow{Fe_2(CO)_9} \text{(cycloheptenone)}$$

144

145   *146*

4,5-Homotropones *(147)* and the corresponding homoaromatic cation(hydroxy-homotropylium ion) *(148)* can also be prepared from the seven-membered ketones. [180]

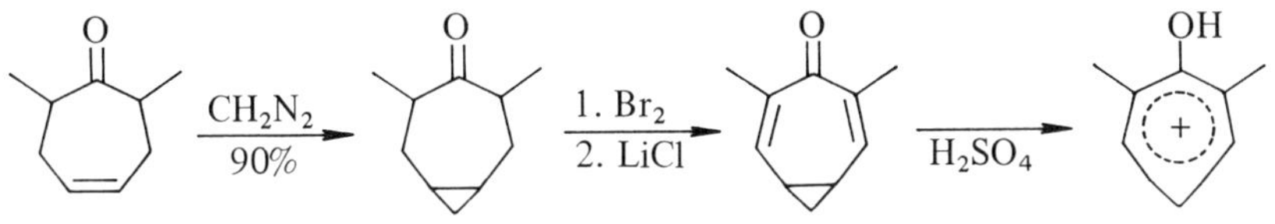

147   148

Furans are exceptionally good receptors of the reactive species generated from dibromo ketones and $Fe_2(CO)_9$. [181] The bicyclic compounds *(149)* thus obtained were transformed into tropone and tropolone derivatives. [182]

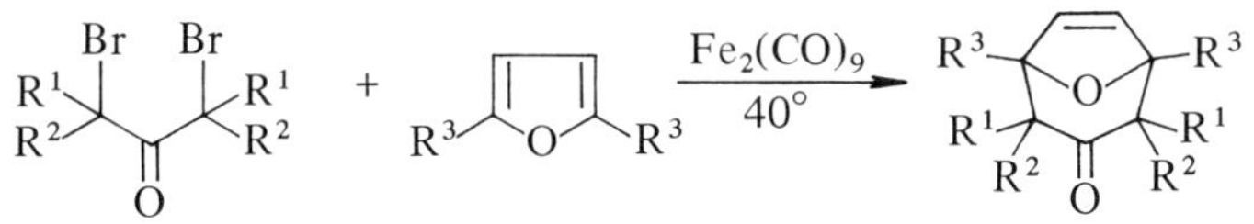

149

The cyclization reaction shown above is not possible with dibromides derived from methyl ketones. This limitation was removed by the use of certain poly-bromo ketones in place of the dibromo ketones. [186] This technique can be applied to the synthesis of tropane alkaloids. Reaction of tetrabromoacetone with N-carbomethoxypyrrole as the receptor gave the bicycloketone *(150)*, which was easily converted into tropine *(151)*.

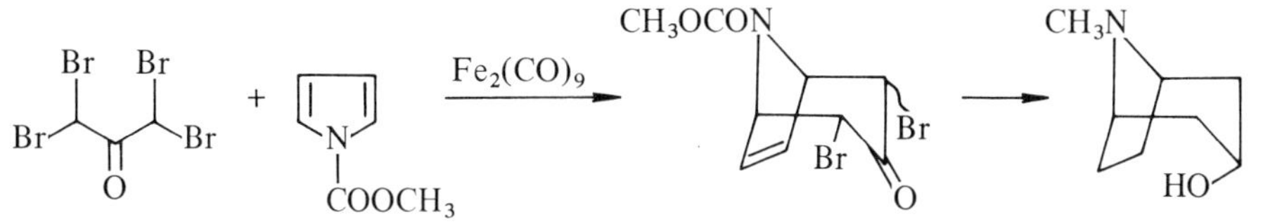

150   151

Similar reaction of dibromo ketones with enamines gave rise to labile β-aminocyclopentanones. Since elimination of the amine moiety therefrom takes place readily, the cyclopentenones *(152)* can be obtained in a high yield by a

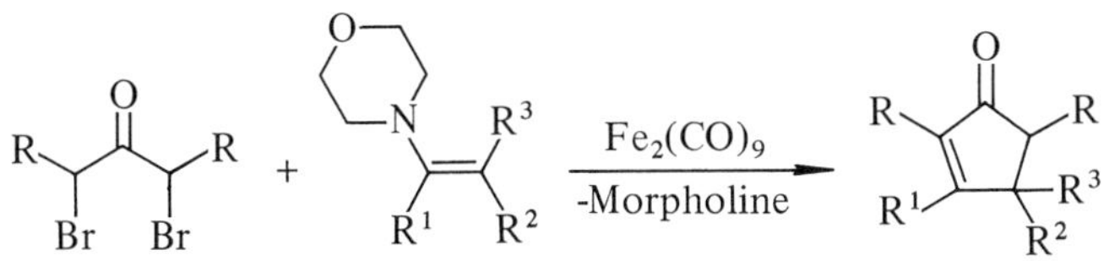

152

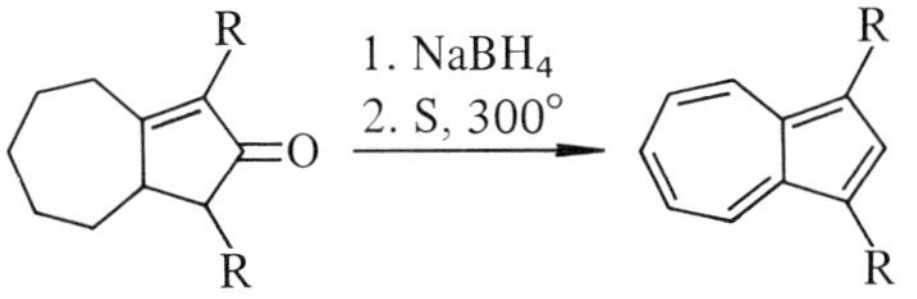

153   154

single-flask procedure. [183] The reaction was extended to the synthesis of azulene *(154)*. The cyclocoupling reaction of dibromo ketones with cyclo-heptanone enamine gave the bicyclo[5,3,0]decenone *(153)*. Reduction of the ketone and the treatment with sulfur effect dehydration and dehydrogenation to give the azulenes *(154)*.

The reaction with aromatic olefins such as styrene, stilbene, and indene also gave five-membered rings. [184] *cis-β*-Deuteriostyrene entered into the cycloaddition in a stereospecific manner.

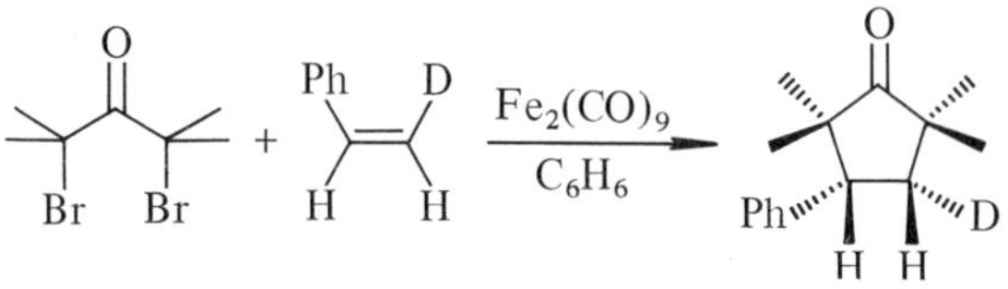

In addition to the above-mentioned carbon-carbon double bonds, a carbon-oxygen double bond of carboxamides reacts with the active three-carbon unit. DMF reacted with dibromo ketones in the presence of $Fe_2(CO)_9$ to give the 3(2H)-furanones *(155)* with elimination of dimethylamine. [185] Dimethyl-acetamide, dimethylbenzamide, and N-methylpyrrolidone can be used in place of DMF. The method was applied to the synthesis of 4-methylmuscarine *(156)*.

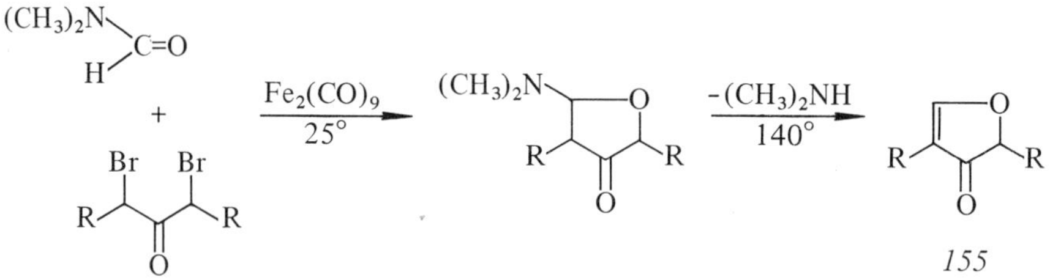

155

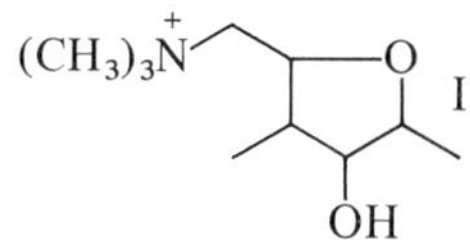

156

Mechanistic studies of the dibromo ketone-$Fe_2(CO)_9$ reaction showed that the first step of the reaction is a two-electron reduction of the dibromo ketone with the zerovalent iron (or oxidative addition of the carbon-bromine bond onto the zerovalent iron complex) forming the iron enolate *(157)*. [187] The second step is an S_N1 type or iron ion assisted elimination of the second bromide ion to give the oxyallyl intermediate *(158)*. This species in turn undergoes the cyclocoupling reactions shown above. Several intramolecular reactions that support this hypothesis have been observed. Treatment of bis(bromobenzyl) ketone with $Fe_2(CO)_9$ gave 1-phenyl-2-indanone *(159)*. The cyclization can be explained by electrocyclization of the oxyallyl intermediate *(158)* followed by aromatization. The oxyallyl species was trapped intermolecularly by added furan to give a high yield of the bicyclic adduct *(160)*.

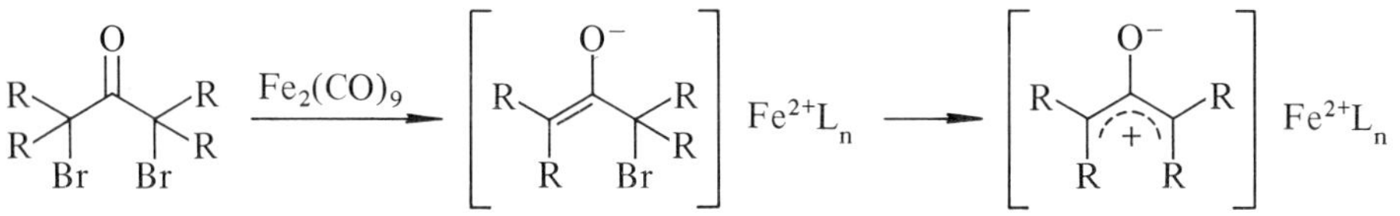

157   *158*

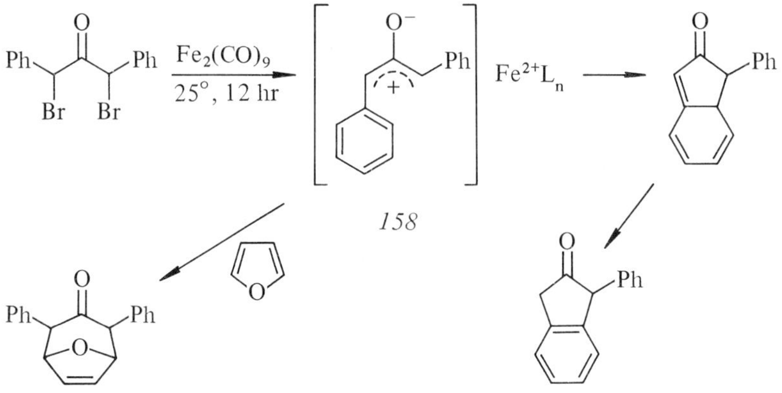

158

160   *159*

Reductive rearrangement of bis(bromoneopentyl) ketone *(161)* afforded 2-tert-butyl-3,3,4-trimethylcyclobutanone *(163)*. The smooth conversion to the four-membered ketone would involve the neopentyl-type rearrangement of di-tert-butyloxyallyl *(162)* and subsequent cyclization.

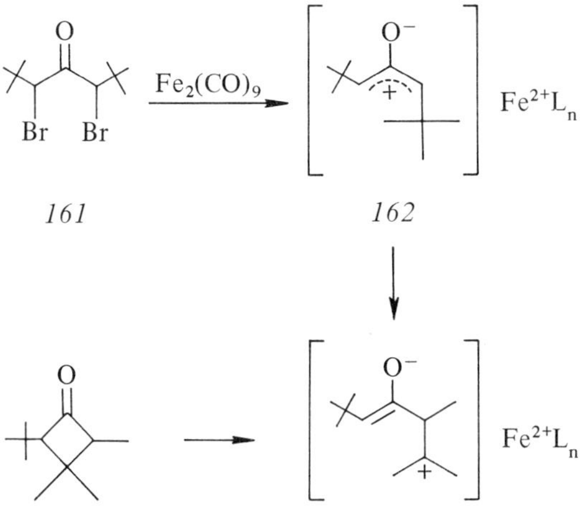

161   *162*

163

The following skeletal change can be explained in terms of the debromination followed by the known cationic [1a,4s] sigmatropic rearrangement. [188]

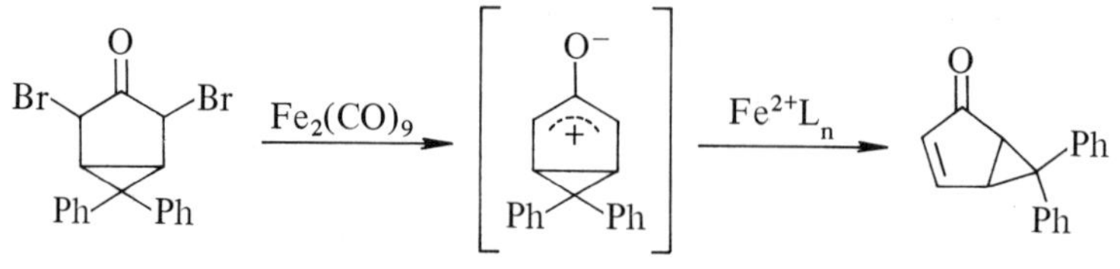

In the reactions giving five- and seven-membered rings described above, one role of the iron complex is to assemble the reactants for an intermolecular coupling reaction.

It is worthwhile, at this time, to compare the reaction of α,α'-dibromo ketones with a carboxamide in the presence of $Fe_2(CO)_9$ and a zinc-copper couple. [189] Debromination of α,α-dibromo ketones with zinc-copper couple in DMF gave 2-dimethylamino-4-methylene-1,3-dioxolane (165) in a high yield. In this reaction, formation of the oxyallylzinc intermediate (164) is assumed which has a 1,3-dipolar nature and reacts with the carbon-oxygen double bond of the carboxamide to give the 1,3-dioxolane (165). Unlike the intermediate of the iron carbonyl reaction, the reaction takes place at the oxygen and carbon termini of the 1,3-dipolar system, rather than at the carbon and carbon termini. As postulated in the Reformatsky reaction, zinc attacks initially at the oxygen of the bromo ketones to form a zinc enolate. The dioxolane (165) is converted to the furanone (155) by acid treatment.

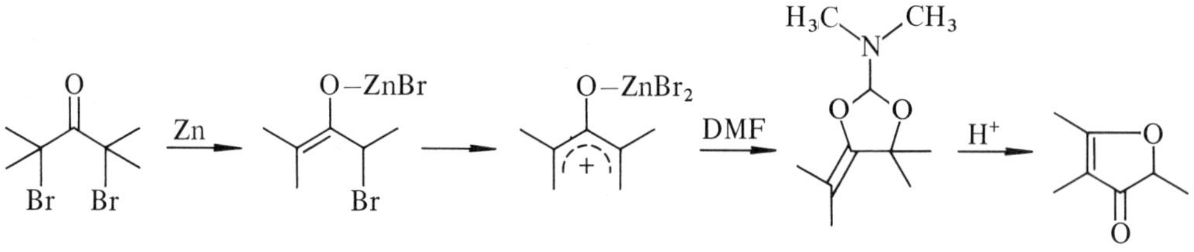

It is worthwhile to note here that a free radical mechanism was proposed for the addition reaction of phenacyl bromide to 1,1-diphenylethylene to give 56% yield of 3,3-diphenylpropyl phenyl ketone in the presence of a zinc-copper couple. [190]

$$PhCOCH_2Br + CH_2\!=\!CPh_2 \xrightarrow[DMSO]{Zn} PhCOCH_2CH_2CHPh_2$$

Certain metal carbonyls are very useful for removing halogens from poly-halogenated compounds. In this reaction, again the oxidative addition step seems to be important, even though there is no direct evidence. Removal of gem-chlorine atoms from polyhalogenated compounds is explained to proceed via unstable carbene transition metal complex formation. [191] The formation of the products shown below was explained via carbene formation followed by coupling.

$$Ph_2CCl_2 + 2Fe(CO)_5 \longrightarrow Ph_2C=CPh_2 + 2FeCl_2 + 10CO$$

$$Ph_2C=CBr_2 + Fe_3(CO)_{12} \longrightarrow (Ph_2C=C=C=CPh_2)Fe_2(CO)_6$$

For the reaction of gem-dihalides with $Fe(CO)_5$, a different mechanism has been proposed by Alper. [178] Treatment of 9,9-dibromofluorene *(166)* with $Fe(CO)_5$ gave 9,9'-dibromobifluorenyl *(167)* and the olefin *(168)*. The formation of these products was explained by the following mechanism which does not involve carbene species.

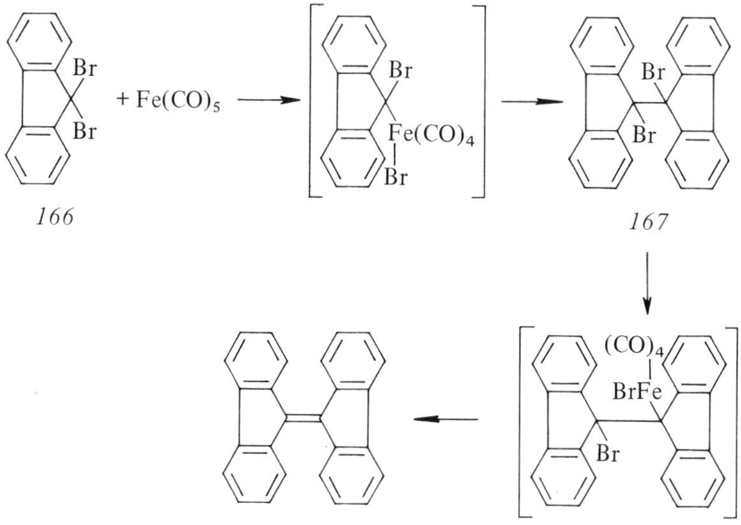

Dehalogenation of dichlorocyclobutene *(169)* with $Fe_2(CO)_9$ gave the cyclobutadiene complex of iron carbonyl *(170)*. [192] The complex is coordinatively saturated and stable. Stabilization of short-lived organic species such as cyclobutadiene and carbene by complex formation is another useful property of transition metal complexes. Isolation of the stable cyclobutadiene complex has opened a unique chemistry of cyclobutadiene, by which several interesting organic compounds have been synthesized.

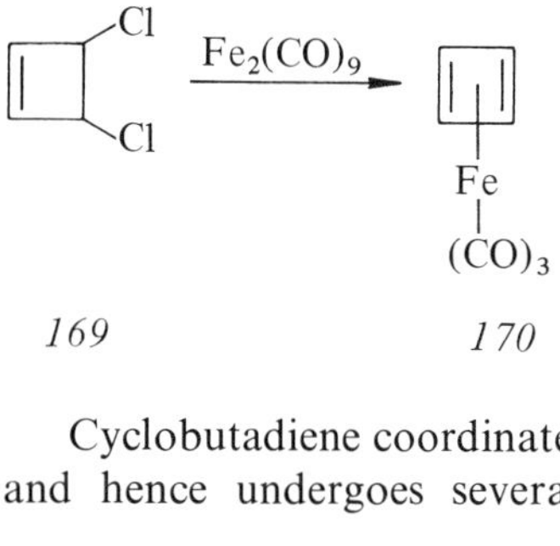

Cyclobutadiene coordinated to the zerovalent iron shows aromatic characters and hence undergoes several typical aromatic substitution reactions. [193]

Examples include Friedel-Crafts acetylation, Vilsmeier formylation, and chloro-methylation. Grignard reagent also reacts with the formylated cyclobutadiene complex. These reactions of the complexed cyclobutadiene proceed without decomposition of the complex itself.

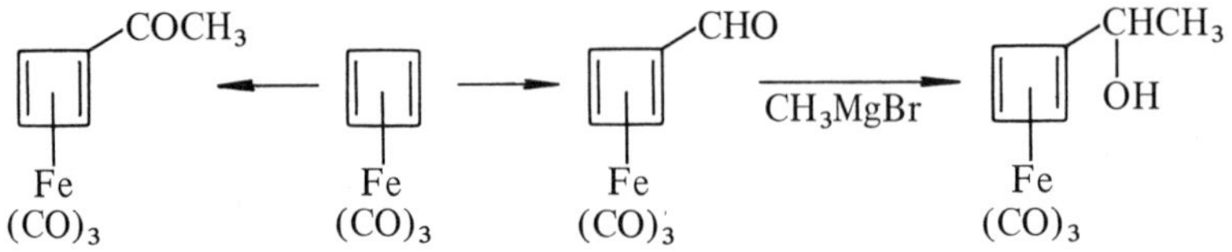

170

These facile substitution reactions are possible by the stabilization of the inter-mediate π-allylic iron carbonyl cation complex *(171)*. The stability of the π-allylic iron cationic complex *(172)* which has a structure similar to the complex *(171)* is known. [194]

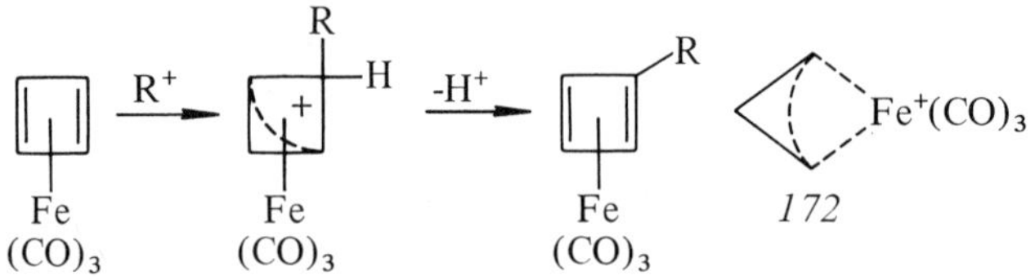

Also hydroxymethylcyclobutadiene can be converted into the stable cationic complex *(173)* by the action of an acid. [195]

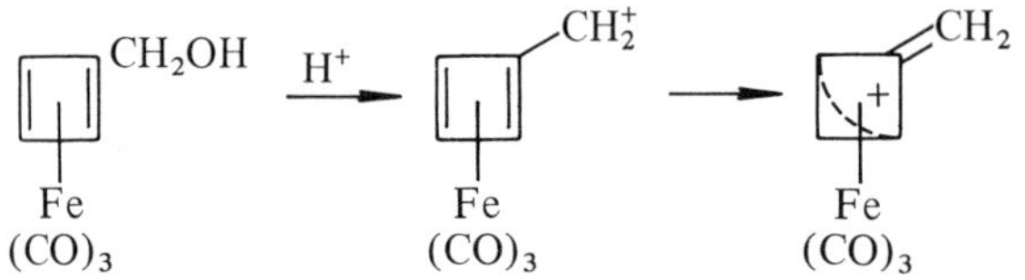

173

When the low valent iron in the complex *(170)* is oxidized with oxidizing agents such as Ce^{4+} ion to a higher valence state, which has no ability to sta-bilize cyclobutadiene, the cyclobutadiene is liberated from the iron and then adds immediately onto other unsaturated compounds. [196, 197] Several inter-esting compounds which are difficult to synthesize by other means were prepared by this method. Oxidation of the optically active (1-ethyl-2-methylcyclobutadiene)-iron tricarbonyl *(174)* with Ce^{4+} ion in the presence of maleic anhydride prod-uced the optically inactive bicyclo[2,2,0]hexene *(175)*. [198] This result of racemization clearly shows that free cyclobutadiene is involved in the cyclo-addition reaction.

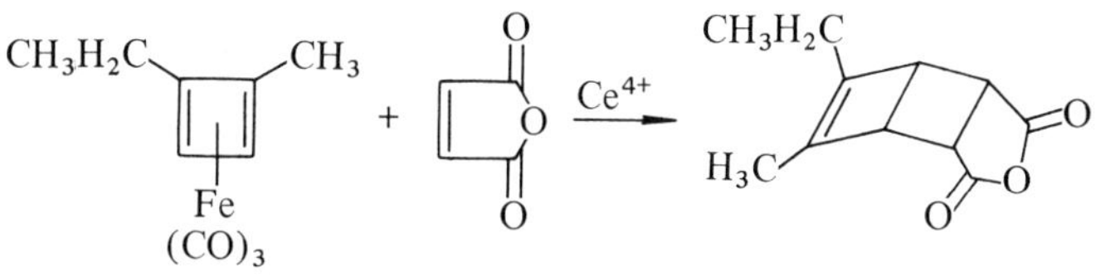

174   *175*

The addition reaction of an acetylenic compound gave the Dewar benzene *(176)*. [199, 200]

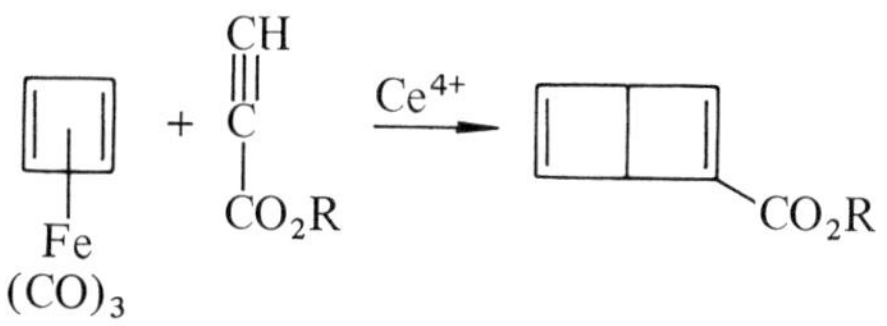

170   *176*

Benzoquinone adds to cyclobutadiene easily. Hypostrophene *(177)* was synthesized from the addition product by the following sequence. [201]

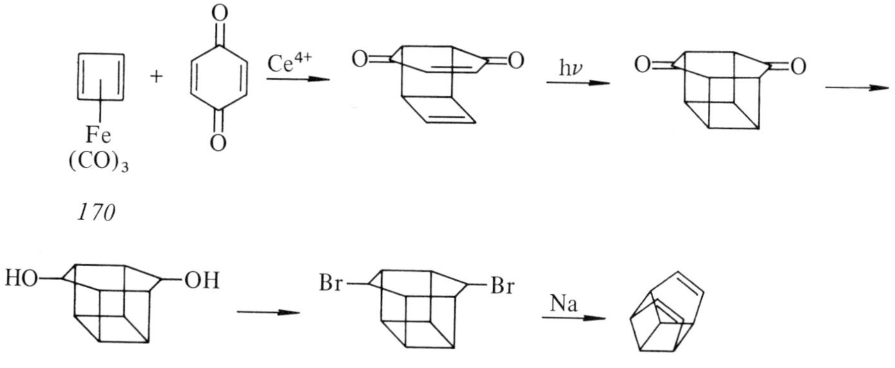

170

177

Similarly, the addition product *(179)* was obtained by the reaction of dibromobenzoquinone *(178)*. From this adduct, cubane *(180)* has been prepared by photo-induced cycloaddition, Favorskii reaction, and decarboxylation. [202]

Liberation of coordinated carbon monoxide by irradiation from (cyclobutadiene)iron tricarbonyl *(170)*, like Fe(CO)$_5$, gave a coordinatively unsaturated complex, which then undergoes various cycloaddition reactions. A ketal of tropone *(181)* reacted with the cyclobutadiene complex *(170)* under irradiation to give the tricyclic compound *(182)*, which is still stabilized by the coordination to the low valent iron. The iron was removed by the oxidation with Ce^{4+} ion. Photo-induced intramolecular cycloaddition reaction of the liberated

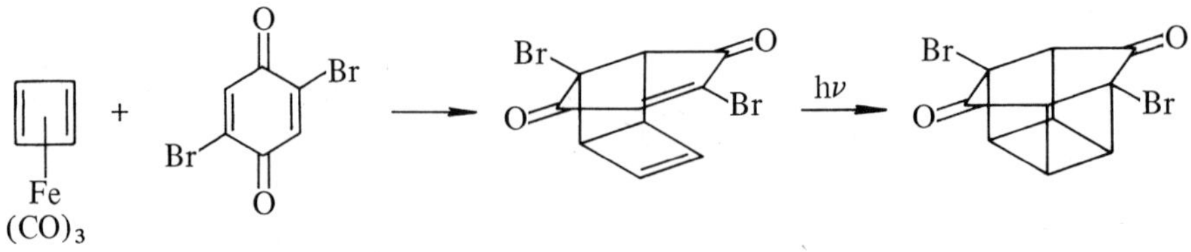

170 178 179

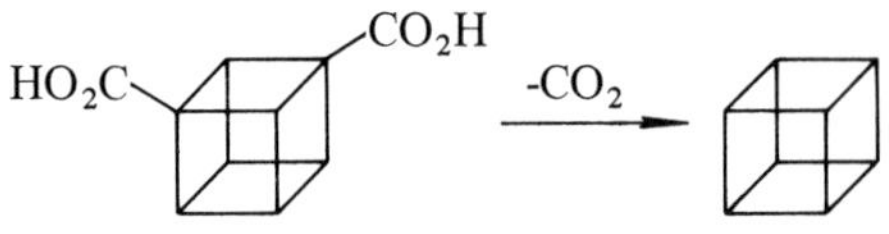

180

tricyclic compound and subsequent acid treatment afforded homopentaprisma-none *(183)*. [203]

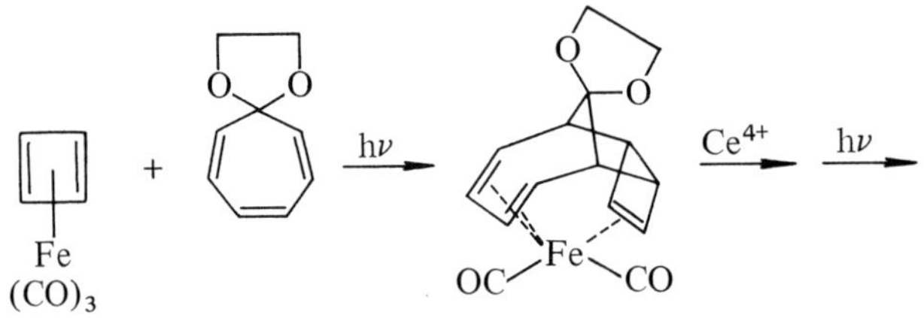

170 181 182

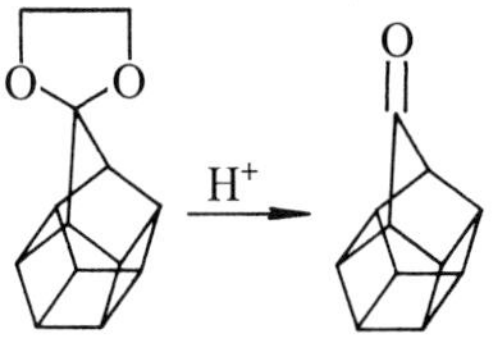

183

Addition to diazanorcaradiene *(184)* gave the tetracyclic azo compound *(185)*, which was then converted to bicyclo[5,2,0]nonatriene *(186)*. [204]

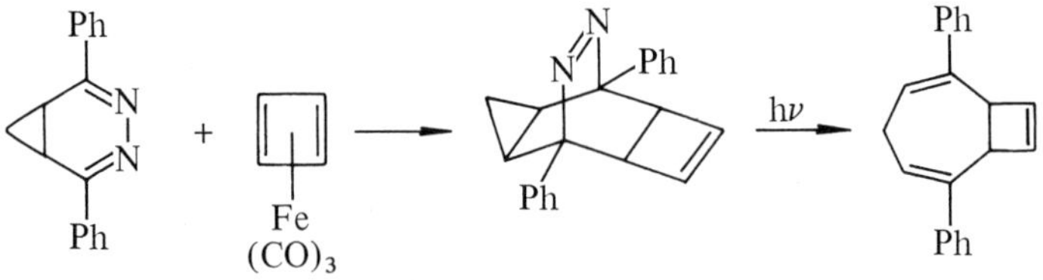

184 170 185 186

6. Miscellaneous Examples of Oxidative Addition Reactions

In the preceding sections, the most widely observed oxidative addition reactions involving various bonds containing carbon and hydrogen atoms are described. In this section, miscellaneous examples involving other bonds will be described briefly, although their use in organic synthesis is rather limited.

Similar to acyl halides, oxidative addition of sulfonyl chlorides is known. Treatment of sulfonyl chloride with $Fe(CO)_5$ at temperatures below $-20°$ gave the complex *(187)* with sulfur-chlorine bond cleavage. Further reaction of the complex produced the α-disulfone *(188)* by coupling. [178]

$$RSO_2Cl + Fe(CO)_5 \longrightarrow RSO_2\underset{\underset{Cl}{|}}{Fe}(CO)_4 \longrightarrow RSO_2SO_2R$$

$$\qquad\qquad\qquad\qquad 187 \qquad\qquad\qquad 188$$

Cationic metal-alkyl complexes *(189)* can be prepared by the oxidative addition reaction of unsaturated complexes of iridium, rhodium, platinum, and molybdenum with carbocation reagents such as methylfluorosulfonate or trimethyloxonium hexafluorophosphate. [205, 206]

$$IrCl(CO)(PPh_3)_2 + CH_3{-}SO_3F \longrightarrow [IrCl(CO)(PPh_3)_2(CH_3)]^+(SO_3F)^-$$

$$\qquad\qquad\qquad\qquad\qquad\qquad\qquad\qquad\qquad 189$$

SO_3F anion has a low tendency to coordinate to metals and can be displaced at room temperature with another anion.

$$[IrCl(CO)(PPh_3)_2(CH_3)](SO_3F) + Et_4NCl$$
$$\longrightarrow IrCl_2(CO)(PPh_3)_2(CH_3) + (Et_4N)(SO_3F)$$

Oxidative addition of nitrosyl chloride to an arylazo complexes of molybdenum gave a nitrosyl complex. [207]

$$(\pi\text{-}C_5H_5)Mo(CO)_2(N_2Ph) + NO{-}Cl \longrightarrow (\pi\text{-}C_5H_5)Mo(NO)(N_2Ph)Cl + 2CO$$

Certain examples of oxidative addition involving the cleavage of bonds containing oxygen atoms are known. The peroxide bond of benzoyl peroxide or dialkyl peroxide is cleaved and adds to a zerovalent nickel complex. [115]

$$RO{-}OR + NiL_n \longrightarrow RO\underset{\underset{L_n}{|}}{-}Ni{-}OR$$

Various *o*-quinones undergo the following addition reactions. [209—211]

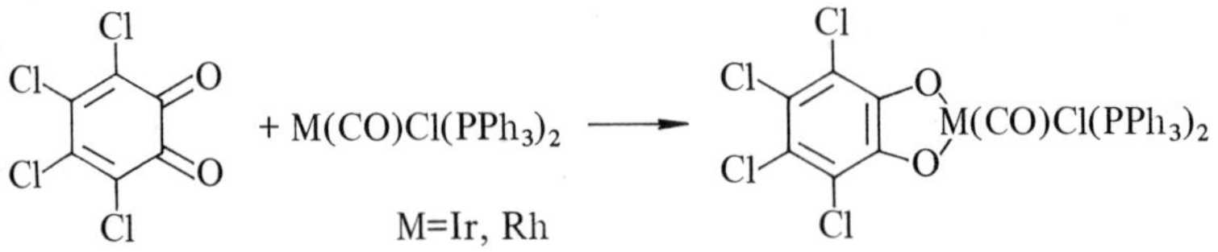

Oxygen molecule adds to the Vaska complex [208] and other complexes. Although the chemistry of the oxygen complexes and their reactions have not been elucidated clearly, the oxygen addition is important and interesting in connection with oxygen activation with metal catalysts.

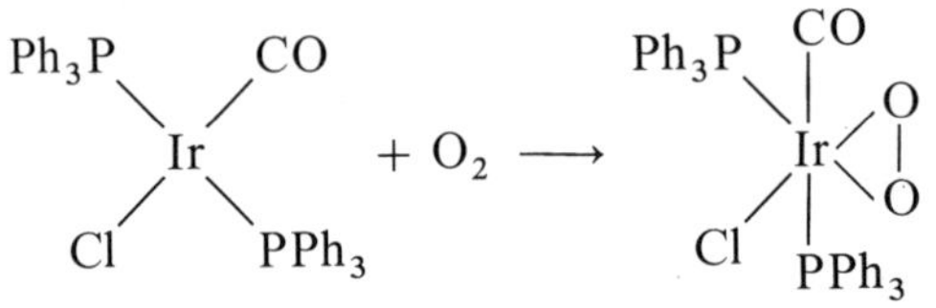

The high reactivity of the coordinated oxygen was shown by the facile formation of benzoyl peroxide in a good yield by the reaction of a dioxygen complex of nickel *(190)* with benzoyl bromide at $-70°$. [213]

$$NiO_2(t\text{-BuNC})_2 + 2\,PhCOBr \longrightarrow Ph-\underset{\underset{O}{\|}}{C}-O-O-\underset{\underset{O}{\|}}{C}-Ph + NiBr_2(t\text{-BuNC})_2$$
$$190$$

Further studies are necessary on the chemistry of oxygen complexes.

Certain diaryl disulfide undergo oxidative addition by sulfur-sulfur bond cleavage with the Vaska complex to form the binuclear product *(191)* by the loss of triphenylphosphine. [212]

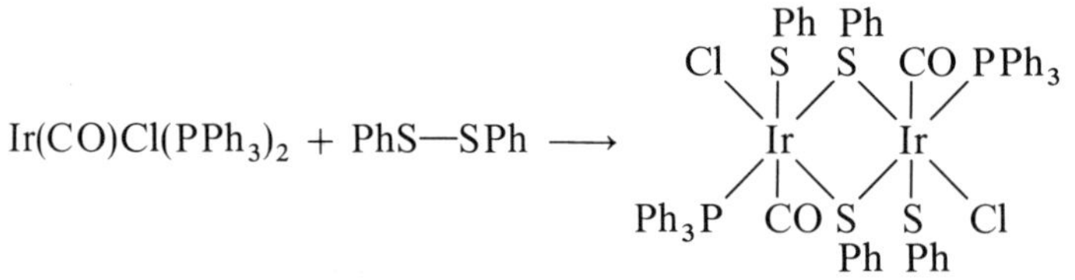

$$Ir(CO)Cl(PPh_3)_2 + PhS-SPh \longrightarrow$$

191

σ-Bond formation by an oxidation reaction takes place with a few complexes at higher valence state. The five-coordinate d^7 complex of bivalent cobalt, $Co(CN)_5^{3-}$, undergoes the following type of oxidation reaction with various covalent molecules [δ]. A free radical nature of the oxidation was proposed.

$$2\,Co(CN)_5^{3-} + X-Y \longrightarrow Co(CN)_5X^{3-} + Co(CN)_5Y^{3-}$$
$$X-Y = H-H, Br-Br, HO-OH, CH_3I, I-CN$$

Bivalent chromium compounds also undergo similar oxidation reactions with organic halides to give σ-bonded compounds. [214—215]

7. Oxidative Metal-Metal Bond Cleavage

Transition metal complexes are classified into mononuclear and polynuclear (di-, trinuclear etc.) complexes. Depending on the effective atomic number of metal and the kind of ligands, the complexes can be mononuclear like $Fe(CO)_5$ or dinuclear like $Co_2(CO)_8$. The characteristic feature of certain dinuclear metal complexes is that they have a metal-metal bond, which is cleaved oxidatively by the reaction of some covalent molecules to form σ-bonded complexes. The reaction offers another way of σ-bond formation and is useful for organic synthesis.

$$-M-M- + X-Y \longrightarrow -M-X + -M-Y$$

The most well-established example is the reaction of $Co_2(CO)_8$ with hydrogen. This reaction is the first step of the catalysis in the oxo reaction to form cobalt hydride *(192)*. A similar reaction is the cobalt carbonyl-catalyzed hydrosilylation reaction, in which hydrosilanes behave similar to hydrogen in hydrogenation reaction.

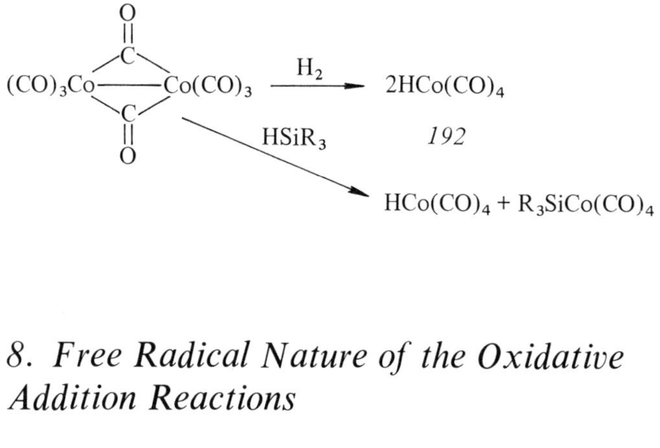

8. Free Radical Nature of the Oxidative Addition Reactions

The elucidation of the nature of the metal bonds formed by various oxidative addition reactions discussed so far is an important problem. In some oxidative addition reactions, the metals behave as nucleophiles and the reactions are explained by a S_N2 mechanism (p. 14). Also a free radical mechanism has been proposed in a few cases. [216—219] The mechanism is still controversial and needs further critical investigation. In the following, studies on the oxidative addition are surveyed from a viewpoint of free radicals.

Osborn showed that the oxidative addition to a univalent iridium complex proceeds by a free radical mechanism. [216] Many alkyl halides were found to undergo addition to *trans*-$Ir(CO)Cl[P(CH_3)_3]_2$ at greatly enhanced rates if a small quantity of molecular oxygen or a radical initiator was present. The presence of a small quantity of a radical scavenger such as duroquinone retarded the addition. Oxidative addition of optically active ethyl α-bromopropionate to the complex proceeded with racemization and was inhibited by galvinoxyl. [217] Based on these experimental results, the following radical mechanism was proposed for the oxidative addition.

$$Ir^{1+} + Q\bullet \longrightarrow Ir^{2+}-Q$$
$$Ir^{2+}-Q + R-Br \longrightarrow Br-Ir^{3+}-Q + R\bullet$$
$$Ir^{1+} + R\bullet \longrightarrow Ir^{2+}-R$$
$$Ir^{2+}-R + R-Br \longrightarrow Br-Ir^{3+}-R + R\bullet$$
$$Q = \text{initiator}$$

However, methyl iodide reacted extremely rapidly with the iridium complex even in the presence of the radical inhibitor, indicating that an alternative pathway, probably an S_N2 type process, may be operative for certain substrates. The formation of a free radical as an intermediate in the oxidative addition of alkyl halides to a zerovalent platinum complex was reported. [218]

$$R-X + Pt(PPh_3)_3 \longrightarrow [PtX(PPh_3)_2]\bullet + R\bullet + PPh_3$$

Also liberation of an alkyl free radical in reactions of dialkylplatinum complexes with olefins was observed by ESR. [219] Hopgood showed that the addition of iodine to platinum acetylacetonate is chain propagated in nature. [221]

$$Pt(acac)_2 + I_2 \longrightarrow \textit{trans-}Pt^{4+}(acac)_2 I_2$$

$Fe(CO)_5$ can be used for the selective conversion of enol acetates, vinyl chlorides, and α,β-unsaturated aldehydes to the corresponding olefins. [220] Also, α-acetoxy ketones are reduced to the respective ketones.

X=OCOCH₃, CHO, Cl

These reactions proceed smoothly in solvents like dibutyl ether or cumene suggesting that the hydrogen donating ability of the solvents is an important factor. Also tri-n-butyltin hydride can be used as a source of hydrogen. Reduction under anhydrous conditions in dibutyl ether, followed by work-up with D_2O did not lead to deuterium incorporation. Based on this evidence, the reaction was explained by the oxidative addition followed by iron-carbon bond homolysis to form vinyl radicals. Abstraction of hydrogen from the solvent results in the final reduced products.

$$\text{>C=C<}_{X} + Fe(CO)_5 \longrightarrow \text{>C=C<}_{\underset{X}{\overset{|}{Fe(CO)_3}}} \longrightarrow \text{>C=C<}_{\underset{Fe(CO)_3X}{\bullet}} \xrightarrow{\text{solvent}} \text{>C=C<}_H$$

The free radical nature of the oxidative addition to transition metal compounds has been shown more clearly in the reactions of organic polyhalides with metal carbonyls. These reactions are not stoichiometric, but catalytic with regard to the metal carbonyls. Free radical reactions of the polyhalides are carried out usually by using radical initiators such as peroxides. Some reactions catalyzed by metal carbonyls are similar to those catalyzed by organic peroxides, but there are a few unique reactions of the metal carbonyls.

Addition of carbon tetrachloride to olefins to form 1,1,1,3-tetrachloro-alkanes *(193)* is carried out by using metal carbonyls such as $Fe(CO)_5$, $Mo(CO)_6$, $Co_2(CO)_8$, $[\pi\text{-}C_5H_5Fe(CO)_2]_2$ and $[\pi\text{-}C_5H_5Mo(CO)_3]_2$ as a catalyst. [222—225]

$$CCl_4 + RCH{=}CH_2 \longrightarrow RCHClCH_2CCl_3$$
$$\textit{193}$$

The reaction is also catalyzed by ferric chloride [226] and copper oxide. [227, 228] The mechanism involving oxidative addition or oxidative cleavage was proposed.

$$M(CO)_n + CCl_4 \longrightarrow \underset{Cl\bullet}{\overset{Cl_3C\bullet}{\cdots}} M(CO)_{n-1} + CO$$

$$\bullet CCl_3 + RCH{=}CH_2 \longrightarrow \underset{\bullet}{R}CH{-}CH_2CCl_3 \xrightarrow{CCl_4}$$

$$\underset{\overset{|}{Cl}}{R}CHCH_2CCl_3 + \bullet CCl_3$$

The reaction of chloroform with olefins catalyzed by $Fe(CO)_5$ takes a different course depending on the solvents used. [229, 230] When the reaction is carried out without solvent, carbon-hydrogen bond fission takes place. Similarly the carbon-hydrogen bond fission is observed in peroxide-catalyzed reactions. On the other hand, the reaction in the presence of solvents such as alcohols or acetonitrile, or in the presence of amines, causes carbon-chlorine bond fission.

$$\underset{\overset{|}{Cl}}{\overset{\overset{\displaystyle Cl}{|}}{H{-}C{-}Cl}} + R{-}CH{=}CH_2 \underset{\searrow RCHCH_2CHCl_2}{\overset{\nearrow RCH_2CH_2CCl_3}{\big<}} \quad \underset{\overset{|}{Cl}}{}$$

Addition reaction of trichloroacetate to olefins is also catalyzed by metal carbonyls. [231, 232] Unlike the reaction catalyzed by organic peroxides, lactone formation was observed with some metal carbonyls. $Co_2(CO)_8$ catalyzed the simple addition reaction to give 2,2,4-trichlorocarboxylates *(194)*. On the other hand, when $[\pi\text{-}C_5H_5Mo(CO)_3]_2$ was used as the catalyst, α,α-dichloro-γ-alkyl-γ-butyrolactones *(195)* were obtained. This result shows that reaction course of the free radical type reactions catalyzed by the metal carbonyls are influenced by the type of metal carbonyls. This difference can be explained partly by the different interactions of the metal carbonyls with the free radical species formed.

$$RCH{=}CH_2 + Cl_3CCO_2R \longrightarrow RCHClCH_2CCl_2CO_2R$$

194

195

The above examples show that free radicals formed by the reaction of polyhalides with the metal carbonyls are different from the usual free radicals due to coordination or interaction with the metallic species. Free radical polymerization of vinyl monomers such as acrylate or methacrylate, is initiated by a catalytic system formed from organic halides and metal carbonyls. The polymerization reactions were studied extensively by Bamford and his coworkers using various metal carbonyls. [233]

When carbon tetrachloride, carbon monoxide, and olefins reacted in the presence of a catalytic amount of metal carbonyl, carbonylation took place producing a high yield of 2-alkyl-4,4,4-trichlorobutyryl chloride *(196)*. [224] The reaction was catalyzed by the dinuclear metal carbonyls such as $Co_2(CO)_8$, $[\pi\text{-}C_5H_5Mo(CO)_3]_2$, and $[\pi\text{-}C_5H_5Fe(CO)_2]_2$. The latter was the most active catalyst and the reaction proceeds even in the presence of a radical inhibitor. The following mechanism was proposed for this free radical carbonylation reaction of olefins.

$$[M(CO)_nL]_2 + CCl_4 \longrightarrow Cl_3C{\bullet}\,M(CO)_nL + Cl{\bullet}\,M(CO)_nL$$

$$Cl_3C{\bullet}\,M(CO)_nL + RCH{=}CH_2 \longrightarrow Cl_3CCH_2CHR{\bullet}\,M(CO)_nL \xrightarrow{CO}$$

$$Cl_3CCH_2CHRCO{\bullet}\,M(CO)_nL \xrightarrow{CCl_4} Cl_3CCH_2CHRCOCl + Cl_3C{\bullet}\,M(CO)_nL$$

196

Synthetic utility of the free radical formed from carbon tetrachloride and metal carbonyls has been extended to the unique reaction of amines. The reaction of aniline and carbon tetrachloride in the presence of a catalytic amount of metal carbonyl such as $Co_2(CO)_8$ and $Mo(CO)_6$, gave 90% yield of p-amino-N,N-diphenylbenzamidine *(197)*. [234] Hydrolysis of the latter gave p-aminobenzoic acid *(198)*. This reaction of aniline offers a good method for introducing a carboxylic acid group into aromatic amines.

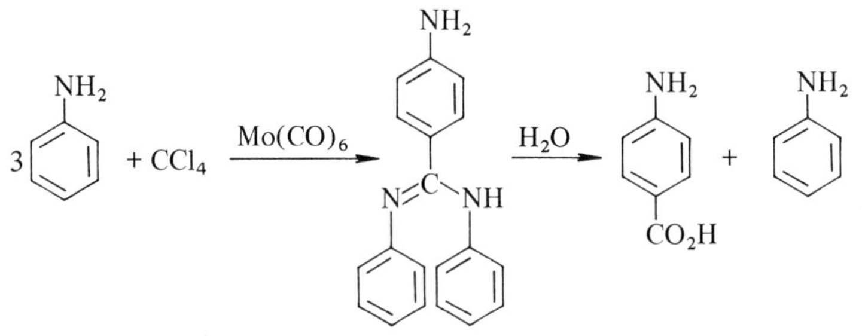

197   198

At the first stage of this reaction, trichloromethyl radical is formed from carbon tetrachloride and the metal carbonyl. This radical abstracts hydrogen from aniline. The resulting amino radical reacts at the *para*-position with the metal-bonded trichloromethyl moiety to give *p*-trichloromethylaniline. Finally the trichloromethyl group reacts with two moles of aniline to give the amidine *(197)*.

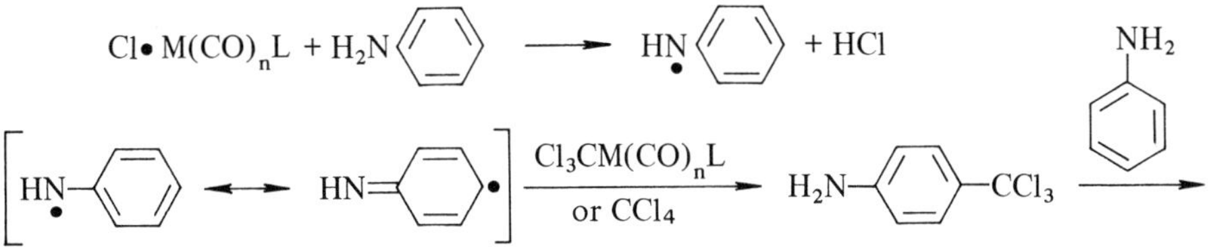

When substituted anilines were treated with carbon tetrachloride under carbon monoxide pressure in the presence of the same metal carbonyl catalyst, a cyclization reaction took place. Thus 7-chloro-3-(*m*-chlorophenyl)-4(3H)quinazolinone *(199)* was obtained from *m*-chloroaniline in 80 % yield.

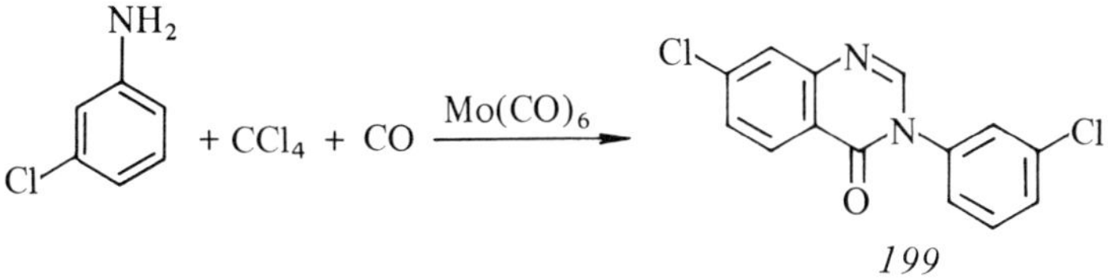

199

When the carbon monoxide pressure was increased to above 150 atm, two moles of carbon monoxide took part in the reaction. The reaction of *p*-chloroaniline with carbon tetrachloride and carbon monoxide gave 7-chloro-1-(*p*-chloro-phenyl)-2H(1H)quinoxalinone *(200)* and 6-chloro-3-(*p*-chlorophenyl)-2,4-(1H, 3H)quinazolinedione *(201)*.

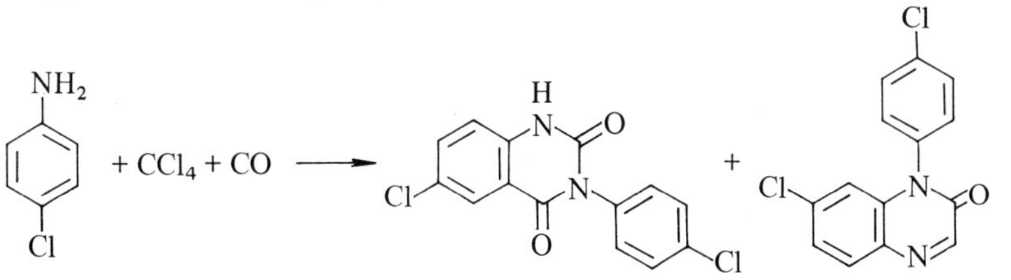

201   200

Formation of the quinazolinone *(199)* can be explained by the following mechanism. Carbon tetrachloride reacts with the two amino groups and then the cylization takes place by the carbon monoxide attack on the aromatic ring.

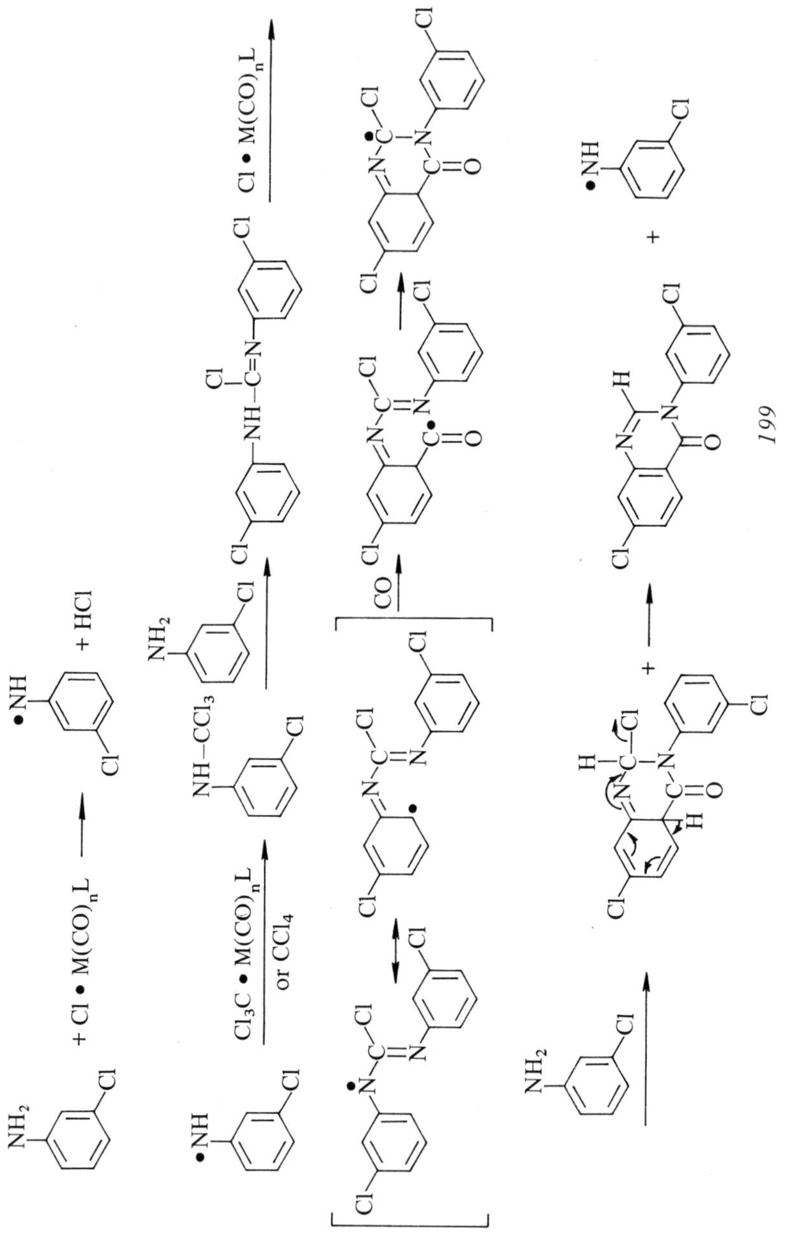

When benzylamine was allowed to react with carbon tetrachloride in the presence of $Co_2(CO)_8$, triphenylimidazole *(202)* and triphenylimidazoline *(203)* were obtained. [235]

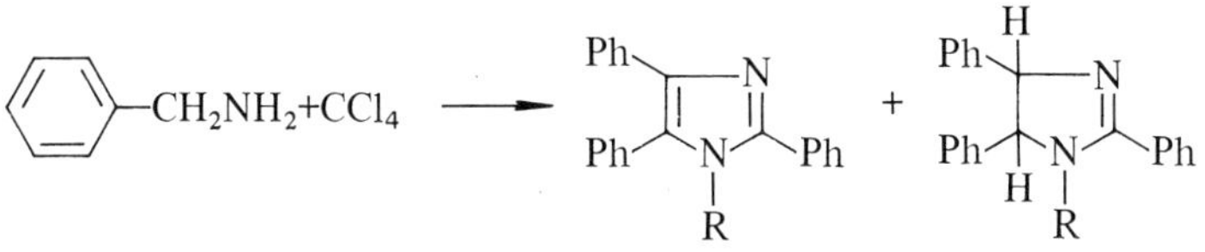

CH_2NH_2+CCl_4
Ph
N
Ph
N
Ph
R
+
H
Ph
N
Ph
N
Ph
H
R

202 *203*

R=H, $PhCH_2$

When the reaction was carried out for a short time or under mild conditions, the yield of the imidazoline increased. Furthermore, from sterically hindered *o*-chlorobenzylamine, 1-(*o*-chlorobenzyl)-2,4,5-tris(*o*-chlorophenyl)imidazolidine *(204)* was obtained as the only product.

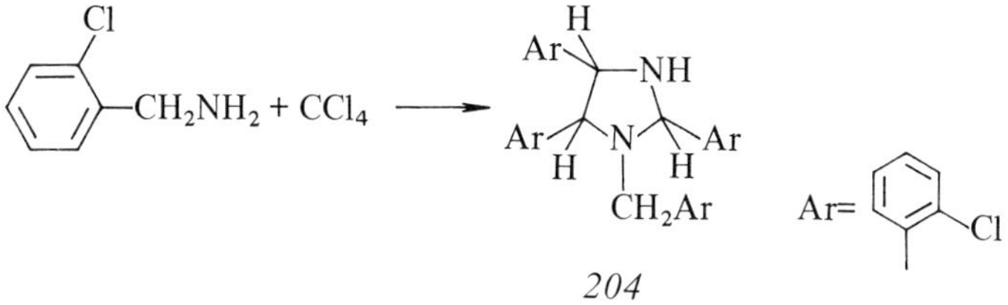

Cl
CH_2NH_2 + CCl_4
H
Ar
NH
Ar
N
Ar
H
H
CH_2Ar
Ar=
Cl

204

From this evidence, it is clear that the imidazolidine *(205)* is formed as the primary product, which is then dehydrogenated to give the imidazoline *(203)*. Further reaction proceeds to form the imidazole *(202)* by hydrogen abstraction.

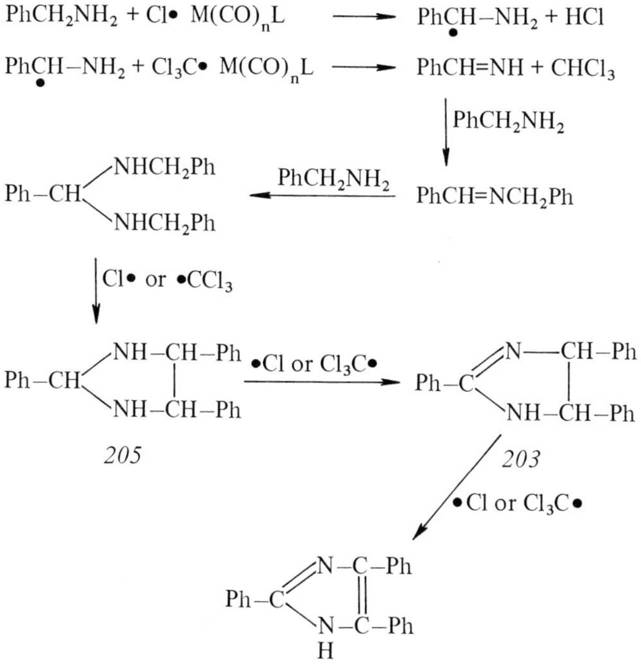

$PhCH_2NH_2$ + Cl• $M(CO)_nL$ → $Ph\dot{C}H-NH_2$ + HCl
$Ph\dot{C}H-NH_2$ + Cl_3C• $M(CO)_nL$ → PhCH=NH + $CHCl_3$
$PhCH_2NH_2$
PhCH=NCH_2Ph
$PhCH_2NH_2$
Ph–CH($NHCH_2Ph$)$_2$
Cl• or •CCl_3
Ph–CH
NH–CH–Ph
NH–CH–Ph
205
•Cl or Cl_3C•
Ph–C
N—CH–Ph
NH–CH–Ph
203
•Cl or Cl_3C•
Ph–C
N–C–Ph
N–C–Ph
H

202

9. Reactions of Transition Metal Compounds with Organo-Nontransition Metal Reagents

A common method for the synthesis of σ-bonded alkyl transition metal complexes is the reaction of transition metal compounds, such as halides or acetyl-acetonates, with organometallic anionic alkylating agents, such as alkylmagnesium, alkyllithium, alkylaluminum, organomercury, and organozinc compounds. [236]

$$L_nMX + R-M' \longrightarrow L_nM-R + M'-X$$

The most famous example of this type of reaction is the *in situ* formation of the Ziegler catalyst, useful for oligomerization and polymerization of olefins, from transition metal compounds and alkylaluminum. In most cases, unstable alkyl or hydride complexes are considered to be formed, and highly active catalysts are used without isolation. The reaction in the presence of certain ligands, such as carbon monoxide, pyridine, bipyridyl, triaryl- or trialkylphosphines, could lead, in some instances, to isolable σ-bonded alkyl transition metal complexes. One example is the reaction of iron acetylacetonate with diethylaluminum monoethoxide in the presence of bipyridyl to form the diethyliron complex *(206)*. [237—239]

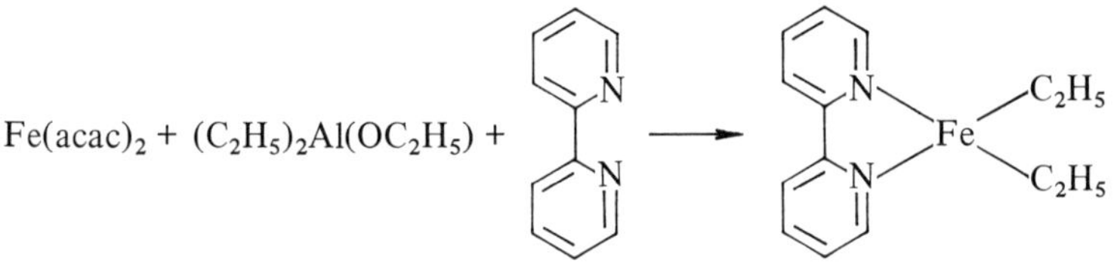

$$Fe(acac)_2 + (C_2H_5)_2Al(OC_2H_5) + $$

206

The stable alkyl complexes *(208)* were obtained by the reaction of transition metal halides with trimethylsilylmethyllithium *(207)*. [240] Since there is no possibility of β-elimination reaction with this alkyl group, the complex can be isolated as a stable one.

$$(CH_3)_3SiCH_2-Li + M-X \longrightarrow (CH_3)_3SiCH_2-M + LiX$$
$$\quad\quad 207 \quad\quad\quad\quad\quad\quad\quad\quad\quad\quad 208$$
$$M = Mo,\ Cr,\ Ti,\ Mn.$$

The reaction of Grignard or organolithium reagents with halo complexes of bivalent platinum or palladium has been widely used for the preparation of alkyl complexes. In certain cases, quite stable alkyl complexes of noble metals can be prepared. [241]

$$[(C_2H_5)_3P]_2Pt(CH_3)I + CH_3MgI \longrightarrow [(C_2H_5)_3P]_2Pt(CH_3)_2 + MgI_2$$

A useful catalytic process involving the reaction of Grignard reagent with transition metal complexes, oxidative addition, and reductive elimination is exemplified by the selective cross-coupling reactions of Grignard reagents with various organic compounds. The cross-coupling of organic groups by the reactions of Grignard reagents with organic halides has been known for a long time. Soluble catalysts consisting of silver, copper, or iron in THF are very effective for coupling Grignard reagents with alkyl halides. Silver is useful for homo-coupling, whereas copper and iron are suitable for cross-coupling of vinyl halides. [242, 243]

The usefulness of the cross-coupling reaction of Grignard reagents has been further enhanced by using a nickel-phosphine complex as a catalyst. The cross-coupling of allylic alcohols and Grignard reagents was carried out in the presence of a catalytic amount of dichlorobis(triphenylphosphine)nickel. [244, 245]

$$R-MgX + \underset{\underset{\displaystyle OH}{\displaystyle |}}{R'CH=CHCHR''} \xrightarrow{NiCl_2(PPh_3)_2} \left\{ \begin{array}{l} \underset{\underset{\displaystyle R}{\displaystyle |}}{R'CH=CHCHR''} \\[2em] \underset{\underset{\displaystyle R}{\displaystyle |}}{R'CHCH=CHR''} \end{array} \right.$$

A mixture of isomeric olefins was obtained. The reaction proceeds *via* a π-allylic nickel complex formed by the oxidative addition of allylic alcohols to a zero-valent nickel complex.

Aryl and vinyl halides were also coupled with Grignard reagents in the presence of a catalytic amount of various nickel compounds such as nickel acetylacetonate or dichlorobis(triphenylphosphine)nickel. [246—248]

$$Ph-X + R-MgX \xrightarrow{Ni \ complex} Ph-R + MgX_2$$

The following two well-established elemental reactions are the basis of this novel catalytic reaction. The first reaction is the coupling of two organic groups on the same nickel atom induced by the action of an organic halide, [249] while the complex itself is converted into the corresponding organonickel halide complex *(209)*.

        206   *209*

The second one is the well-known reaction between the organonickel halide *(210)* and Grignard reagent forming the corresponding diorganonickel complex *(211, corresponds to 206)*. [250, 251]

$$\underset{210}{\overset{\displaystyle L}{\underset{\displaystyle L}{\diagdown}}\underset{\displaystyle Cl}{\overset{\displaystyle Ph}{\diagup}}Ni} \quad + \; R\!-\!MgX \; \longrightarrow \; \underset{211}{\overset{\displaystyle L}{\underset{\displaystyle L}{\diagdown}}\underset{\displaystyle R}{\overset{\displaystyle Ph}{\diagup}}Ni} \quad + \; MgX_2$$

The combination of these two elemental processes constitutes the novel catalytic process of the coupling. The following mechanism has been proposed for this process. Catalysis is initiated by the reaction of dihalobis(triphenylphosphine)-nickel with Grignard reagent to form the intermediate diorganonickel complex *(212)*, which undergoes reductive elimination of the two organo groups to give the coupled product *(213)* by the addition of aryl halide (R'X). The re-generated zerovalent nickel-phosphine complex *(214)* again undergoes oxidative addition of the aryl halide to form the organonickel-halide complex *(215)*. Successive reaction of this complex with Grignard reagent forms a new diorgano complex *(216)*, from which the cross-coupling product *(217)* is released again by the attack of the aryl halide to complete the catalytic cycle. Thus the reaction can be carried out simply by the addition of Grignard reagent to an organic halide in the presence of a catalytic amount of dihalobis(phosphine)nickel. The cross-coupling products are obtained in high yields.

$$\underset{212}{L_2NiX_2 + 2RMgX \; \longrightarrow \; L_2NiR_2} + 2MgX_2$$

$$\underset{212}{L_2NiR_2} \; \longrightarrow \; \underset{214}{(L_2Ni)} + \underset{213}{R\!-\!R}$$

$$L_2Ni + R'X \; \longrightarrow \; \underset{215}{L_2Ni(R')(X)}$$

$$\underset{215}{\overset{\displaystyle X}{\underset{\displaystyle R'}{L_2Ni}}} \; \underset{\underset{217}{R\!-\!R'}}{\overset{RMgX}{\rightleftharpoons}} \; \underset{R'X}{\overset{MgX_2}{}} \; \underset{216}{\overset{\displaystyle R}{\underset{\displaystyle R'}{L_2Ni}}}$$

Dichloro[1,2-bis(diphenylphosphino)ethane]nickel, which is easily prepared, seems to be the best catalyst for the reaction. [252] The organic halides, such as vinyl and aryl halides, are used. The yields of the coupled products are generally higher than 90%.

A coupling reaction of a similar nature is also possible with a rhodium complex, although it is less useful than the nickel-catalyzed reaction. RhCl(CO)(PPh$_3$)$_2$ *(5)* is inert for oxidative addition reaction under normal conditions. As described previously (p. 12), displacement of triphenylphosphine in RhCl(PPh$_3$)$_3$ by carbon monoxide makes the complex much less reactive for oxidative addition. The complex *(5)* can be used for unsymmetrical ketone formation by treatment with an organolithium compound. [253] Organolithium or Grignard reagent reacted with this complex at $-78°$ to form the alkylrhodium complex *(218)*, which underwent a facile oxidative addition reaction with acyl halides at $-78°$ to give the six-coordinated *acylalkyl*rhodium complex *(219)*. Then the coupling of the alkyl and acyl groups by reductive elimination took place to give a ketone *(220)*. The regenerated halorhodium complex can be isolated. The intermediary alkylrhodium complex *(218)* does not react with other functional groups such as nitriles and carbonyl groups. Thus the method allows the ready conversion of acyl halides to ketones in high yields under mild conditions. The reaction is not truly catalytic with regard to the rhodium complex, but the rhodium complex can be reused after recovery.

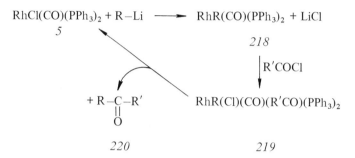

The enhanced reactivity toward the oxidative addition of acyl chloride to the alkylrhodium complex *(218)*, compared with $RhCl(CO)(PPh_3)_2$, is possibly a result of the increased electron density at the central metal caused by replacement of the chlorine with the alkyl group.

Similarly methyltris(triphenylphosphine)rhodium *(221)* readily accepts alkyl, aryl, and alkenyl halides under mild conditions in coordinating solvents, and carbon-carbon bond formation takes place. [254] The reaction proceeds *via* oxidative addition and reductive elimination.

$RhI(PPh_3)_3$ can be reconverted to the alkylrhodium complex *(221)* by the reaction of Grignard reagent. Thus the overall reaction is the coupling of organic halides and Grignard reagents.

In addition to the alkyl complex formation mentioned above, metal hydrides are prepared by the use of group III metal hydrides. Treatment of $NiX_2(PR_3)_2$ with sodium borohydride leads directly to nickel hydride complexes. The selective metal-hydride reducing agent, $Na[HB(CH_3)_3]$, has been used for the synthesis of π-allylnickel hydride. [255]

$$NiCl_2(PR_3)_2 + NaBH_4 \xrightarrow{\hspace{3cm}} HNiCl(PR_3)_2$$

$$(\pi\text{-}C_3H_5)NiBr(PR_3) + Na[HB(CH_3)_3] \longrightarrow (\pi\text{-}C_3H_5)NiH(PR_3)$$

$$R = cyclohexyl$$

Organomercury compounds, which are easy to prepare and have stable alkyl-metal bonds, can be used for the synthesis of transition metal alkyl complexes. Their alkyl groups can be transferred to transition metal complexes to form stable organo-transition metal complexes. Diphenylmercury reacts with one equivalent of the dichloroplatinum complex *(222)* to produce phenylmercuric chloride and the phenylated platinum complex *(223)* in high yields. [241]

$$PtCl_2[PPh(CH_3)_2] + HgPh_2 \longrightarrow PtClPh[PPh(CH_3)_2] + HgPhCl$$
$$222223$$

Sometimes, alkylation of carbonyl complexes with alkylmercury compounds proceeds with insertion of carbon monoxide to form acyl complexes.

$$2\,Pt(CO)Cl_2(PPh_3) + 2\,Hg(CH_3)_2 \longrightarrow [Pt(CH_3CO)Cl(PPh_3)]_2 + 2CH_3HgCl$$

Application of this alkylation reaction as a method of σ-bond formation to organic synthesis has been made by Heck, using palladium salts. [256—263] Alkyl and aryl groups of organomercury compounds are transferred to the palladium salt to form very reactive alkyl- or arylpalladium complexes, which are then used *in situ* for organic synthesis. A homogeneous solution of palladium salt and alkyl- or arylmercury compounds in a polar solvent reacts to form the solvated alkyl- or arylpalladium salts *(224)*. These species are unstable and generally decompose even at room temperature to form coupled products or olefins if β-hydrogen is present in the alkyl group. In the presence of an olefin, however, the alkyl groups without β-hydrogen, as well as aryl groups, add to the olefin forming the alkylethyl- or arylethylpalladium salts *(225)*, which then rapidly decompose to palladium hydride and alkylated or arylated olefins *(226)*. [257]

$$PdX_2 + RHgX \xrightarrow{\hspace{2cm}} R\text{—}PdX + HgX_2$$
$$224$$

$$R\text{—}PdX \;|\; R'CH\text{=}CH_2 \longrightarrow \underset{R'}{RCH_2CH\text{—}PdX} + \underset{R'}{RCHCH_2\text{—}PdX}$$

$$225$$

$$RCH\text{=}CHR' \quad + \quad \underset{R'}{\overset{R}{>}}C\text{=}CH_2 + H\text{—}PdX$$

$$226$$

The scope of the reaction which effects alkylation and arylation of olefins, has been investigated. [263] The reaction of propylene with phenylpalladium acetate *(227)* in methanol at 30° gave a 66% yield of a mixture of olefins containing 60% *trans*-propenylbenzene *(228)*, 9% *cis*-propenylbenzene *(229)*, 15% allylbenzene *(230)* and 16% α-methylstyrene *(231)*.

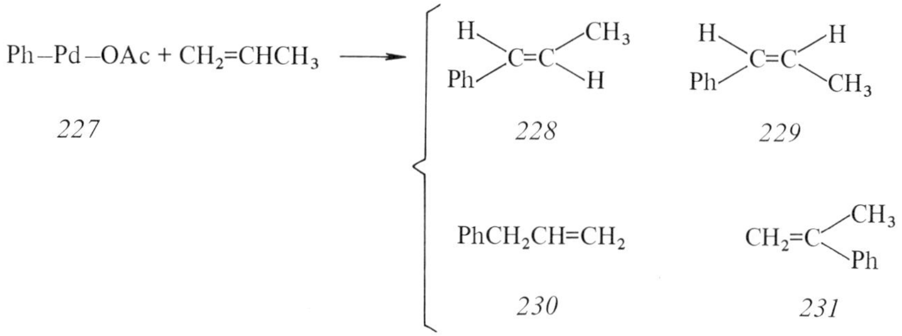

$$Ph-Pd-OAc + CH_2=CHCH_3 \longrightarrow$$

227

228

229

230

231

The arylpalladium salts *(232)*, prepared similarly *in situ*, reacted with allylic halides to give allyl derivatives of aromatic compounds by the following addition and elimination reaction. [258]

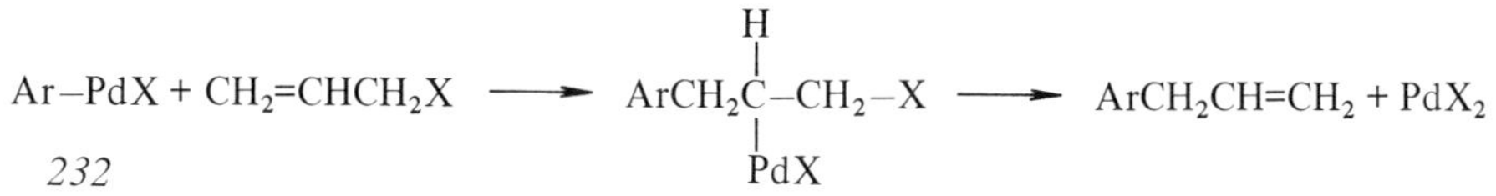

$$Ar-PdX + CH_2=CHCH_2X \longrightarrow ArCH_2\overset{\underset{\displaystyle |}{H}}{\underset{\underset{\displaystyle PdX}{|}}{C}}-CH_2-X \longrightarrow ArCH_2CH=CH_2 + PdX_2$$

232

Reaction with primary and secondary allylic alcohols provides a useful new route to a variety of the 3-aryl aldehydes or 3-aryl ketones *(233)*. [257]

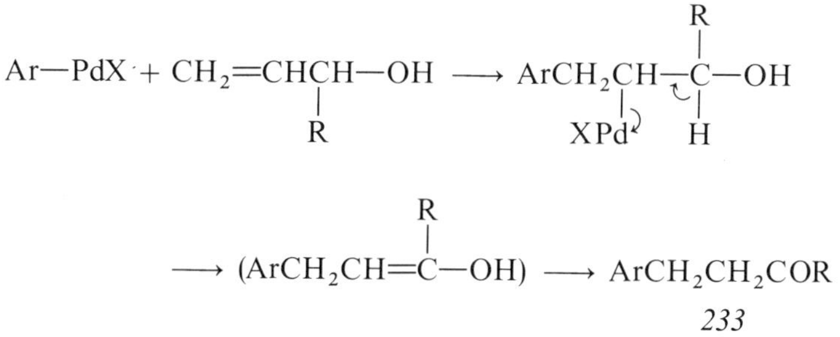

$$Ar-PdX + CH_2=\overset{}{C}HCH-OH \longrightarrow ArCH_2\overset{}{C}H\overset{\underset{\displaystyle |}{R}}{\underset{}{-C}}-OH$$

$$\longrightarrow (ArCH_2CH=\overset{\underset{\displaystyle |}{R}}{C}-OH) \longrightarrow ArCH_2CH_2COR$$

233

Reaction of carbomethoxymercuric acetate *(234)* with palladium acetate gives "carbomethoxypalladium salt" *(235)*, which can be used for the carbomethoxylation of olefins to form unsaturated esters. The reaction with α-methylstyrene produced 86% yield of unsaturated esters, 96% of this being the nonconjugated methyl 3-phenyl-3-butenoate *(236)* and only 4% the conjugated product *(237)*. [263]

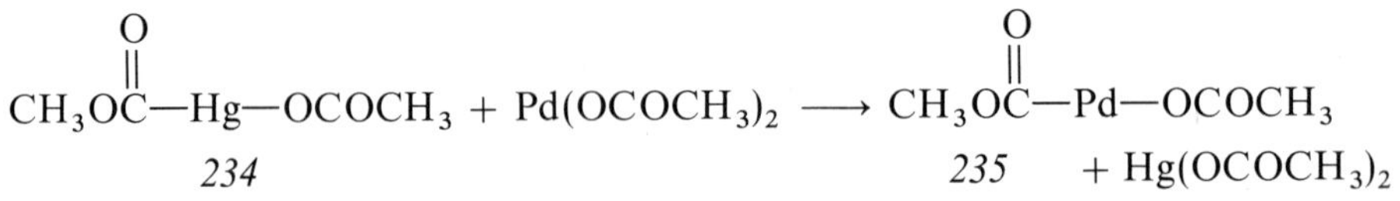

$$CH_3O\overset{\overset{\displaystyle O}{\|}}{C}-Hg-OCOCH_3 + Pd(OCOCH_3)_2 \longrightarrow CH_3O\overset{\overset{\displaystyle O}{\|}}{C}-Pd-OCOCH_3$$

234 $\qquad\qquad\qquad\qquad$ *235* $\quad$ + $Hg(OCOCH_3)_2$

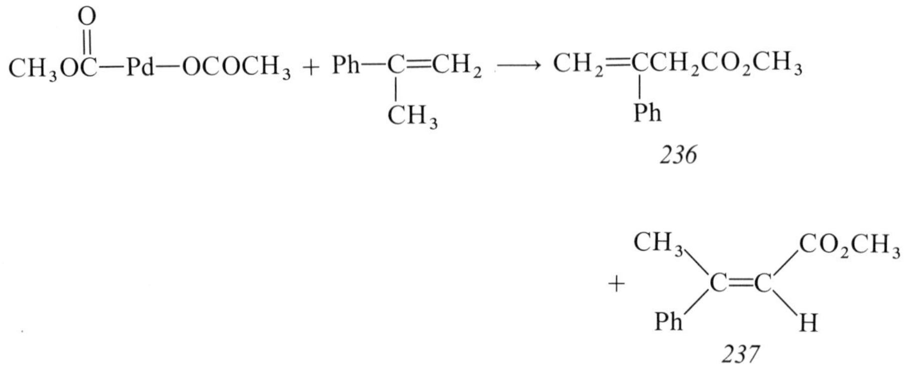

$$CH_3O\overset{\overset{\displaystyle O}{\|}}{C}-Pd-OCOCH_3 + Ph-\underset{\underset{\displaystyle CH_3}{|}}{C}=CH_2 \longrightarrow CH_2=\underset{\underset{\displaystyle Ph}{|}}{C}CH_2CO_2CH_3$$

236

$$+ \quad \underset{Ph}{\overset{CH_3}{\diagdown}}C=C\underset{H}{\overset{CO_2CH_3}{\diagup}}$$

237

The arylpalladium salts *(232)* also reacted with carbon monoxide in hydroxylic solvents to form aryl carboxylic acids or their derivatives depending on the solvents. [264]

$$Ph-PdX + CO \longrightarrow PhCOX$$

232

Under certain conditions, diaryl ketones are formed in a moderate yield from arylmercuric salts and carbon monoxide using palladium salts as a catalyst. [262]

Organomercury compounds react with $Co_2(CO)_8$ in THF to give ketones in moderate yields. [265]

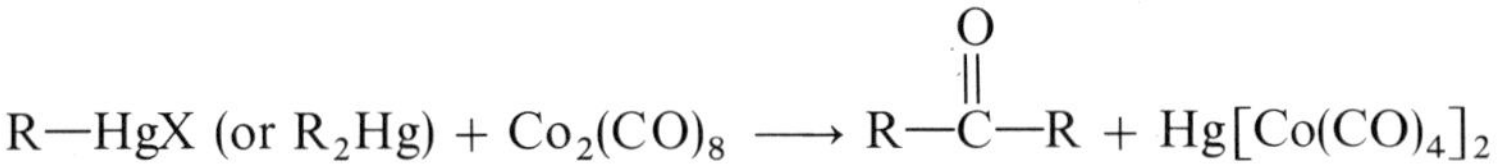

$$R-HgX \text{ (or } R_2Hg) + Co_2(CO)_8 \longrightarrow R-\overset{\overset{\displaystyle O}{\|}}{C}-R + Hg[Co(CO)_4]_2$$

In this reaction, the organomercury compound behaves as an alkylating agent of the cobalt complex. The carbonylation of the mercury compounds to form ketones was improved by carrying out the reaction in the presence of a catalytic amount of $Co_2(CO)_8$ or $Hg[Co(CO)_4]_2$ under uv irradiation. [266]

It should be noted here that alkylation of other metals by transfer of an alkyl group of organoaquocobaloxime complexes proceeds with the cleavage of the cobalt-carbon σ-bond; [267] organocobaloxime reacted with Cr^{2+} with essentially quantitative transfer of the alkyl group to the chromium.

$$RCo(Hdmg)_2H_2O + Cr^{2+} + 2H^+ \longrightarrow Co^{2+} + (H_2O)CrR^{2+} + 2H_2dmg$$

$$dmg = \text{dimethylglyoximato}$$

10. Carbonylmetallate Anions

a) Introduction

Alkylation of metal carbonyls treated in this chapter is related to the reactions discussed in the preceding chapter. But the reaction of metal carbonyls is treated in an independent chapter because of its special importance and wide applicability in organic synthesis.

Reactive organic halides add oxidatively to metal carbonyls affording useful intermediates for organic synthesis (p. 38). The reactivity of metal carbonyls can be enhanced further by converting them into carbonylmetallate anions by treating them with certain alkyl and aryl metal compounds such as organolithium or certain bases. The reactions of carbonyls of nickel, iron, molybdenum, tungsten, and chromium have been studied extensively. By converting metal carbonyls into carbonylmetallate anions, electron density of the metal is increased. Higher electron density of the metal facilitates oxidative addition reactions and hence the reactivity of the metal carbonyls is enhanced.

Organolithium compounds are very reactive toward metal carbonyls. The anionic organic moiety attacks the electropositive carbon of one of the carbon monoxide ligands of the metal carbonyl, even at low temperatures, to form rather stable acylcarbonylmetallate *(238)*.

$$R-Li + M(CO)_n \longrightarrow Li[RCO-M(CO)_{n-1}]$$

238

The products of the reaction with $Ni(CO)_4$ and $Fe(CO)_5$ are most useful for organic synthesis. They are called lithium acylcarbonylmetallate, or lithium acylmetalcarbonylate. [148, 149] Similar reactions with molybdenum, tungsten, and chromium carbonyls studied by Fischer opened the interesting chemistry of carbene species coordinated to metal.

b) Nickel and Iron Complexes

Lithium acylcarbonylmetallates are reactive species useful for organic synthesis in that they can be regarded as an equivalent to the acyl anion. The most useful property of the acyl- and aroylcarbonylmetallates is their nucleophilic character of the coordinated ligands. In the usual organic chemistry, the acyl group is electrophilic and no example of a nucleophilic reaction of the acyl group is known, [268] except for the acyl anion postulated in the reaction of phenyllithium with carbon monoxide. [269]

$$PhLi + CO \longrightarrow \left(\begin{matrix} & O \\ & \parallel \\ & C \\ Ph^\diagup & \diagdown Li \end{matrix} \right)$$

Hydrolysis of lithium aroyltetracarbonylferrate *(239)* gives aldehydes. This reaction is especially useful for the synthesis of aldehydes which contain unsaturated functional groups. [270, 271]

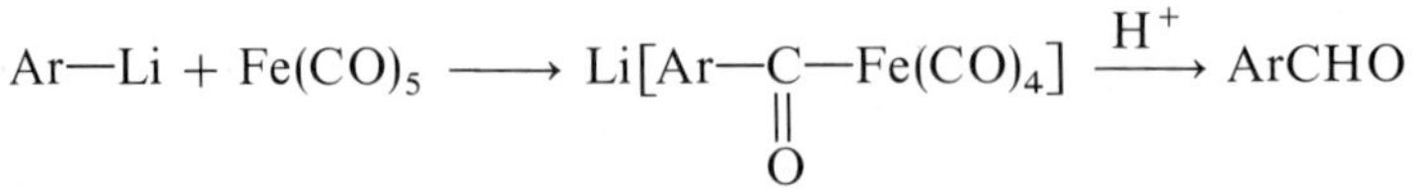

$$\text{Ar—Li} + \text{Fe(CO)}_5 \longrightarrow \text{Li}[\text{Ar—}\underset{\underset{O}{\|}}{C}\text{—Fe(CO)}_4] \xrightarrow{\text{H}^+} \text{ArCHO}$$

239

Unlike the iron complex, the corresponding lithium aroyltricarbonylnickelate *(240)* gives acyloins *(241)* instead of aldehydes by hydrolysis with aqueous methanol containing hydrochloric acid. When the complex was heated or treated with bromine, the α-diketones *(242)* were formed. [272]

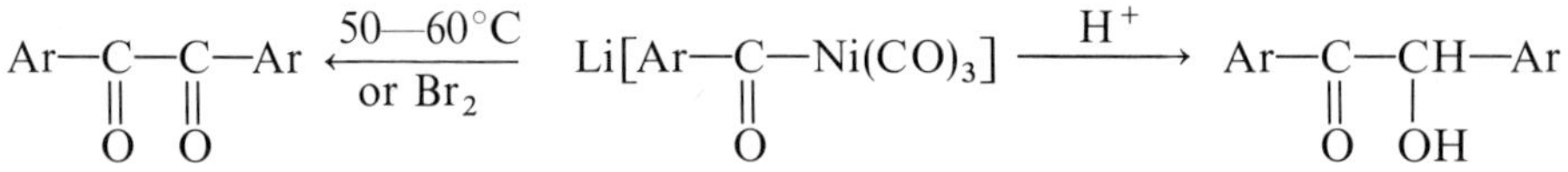

$$\text{Ar—}\underset{\underset{O}{\|}}{C}\text{—}\underset{\underset{O}{\|}}{C}\text{—Ar} \xleftarrow[\text{or Br}_2]{50\text{—}60°C} \text{Li}[\text{Ar—}\underset{\underset{O}{\|}}{C}\text{—Ni(CO)}_3] \xrightarrow{\text{H}^+} \text{Ar—}\underset{\underset{O}{\|}}{C}\text{—}\underset{\underset{OH}{|}}{C}H\text{—Ar}$$

242   *240*   *241*

The nucleophilic character of the acyl group is apparent from the reactions with alkyl halides or acyl halides. With the iron complex *(239)*, unsymmetrical ketones were formed. [273]

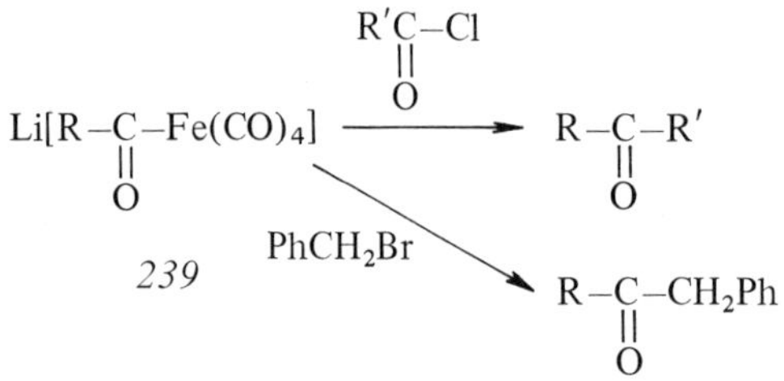

$$\text{Li}[\text{R—}\underset{\underset{O}{\|}}{C}\text{—Fe(CO)}_4] \underset{\underset{PhCH_2Br}{}}{\overset{\overset{R'C-Cl (O)}{}}{\longrightarrow}} \begin{array}{l} \text{R—}\underset{\underset{O}{\|}}{C}\text{—R'} \\ \\ \text{R—}\underset{\underset{O}{\|}}{C}\text{—CH}_2\text{Ph} \end{array}$$

239

On the other hand, the corresponding nickel complexes give acyloins *(243)* or en-diol diesters *(244)*.

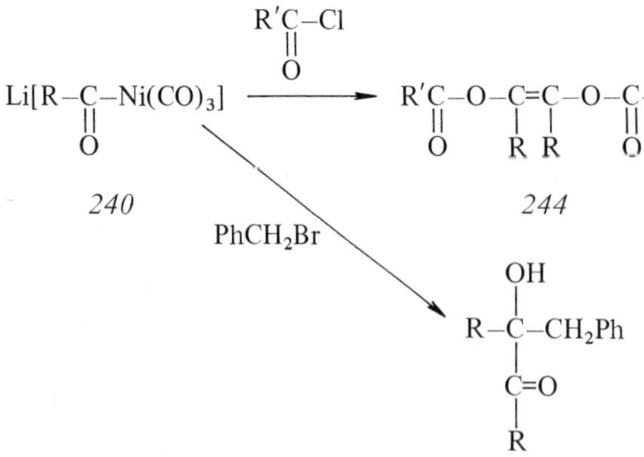

$$\text{Li}[\text{R—}\underset{\underset{O}{\|}}{C}\text{—Ni(CO)}_3]$$

240

R'C–Cl (O) → R'C—O—C=C—O—C—R' (244)

PhCH₂Br →

OH | R—C—CH₂Ph | C=O | R

243

As expected, the nucleophilic, Michael-type addition reactions were capable of forming 1,4-dicarbonyl compounds. [274]

$$Li[R-\underset{\underset{O}{\|}}{C}-Ni(CO)_3] + CH_2=CH-\underset{\underset{O}{\|}}{C}-R' \longrightarrow R-\underset{\underset{O}{\|}}{C}-CH_2CH_2-\underset{\underset{O}{\|}}{C}-R'$$

1,4-Dicarbonyl compounds were also formed by the reaction of acetylenic compounds with 2 moles of the acylnickelate. [275] No reaction took place with the corresponding iron complex.

$$Li[R-\underset{\underset{O}{\|}}{C}-Ni(CO)_3] + R'C\equiv CH \xrightarrow[H^+]{-70°} R-\underset{\underset{O}{\|}}{C}-\overset{\overset{R'}{|}}{C}HCH_2\underset{\underset{O}{\|}}{C}-R$$

Lithium carbamoyltricarbonylnickelate (245) is formed from the reaction of lithium amide and $Ni(CO)_4$. [276, 277] This complex reacted with phenylacetylene to give 2-phenyl-N,N,N',N'-tetramethylsuccinamide (246) as a main product and N,N-dimethylcinnamamide (247) as a minor product. [276]

$$LiN(CH_3)_2 + Ni(CO)_4 \longrightarrow Li[(CH_3)_2N\underset{\underset{O}{\|}}{C}Ni(CO)_3]$$

245

$$\xrightarrow{PhC\equiv CH} (CH_3)_2N-\underset{\underset{O}{\|}}{C}-\overset{\overset{|}{Ph}}{C}HCH_2\underset{\underset{O}{\|}}{C}-N(CH_3)_2 + PhCH=CH\underset{\underset{O}{\|}}{C}-N(CH_3)_2$$

246   247

When the complex (245) was treated with organic halides or acyl halides, dimethylaminocarbonylation proceeded smoothly; the reaction of benzyl bromide gave N,N-dimethylphenylacetamide (248) in a yield of 64.5%.

$$Li[(CH_3)_2N\underset{\underset{O}{\|}}{C}Ni(CO)_3] + PhCH_2Br \longrightarrow PhCH_2CON(CH_3)_2$$

248

245

An interesting reaction of lithium acylmetallates is the conversion to carbene complexes; acylcarbonylferrate was converted to ethoxyalkylcarbene tetracarbonyliron complex (249) by treatment with oxonium salts or trityl chloride. [278, 279]

More comprehensive studies of the carbene complexes of transition metals prepared by the similar reaction of $W(CO)_6$, $Mo(CO)_6$, and $Cr(CO)_6$ have been carried out by Fischer. This topic is treated separately.

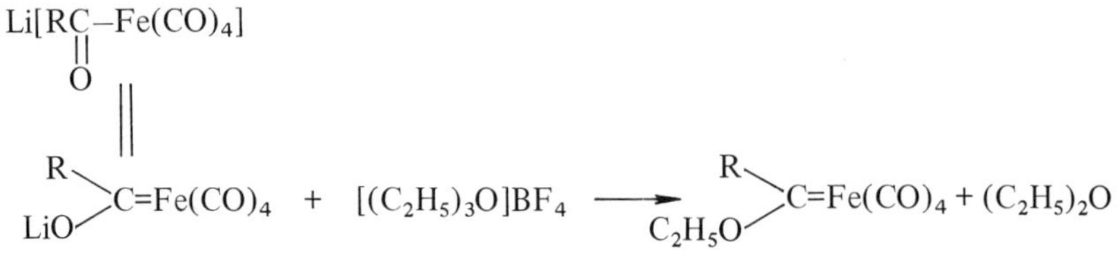

$$Li[R\underset{\underset{O}{\|}}{C}{-}Fe(CO)_4]$$

$$\underset{LiO}{\overset{R}{\diagdown}}C{=}Fe(CO)_4 \ + \ [(C_2H_5)_3O]BF_4 \ \xrightarrow{} \ \underset{C_2H_5O}{\overset{R}{\diagdown}}C{=}Fe(CO)_4 + (C_2H_5)_2O$$

$$\qquad\quad 239 \qquad\qquad\qquad\qquad\qquad\qquad\qquad\qquad 249$$
$$\qquad\qquad\qquad\qquad\qquad\qquad\qquad\qquad\qquad\qquad +LiBF_4$$

Metal carbonyls also react with various bases. Alkali metals, alkali amalgam, alkali alkoxide, and even alkali hydroxide are used for the preparation of carbonylmetallate anions, which have a strong nucleophilicity. The degree of the nucleophilicity of the carbonylmetallate anions varies depending on the identity of metals and ligands. Strongly nucleophilic carbonylmetallate anions are formed by the coordination of a cyclopentadienyl group. Thus σ-bond formation takes place easily with less reactive alkyl halides, which are not active enough to participate in the oxidative addition toward simple, neutral metal carbonyls. Then insertion of carbon monoxide and other molecules takes place to give acyl metal complexes or organic products. The reaction of $[\pi\text{-}C_5H_5Fe(CO)_2]^-$ and other cyclopentadienyl substituted anions has been studied mostly from the standpoint of coordination chemistry. [280, 281] Synthesis of alkyl and acyl complexes has been carried out with these strongly nucleophilic cyclopentadienyl complexes of iron, molybdenum, tungsten, and other metals, but they have not been used for organic synthesis.

Studies of the reactions of carbonylmetallate anions from the synthetic point of view have been carried out with the iron and nickel complexes which can be synthesized easily from $Fe(CO)_5$ and $Ni(CO)_4$ and certain bases. No reaction takes place with methyl iodide and $Fe(CO)_5$; the reaction of iodobenzene with $Fe(CO)_5$ at room temperature is negligible whereas these iodides react readily with carbonylferrate. [282] Furthermore, carbonylation of halides of trigonal and tetragonal carbons is possible with $Ni(CO)_4$ in the presence of alkali alkoxide, especially potassium t-butoxide. The carbonylmetallate complex thus produced effects butoxycarbonylation of moderately active organic halides to give t-butyl esters. [283]

$$R{-}X + Ni(CO)_4 \xrightarrow{\text{t-BuOK}} RCO_2{-}\text{t-Bu}$$

The carbonylation of a vinyl iodide with $Ni(CO)_4$ in the presence of sodium methoxide was applied to the stereospecific synthesis of α-santalol (250). In the key step, the stereospecific carbonylation was involved as shown below. [284]

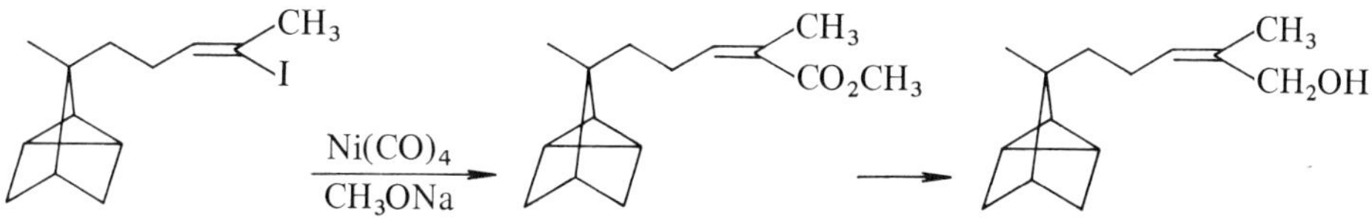

250

The reaction of 1-iodo-5-decyne *(251)* with Ni(CO)$_4$ and potassium t-butoxide gave a mixture of t-butyl 2-(1-cyclopentyl)hexanoate *(252)*, t-butyl 2-cyclo-
pentylidenehexanoate *(253)*, and t-butyl 6-undecynoate *(254)*. The cyclic products were formed by intramolecular insertion of the triple bond before carbon
monoxide insertion took place. [285]

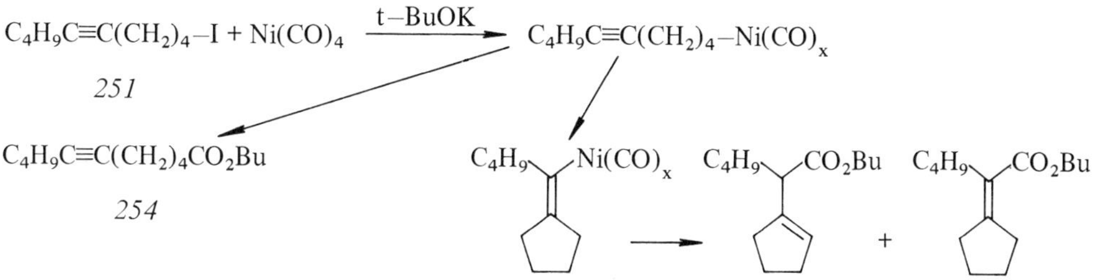

Alkali tetracarbonylferrates are useful reagents for organic synthesis. They
are prepared by reducing Fe(CO)$_5$ with alkali hydroxide or amalgam. The reaction of Fe(CO)$_5$ with potassium hydroxide proceeds as shown below. [286, 287]

$$Fe(CO)_5 + 3\,KOH \longrightarrow KHFe(CO)_4 + K_2CO_3 + H_2$$

$$Fe(CO)_5 + 4\,KOH \longrightarrow K_2Fe(CO)_4 + K_2CO_3 + 2\,H_2$$

Aldehydes, esters, and other carbonyl compounds were synthesized by these
carbonylferrates.

Carbon-iron bonds are formed by the reaction of the carbonylferrate with
alkyl halides. The reaction can be viewed either as an S_N2 displacement at the
carbon or to proceed probably *via* an oxidative addition to the d^{10} iron(-II).
KHFe(CO)$_4$ and Na$_2$Fe(CO)$_4$ react with alkyl bromides, iodides, and tosylates
to give the anionic alkyltetracarbonyliron complex *(255)*. In the presence of
carbon monoxide, alkyl migration takes place to produce the anionic acyl complexes *(256)*. Addition of triphenylphosphine also assists the alkyl-acyl migration.
Protonation of the anion results in the formation of the acyliron hydride *(257)*,
which undergoes reductive elimination to yield aldehydes. [288, 289]

$$\text{Fe(CO)}_5 \xrightarrow{\text{Na–Hg}} \text{Na}_2\text{Fe(CO)}_4 \xrightarrow{\text{R–Br}} \left[\begin{array}{c} \text{R} \\ \text{OC} \diagdown \overset{|}{\underset{\diagup}{\text{Fe}}} \diagup \text{CO} \\ \text{OC} \diagup \diagdown \text{CO} \end{array} \right]^{-} \rightleftharpoons$$

255

$$\left[\begin{array}{c} \overset{\text{O}}{\overset{||}{}} \\ \text{OC} \diagdown \quad \text{C–R} \\ \quad \text{Fe} \\ \text{OC} \diagup \diagdown \text{CO} \end{array} \right]^{-} \xrightarrow{\text{PPh}_3} \left[\begin{array}{c} \text{CO R} \\ \text{OC} \diagdown \overset{|}{\underset{\diagup}{\text{Fe}}} - \overset{|}{\text{C}}\!\!=\!\!\text{O} \\ \text{Ph}_3\text{P} \diagup \diagdown \text{CO} \end{array} \right]^{-} \xrightarrow{\text{H}^+}$$

256

$$\begin{array}{c} \text{R} \\ \text{OC} \diagdown \overset{\text{H}}{\underset{|}{}} \overset{|}{\text{C}}\!\!=\!\!\text{O} \\ \quad \text{Fe} \\ \text{OC} \diagup \overset{|}{\underset{\text{PPh}_3}{}} \diagdown \text{CO} \end{array} \xrightarrow{\text{CO}} \text{Fe(CO)}_4\text{PPh}_3 + \text{R–CHO}$$

257

Reduction of acyl halides to aldehydes is possible by using the same iron complex. [290] Reaction of acyl halides with $\text{Na}_2\text{Fe(CO)}_4$ formed an acyliron complex, which was converted into aldehyde by treatment with acetic acid.

$$\overset{\text{O}}{\overset{||}{\text{R–C–X}}} + \text{Na}_2\text{Fe(CO)}_4 \longrightarrow \text{NaFe(CO)}_4\overset{\text{O}}{\overset{||}{\text{C–R}}} + \text{NaCl}$$

$$\Big\downarrow \text{CH}_3\text{CO}_2\text{H}$$

$$\text{R–CHO}$$

Aldehydes were also prepared from acid anhydrides. [291] The reaction of an acid anhydride with $\text{Na}_2\text{Fe(CO)}_4$ in THF at room temperature followed by acid treatment produced aldehyde in a high yield. The same reaction of anhydrides of dicarboxylic acids gave aldehydic acids. The reaction can be explained by assuming the formation of an acyltetracarbonylferrate as an intermediate.

$$\begin{array}{c} \overset{\text{O}}{\overset{||}{\text{R–C}}} \\ \diagdown \\ \quad \text{O} \quad + \text{Na}_2\text{Fe(CO)}_4 \longrightarrow \left[\text{R–}\overset{\text{O}}{\overset{||}{\text{C}}}\text{–Fe(CO)}_4 \right]^{-} \quad \xrightarrow{\text{H}^+} \\ \text{R'–C} \diagup \\ \overset{||}{\text{O}} \end{array}$$

RCHO

$$+$$

$$\text{R'CO}_2\text{Na} \qquad \text{RCO}_2\text{H}$$

Reduction of acid anhydrides to aldehydes was also carried out with $Co_2(CO)_8$ as the catalyst. [292]

The decomposition of the acyliron complex shown above with iodine in alcohol produces the corresponding ester. α-Bromopropionate was converted into methylmalonate which was the main product accompanied by succinate. [293]

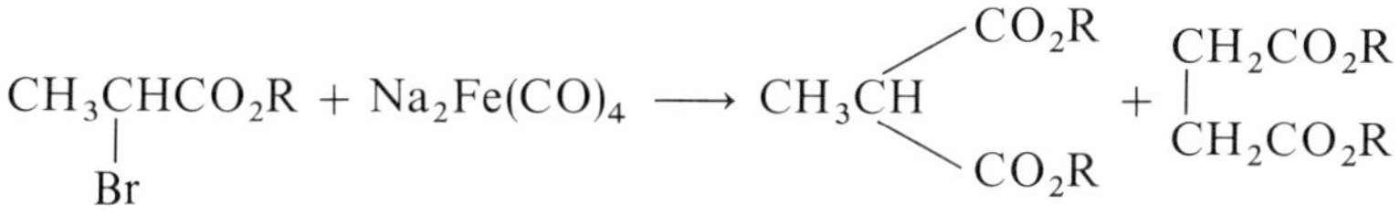

$$CH_3CHCO_2R + Na_2Fe(CO)_4 \longrightarrow CH_3CH \begin{array}{c} CO_2R \\ \diagdown \\ CO_2R \end{array} + \begin{array}{c} CH_2CO_2R \\ | \\ CH_2CO_2R \end{array}$$
(Br below CH)

The reaction of 1,3-dibromopropane (258) with $Na_2Fe(CO)_4$ in the presence of triphenylphosphine afforded the 2-ferracylcyclopentanone complex (259). [294] The reaction can be explained by the intermediacy of an acyliron complex, which undergoes intramolecular nucleophilic attack of the iron to give the cyclic product in a high yield.

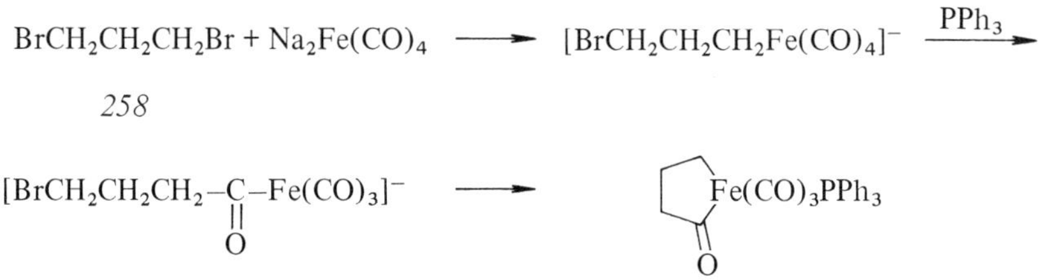

$$BrCH_2CH_2CH_2Br + Na_2Fe(CO)_4 \longrightarrow [BrCH_2CH_2CH_2Fe(CO)_4]^- \xrightarrow{PPh_3}$$

258

$$[BrCH_2CH_2CH_2-\overset{O}{\underset{||}{C}}-Fe(CO)_3]^- \longrightarrow$$

259

Biphthalidylidene (260) was obtained by the reaction of phthaloyl dichloride with $Na_2Fe(CO)_4$. [295] The postulated intermediate of the reaction is the cyclic carbene iron complex (261).

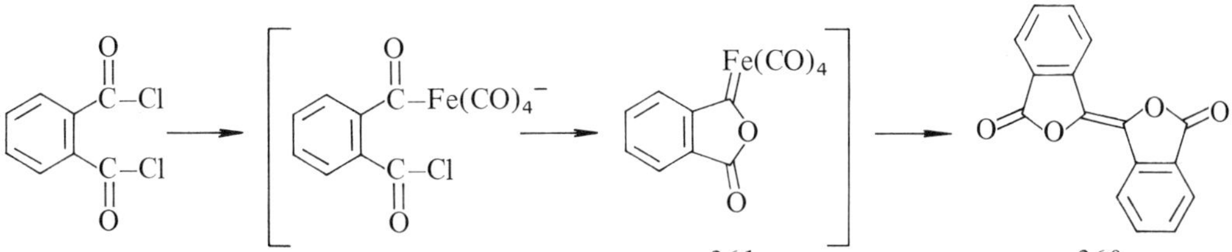

261   260

Halogenation of the acyliron complexes gave acyl halides, which then reacted with water, alcohol, or amine to give the corresponding carboxylic acid derivatives. [296]

$$R-\overset{O}{\underset{||}{C}}-Fe(CO)_4 + X_2 \longrightarrow R-\overset{O}{\underset{||}{C}}-X + FeCl(CO)_4$$

Oxygenation with oxygen or NaOCl also produced carboxylic acids. The reaction proceeded in THF or HMPA with high yields.

The reaction of the tetracarbonylferrate was extended further to ketone synthesis. The synthesis was achieved by the reaction of acyl halides or with $Na_2Fe(CO)_4$ *via* the formation of the acyl complex *(256)*, which reacted with the second alkylating agents. [297] The second alkylating agents should be more reactive than the first alkylating agents. Primary iodides or tosylates, benzyl halides, or α-chloro ethers may also be used. These results can be explained by the following order of decreasing nucleophilicity.

$$Fe(CO)_4^{2-} > RFe(CO)_4^- > RCOFe(CO)_4^-$$

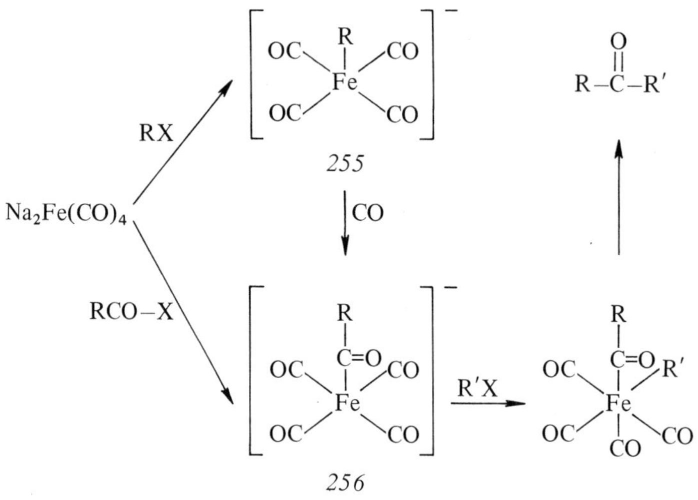

255

256

Thus by this method, unsymmetric ketones can be prepared. [298]

KHFe(CO)$_4$ can be used for the conversion of a terminal olefin oxide to the original olefin. [299, 287]

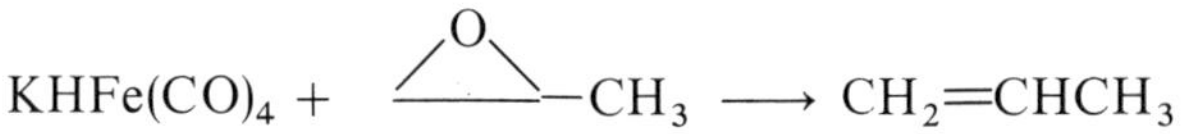

As a competing reaction, the olefin oxide was converted into an acyliron complex, which was decomposed with iodine in alcohol to give a β-hydroxy carboxylate *(262)*. But the yield was not high.

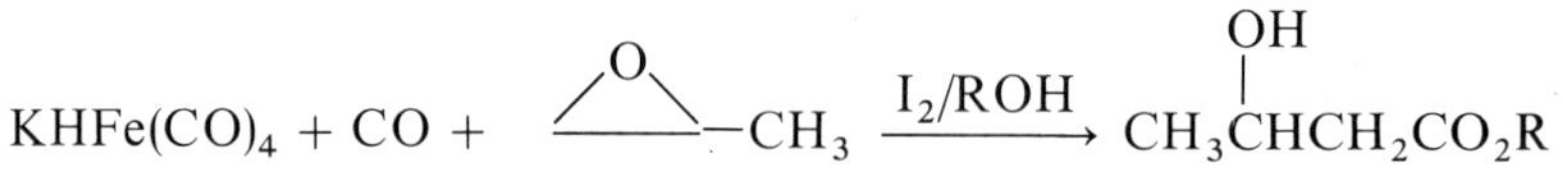

262

Like many other metal hydrides, the iron-hydrogen bond of KHFe(CO)$_4$ is susceptible to olefin insertion. Thus acrylate reacted with the complex in a carbon monoxide atmosphere, and methylmalonate *(263)* was obtained as a main product and succinate *(264)* as a minor product. [300]

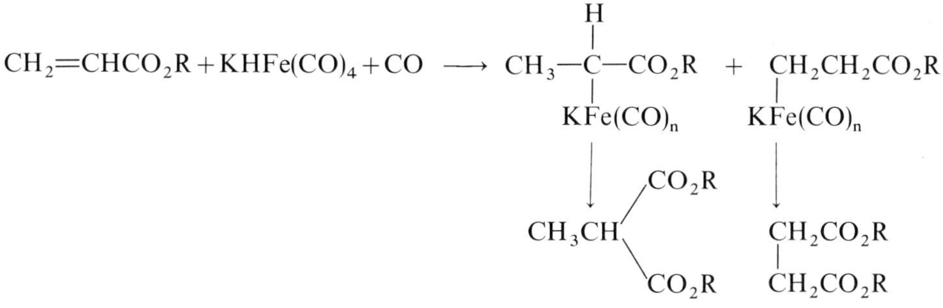

$$CH_2=CHCO_2R + KHFe(CO)_4 + CO \longrightarrow CH_3-\underset{\underset{KFe(CO)_n}{|}}{\overset{\overset{H}{|}}{C}}-CO_2R + CH_2CH_2CO_2R$$

$$\underset{\underset{KFe(CO)_n}{|}}{CH_2CH_2CO_2R}$$

$$CH_3CH\overset{CO_2R}{\underset{CO_2R}{\diagdown}} \qquad \underset{CH_2CO_2R}{\overset{CH_2CO_2R}{|}}$$

263	*264*

A reagent prepared *in situ* from $Fe(CO)_5$ and a small amount of sodium hydroxide in 95% methanol can be used for the reduction of α,β-unsaturated carbonyl compounds. [301]

$$RCH=CHCOR' \longrightarrow RCH_2CH_2-COR'$$

Reaction of bromides or tosylates which have a terminal double bond with $Na_2Fe(CO)_4$ gave cyclic products by intramolecular insertion of the terminal olefins into the iron-carbon bond. [302] Six-membered ketone formation can be explained by the following mechanism.

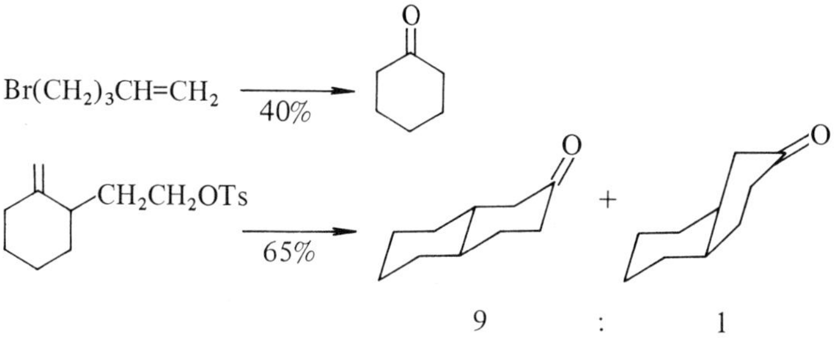

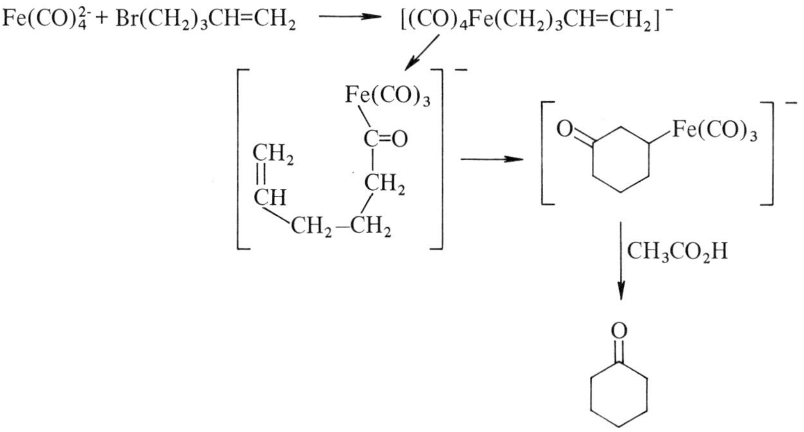

$$9 \quad : \quad 1$$

$$Fe(CO)_4^{2-} + Br(CH_2)_3CH=CH_2 \longrightarrow [(CO)_4Fe(CH_2)_3CH=CH_2]^-$$

5-Bromo-1,2-pentadiene was converted into 2-methylcyclopentenone by a similar intramolecular insertion of the olefinic bond.

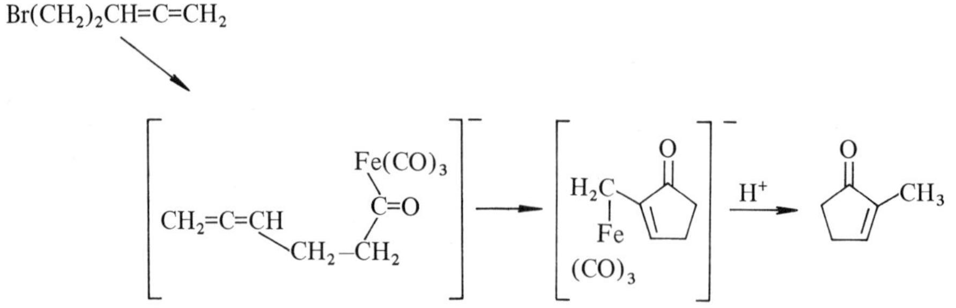

$Br(CH_2)_2CH=C=CH_2$

Reaction of diphenylchlorophosphine with $Na_2Fe(CO)_4$ afforded an anionic phosphide complex *(265)*. The adduct is a potentially ambident nucleophile, but *in situ* alkylation occurred exclusively at phosphorus affording the monomeric phosphine complex *(266)*. [303]

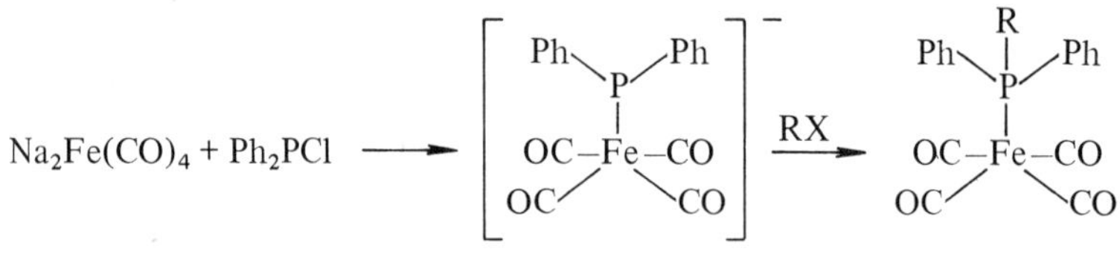

$Na_2Fe(CO)_4 + Ph_2PCl \longrightarrow$

265   *266*

Potassium hexacyanodinickelate $K_4[Ni_2(CN)_6]$ [304] is a complex related to the anionic metal carbonyl. The complex has a dimeric structure with a nickel-nickel bond. [305] The usefulness of this complex is apparent from its facile reaction with a variety of organic halides by splitting the nickel-nickel bond. The complex reacted with benzyl bromide through the formation of a nickel-carbon σ-bond followed by a coupling reaction or carbon monoxide insertion to give bibenzyl or dibenzyl ketone, respectively. [306, 307]

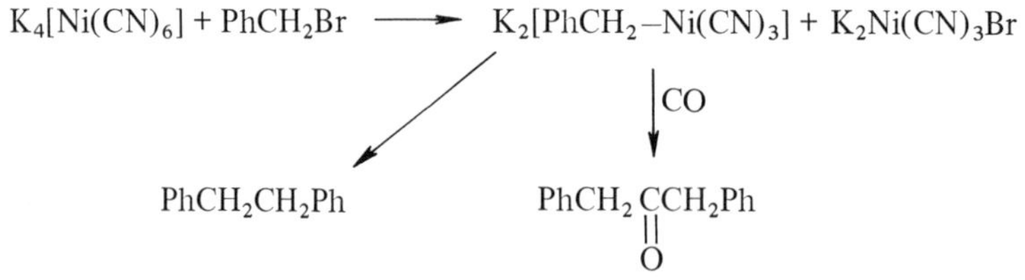

$K_4[Ni(CN)_6] + PhCH_2Br \longrightarrow K_2[PhCH_2-Ni(CN)_3] + K_2Ni(CN)_3Br$

$PhCH_2CH_2Ph$ $PhCH_2 \underset{\underset{O}{||}}{C} CH_2Ph$

Direct cyanation of chemically inactive vinyl halides is also possible with the same complex under mild conditions. [283]

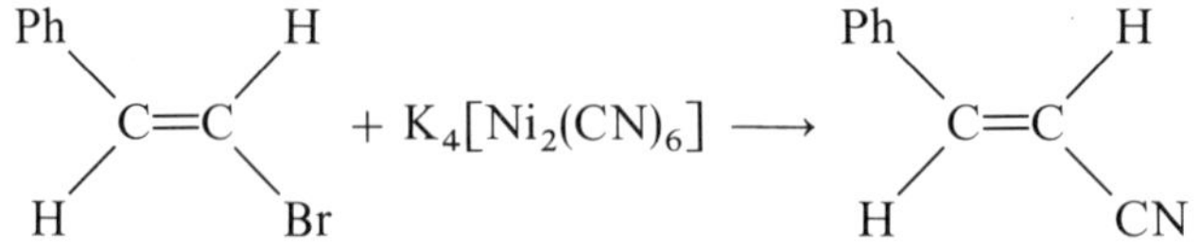

The reaction of *trans*-β-bromostyrene with the same complex in the presence of acrylonitrile in aqueous DMF gave 5-phenyl-4-pentenonitrile *(267)*, 1,4-diphenyl-1,3-butadiene *(268)*, and β-cyanostyrene *(269)*. [308]

$$K_4[Ni(CN)_6] + PhCH{=}CHBr + CH_2{=}CHCN$$

$$\longrightarrow \underset{267}{\underset{56.5\%}{PhCH{=}CHCH_2CH_2CN}} + \underset{268}{\underset{5.2\%}{PhCH{=}CHCH{=}CHPh}}$$

$$+ \underset{269}{\underset{30.7\%}{PhCH{=}CHCN}}$$

Certain dinuclear metal carbonyls give carbonylmetallates. Treatment of $Co_2(CO)_8$ with sodium gives sodium tetracarbonylcobaltate, application of which to organic synthesis will be given later (p. 129).

$$Co_2(CO)_8 + 2Na \longrightarrow 2NaCo(CO)_4$$

c) Chromium, Molybdenum, and Tungsten Complexes

Reactivity of carbonyls of chromium, molybdenum, and tungsten toward organolithium compounds is somewhat lower than that of nickel and iron carbonyls. The most interesting reaction of these Group VIa metal carbonyls, found and studied extensively by Fischer and coworkers, is the formation of stable carbene complexes. [309—313] Phenyl- or methyllithium adds readily to $M(CO)_6$ (M = Cr, Mo, W) at room temperature to form lithium acylmetallate complexes. [314, 271] Treatment of the tetramethylammonium salt *(270)* of these complexes with acid gave the rather unstable hydroxycarbene complexes *(271)*. Hydroxymethyl and hydroxyphenylcarbene complexes were isolated by the treatment of the carbonylate complexes with mineral acid. [315] The treatment of the hydroxycarbene complexes *(271)* with diazomethane gave the neutral methoxycarbene complex *(272)*. [314]

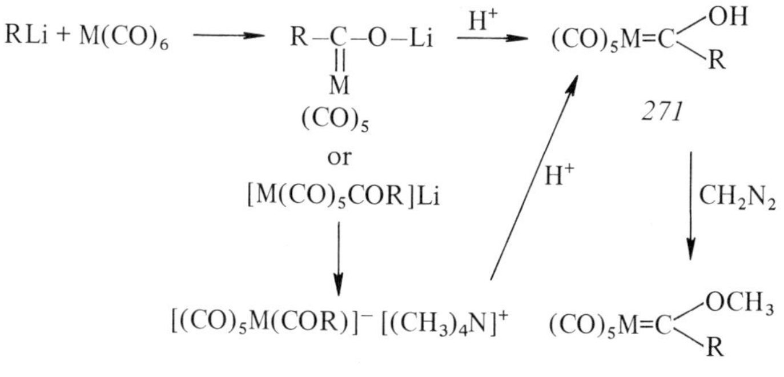

The nature of these carbenoids coordinated transition metal carbonyls is different depending on the identity of metals and ligands. In valence bond description, transition metal complexes of carbenoids exist as a resonance hybrid consisting of "ylene" form *(273)* and two polar forms; the metal-bonded carbanion "ylide" *(274)*, and the metal bonded carbocation "inverse ylide" *(275)*. The relative contribution of these canonical forms is determined by nature of the metals, ligands, and carbene substituents.

$$\left(\; M{=}C{\overset{\displaystyle R}{\underset{\displaystyle R}{\big\backslash}}} \longleftrightarrow M^+{-}C{\overset{\displaystyle R}{\underset{\displaystyle R}{\big\backslash}}} \longleftrightarrow M^-{-}C^+{\overset{\displaystyle R}{\underset{\displaystyle R}{\big\backslash}}} \; \right)$$

273   274   275

The carbene complexes of chromium, molybdenum, and tungsten seem to have the inverse ylide nature *(275)*. The methoxy group serves to stabilize the complexes by electron donation to the electron-deficient carbene carbon atom.

$$\left(\; (CO)_5M{=}C{\overset{\displaystyle Ph}{\underset{\displaystyle OCH_3}{\big\backslash}}} \longleftrightarrow (CO)_5M^-{-}C^+{\overset{\displaystyle Ph}{\underset{\displaystyle OCH_3}{\big\backslash}}} \longleftrightarrow (CO)_5M^-{-}C{\overset{\displaystyle Ph}{\underset{\displaystyle O^+CH_3}{\big\backslash\!\backslash}}} \; \right)$$

In these carbene complexes, the carbenic center is electron deficient due to the presence of the alkoxy group on the carbenic carbon and strong π-accepting carbonyl groups as the ligand on the metal. Extensive studies have been carried out on the reactivities of these carbene complexes.

Thermal decomposition of phenylmethoxycarbene complex gave dimethoxy-stilbene *(276)* by coupling. [316]

$$(CO)_5Cr{=}C{\overset{\displaystyle Ph}{\underset{\displaystyle OCH_3}{\big\backslash}}} \xrightarrow[110°]{C_5H_5N} {\overset{\displaystyle Ph}{\underset{\displaystyle CH_3O}{\big\backslash}}}C{=}C{\overset{\displaystyle OCH_3}{\underset{\displaystyle Ph}{\big\slash}}}$$

276

Attempts to form a cyclopropane by addition to simple olefins of the carbenes generated by thermal-, photochemical-, or pyridine-induced decomposition of these carbene complexes were not always successful. [317] The addition is possible with some activated olefins. Reactions of methoxyphenylcarbene complexes of chromium, tungsten, and molybdenum with methyl *trans*-crotonate *(277)* gave an isomeric mixture of 1-methoxy-1-phenyl-2-carbomethoxy-3-methylcyclopropane *(278)*. [318]

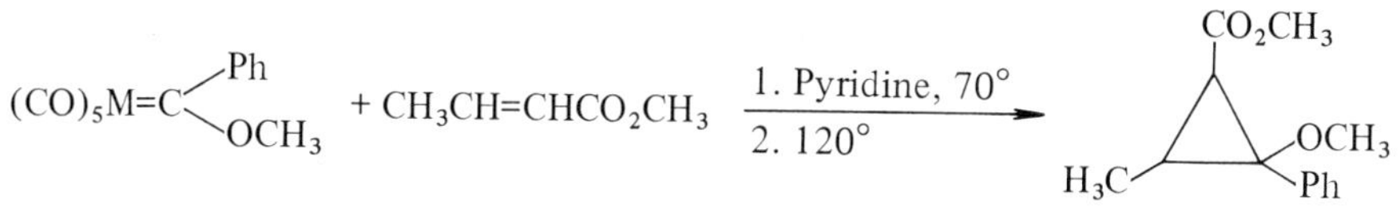
(CO)$_5$M=C(Ph)(OCH$_3$) + CH$_3$CH=CHCO$_2$CH$_3$
1. Pyridine, 70°
2. 120°
CO$_2$CH$_3$
OCH$_3$
Ph
H$_3$C

277 *278*

The reaction with diethyl fumarate gave only one isomer of *trans*-1,2-dicarboethoxy-3-methoxy-3-phenylcyclopropane *(279)*.

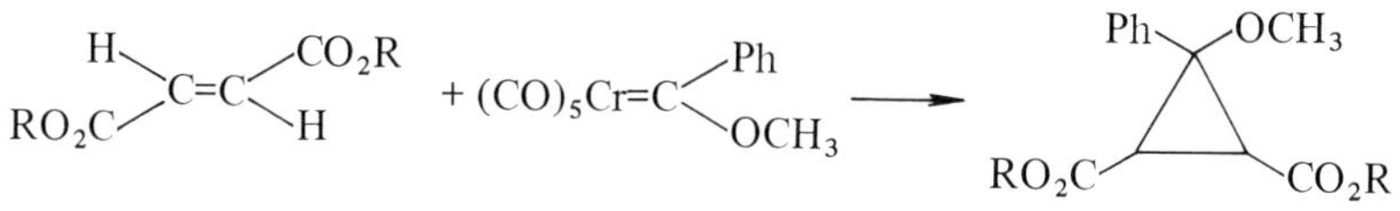
H
CO$_2$R
RO$_2$C
H
+ (CO)$_5$Cr=C(Ph)(OCH$_3$)
Ph
OCH$_3$
RO$_2$C
CO$_2$R

279

The same carbene ligand reacted with ethyl vinyl ether to give the two isomeric 1-methoxy-1-phenyl-2-ethoxycyclopropanes *(280)* and *(281)*, the ratio of which depends on the metals. Free methoxyphenylcarbene is not involved in the reaction, since the metal was observed to influence the ratio of *(280)* and *(281)*. [319]

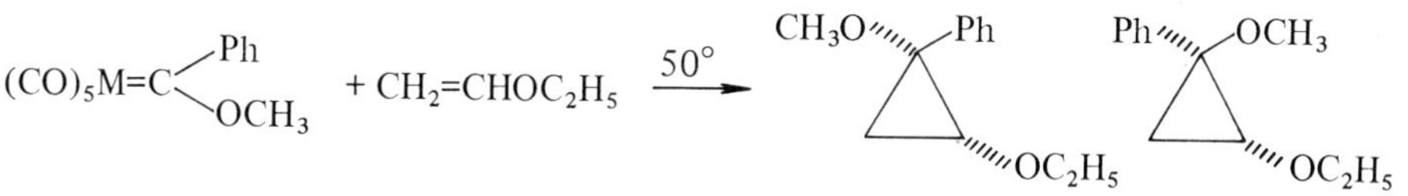
(CO)$_5$M=C(Ph)(OCH$_3$) + CH$_2$=CHOC$_2$H$_5$
50°
CH$_3$O
Ph
OC$_2$H$_5$
Ph
OCH$_3$
OC$_2$H$_5$

	280	*281*
M=Cr	76%	24%
M=W	64%	36%

Further evidence of the absence of a free carbene in the preparation of the cyclopropane from the metal-carbene complexes was obtained by the experiment based on chirality. The chromium-carbene complex *(283)* coordinated by the

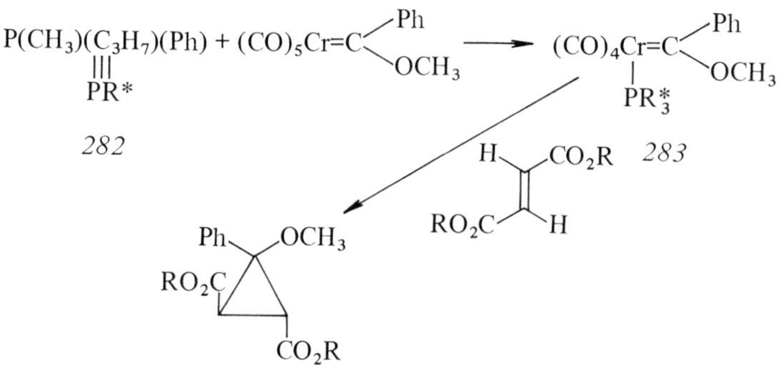
P(CH$_3$)(C$_3$H$_7$)(Ph) + (CO)$_5$Cr=C(Ph)(OCH$_3$)
PR*
282
(CO)$_4$Cr=C(Ph)(OCH$_3$)
PR$_3^*$
283
H
CO$_2$R
RO$_2$C
H
Ph
OCH$_3$
RO$_2$C
CO$_2$R

284

optically active phosphine, (−)(R)-methylphenyl propylphosphine *(282)* was prepared, and the complex was reacted with diethyl fumarate by heating. The cyclopropane derivative *(284)* obtained was optically active. This result indicates clearly that the cyclopropane was formed by the transfer of the carbene within the metal complex, not as the free carbene. [320]

The oxazoline derivatives *(285)* were obtained by cycloaddition reaction of alkoxy(aryl)carbene pentacarbonylchromium with N-acylimines of hexafluoro-acetone *(286)*. [321]

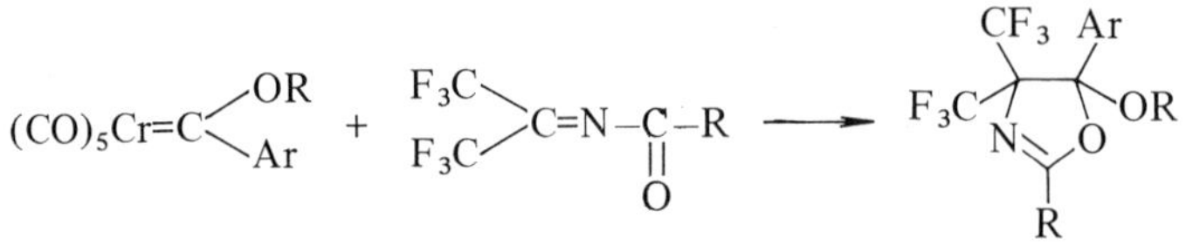

$$(CO)_5Cr=C\diagdown^{OR}_{Ar} \quad + \quad ^{F_3C}_{F_3C}\diagup C=N-\underset{O}{\overset{||}{C}}-R \quad \longrightarrow \quad \text{285}$$

$$286 \qquad\qquad\qquad 285$$

The electrophilic character of the carbenoid carbon is apparent from reactions of the complexes with some nucleophiles such as isocyanate, hydrazine, hydro-xylamine, phenyllithium, and Wittig reagent. [322, 323] The carbene part of the chromium complex *(287)* was converted into vinyl ether by the action of pyridine. [324] Dimethylhydrazine yielded the nitrile complex *(288)*.

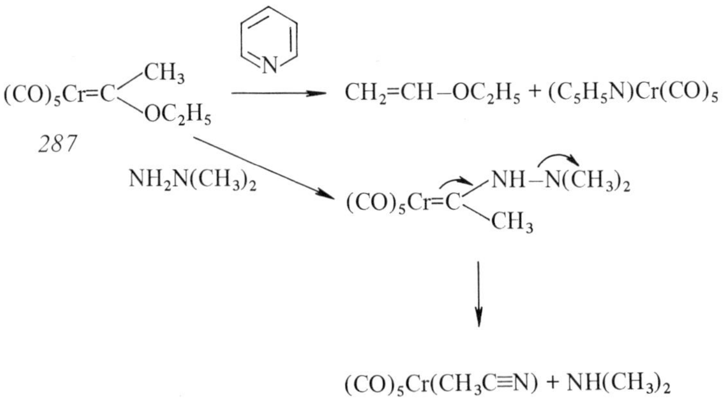

$$(CO)_5Cr=C\diagdown^{CH_3}_{OC_2H_5}$$

$$287$$

$$\longrightarrow CH_2=CH-OC_2H_5 + (C_5H_5N)Cr(CO)_5$$

$$NH_2N(CH_3)_2 \searrow \qquad (CO)_5Cr=C\diagdown^{NH-N(CH_3)_2}_{CH_3}$$

$$(CO)_5Cr(CH_3C≡N) + NH(CH_3)_2$$

$$288$$

Aminolysis of methoxy(phenyl)carbene pentacarbonychromium with amino acid esters yielded the corresponding aminocarbene complexes *(289)*. [325] By this reaction, the amino group can be protected, which may allow peptide synthesis. The ester was hydrolyzed and the reaction of another amino acid ester gave the dipeptide complex *(290)*. The protected amino group of the di-peptide can be cleaved by treatment with trifluoroacetic acid at 20°.

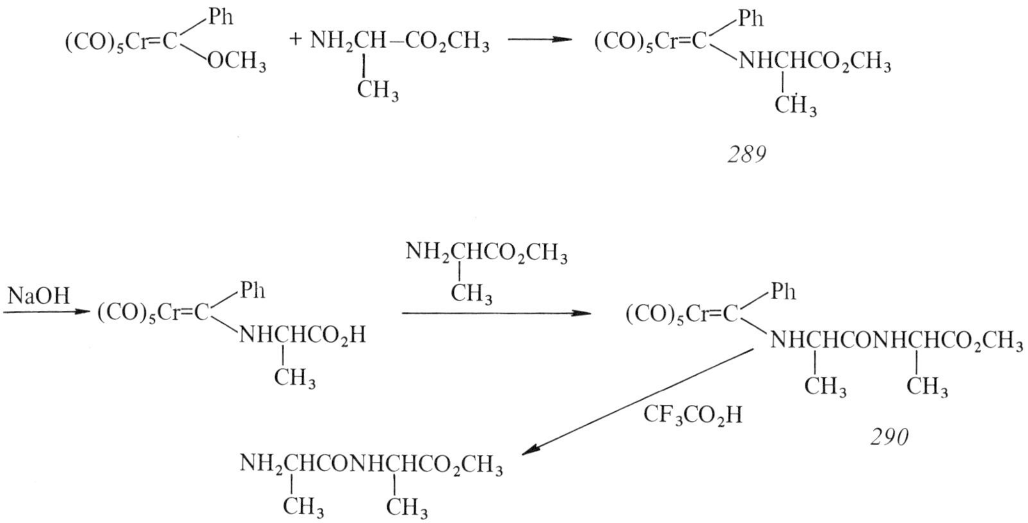

$$(CO)_5Cr{=}C\underset{OCH_3}{\overset{Ph}{\big<}} + NH_2CH{-}CO_2CH_3 \longrightarrow (CO)_5Cr{=}C\underset{NHCHCO_2CH_3}{\overset{Ph}{\big<}}$$

with CH_3 on the second carbon, and CH_3 on the product.

289

$$\xrightarrow{NaOH} (CO)_5Cr{=}C\underset{NHCHCO_2H}{\overset{Ph}{\big<}} \xrightarrow{\underset{CH_3}{\overset{NH_2CHCO_2CH_3}{}}} (CO)_5Cr{=}C\underset{NHCHCONHCHCO_2CH_3}{\overset{Ph}{\big<}}$$

with CH_3 on both carbons.

290

$$\xrightarrow{CF_3CO_2H} NH_2CHCONHCHCO_2CH_3$$

with CH_3 and CH_3.

1,2-Dimethoxy-1,1,2,2-tetraphenylethane *(291)* was obtained among other products by the reaction of phenyllithium. [309]

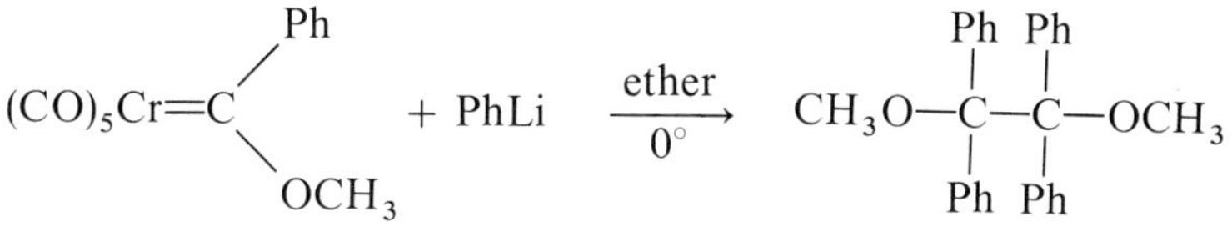

$$(CO)_5Cr{=}C\underset{OCH_3}{\overset{Ph}{\big<}} + PhLi \xrightarrow[0°]{ether} CH_3O{-}\underset{\underset{Ph}{|}}{\overset{\overset{Ph}{|}}{C}}{-}\underset{\underset{Ph}{|}}{\overset{\overset{Ph}{|}}{C}}{-}OCH_3$$

291

When the phenylmethoxycarbene complex of tungsten was treated with phenyllithium and then with hydrogen chloride at $-78°$, diphenylcarbene pentacarbonyltungsten *(292)* was isolated. Due to the lack of the stabilizing effect of the methoxy group, the diphenylcarbene complex is less stable than the original methoxycarbene complex. [326]

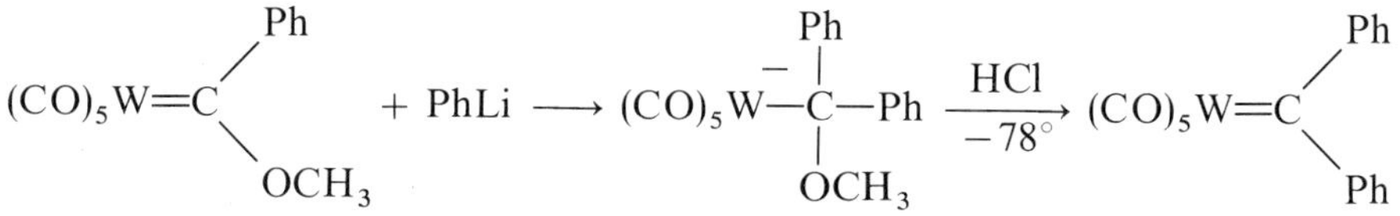

$$(CO)_5W{=}C\underset{OCH_3}{\overset{Ph}{\big<}} + PhLi \longrightarrow (CO)_5\overset{-}{W}{-}\underset{\underset{OCH_3}{|}}{\overset{\overset{Ph}{|}}{C}}{-}Ph \xrightarrow[-78°]{HCl} (CO)_5W{=}C\underset{Ph}{\overset{Ph}{\big<}}$$

292

Another example of nucleophilic attack on the carbene complex was observed with methylidenetriphenylphosphorane (Wittig reagent) *(293)* to give methyl 1-phenylvinyl ether *(294)* and pentacarbonyltriphenylphosphine complex *(295)*. [327]

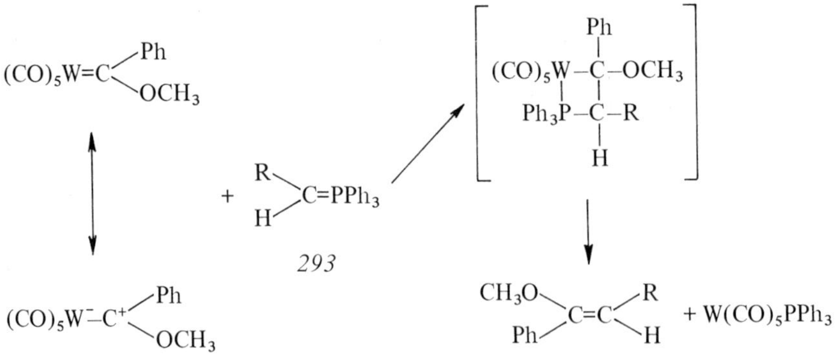

$$(CO)_5W=C\begin{smallmatrix} Ph \\ \\ OCH_3 \end{smallmatrix}$$

$$+ \quad \begin{smallmatrix} R \\ \\ H \end{smallmatrix}C=PPh_3$$

293

$$\left[\begin{smallmatrix} & & Ph \\ & & | \\ (CO)_5W-C-OCH_3 \\ & & | \\ Ph_3P-C-R \\ & & | \\ & & H \end{smallmatrix} \right]$$

$$(CO)_5W-C^+\begin{smallmatrix} Ph \\ \\ OCH_3 \end{smallmatrix}$$

$$\begin{smallmatrix} CH_3O \\ \\ Ph \end{smallmatrix}C=C\begin{smallmatrix} R \\ \\ H \end{smallmatrix} \quad + W(CO)_5PPh_3$$

294   295

However, the reaction of the phosphoranes with alkylmethoxycarbene complex failed to produce the vinyl ether due to abstraction of a highly acidic hydrogen from the carbon α to the carbene carbon atom. For this purpose, more weakly basic diazoalkanes are suitable instead of the phosphoranes. Thus facile reaction took place with diazomethane to give a high yield of the enol ether (296). [328]

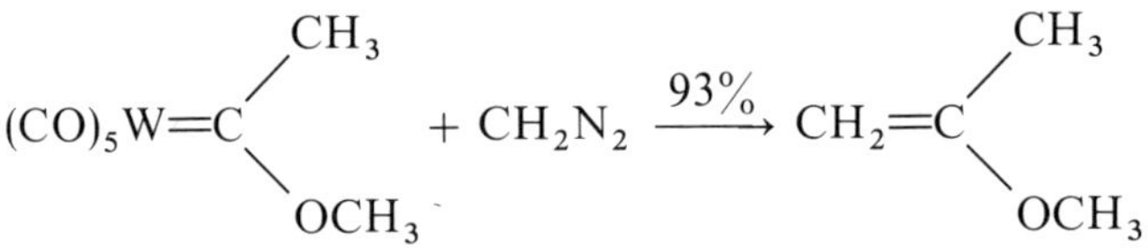

$$(CO)_5W=C\begin{smallmatrix} CH_3 \\ \\ OCH_3 \end{smallmatrix} \quad + CH_2N_2 \xrightarrow{93\%} CH_2=C\begin{smallmatrix} CH_3 \\ \\ OCH_3 \end{smallmatrix}$$

296

The reaction seems to proceed by nucleophilic attack of the diazo carbon atom at the electron-deficient carbene carbon to form the betain-like intermediate (297), which subsequently fragments to form the enol ether (296), nitrogen, and coordinatively unsaturated pentacarbonyltungsten (298).

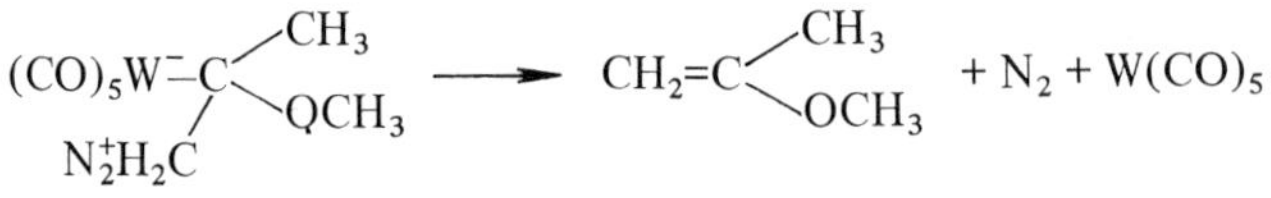

$$(CO)_5W^-\underset{N_2^+H_2C}{\overset{}{C}}\begin{smallmatrix} CH_3 \\ \\ OCH_3 \end{smallmatrix} \longrightarrow CH_2=C\begin{smallmatrix} CH_3 \\ \\ OCH_3 \end{smallmatrix} \quad + N_2 + W(CO)_5$$

297   296   298

The above reactions suggest that the carbon α to the carbene carbon should readily give an anion by the reaction with a strong base. The following reaction proved this hypothesis. The reaction of butyllithium with the tungsten-carbene complex generated the anion (299), which reacted with methyl fluorosulfonate to give ethylmethoxycarbene pentacarbonyltungsten (300). The reaction of the anion with acetyl chloride also gave the enol acetate complex (301). Methyl 3-acetoxy-2-butenoate (302) was obtained by the oxidation of this complex with ceric ammonium nitrate. [329]

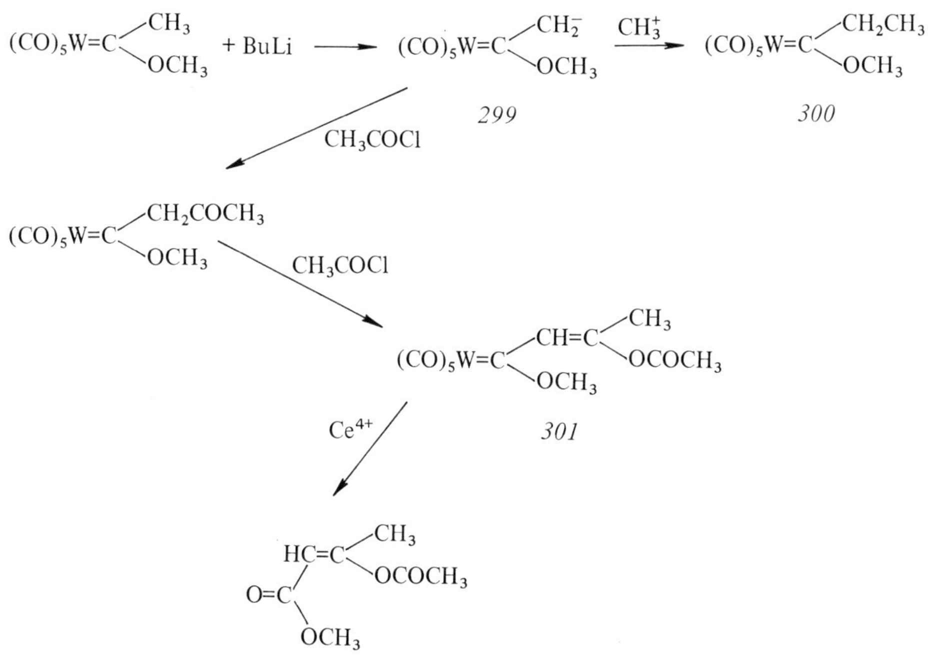

$(CO)_5W=C{\overset{CH_3}{\underset{OCH_3}{}}}$ + BuLi $\longrightarrow$ $(CO)_5W=C{\overset{CH_2^-}{\underset{OCH_3}{}}}$ $\xrightarrow{CH_3^+}$ $(CO)_5W=C{\overset{CH_2CH_3}{\underset{OCH_3}{}}}$

299   *300*

CH₃COCl

$(CO)_5W=C{\overset{CH_2COCH_3}{\underset{OCH_3}{}}}$

CH₃COCl

$(CO)_5W=C{\overset{CH=C{\overset{CH_3}{\underset{OCOCH_3}{}}}}{\underset{OCH_3}{}}}$

301

Ce⁴⁺

$O=C{\overset{HC=C{\overset{CH_3}{\underset{OCOCH_3}{}}}}{\underset{OCH_3}{}}}$

302

In connection with the reactivity of the carbene complexes, it should be noted that there is a structural similarity between transition metal-carbene complexes and the corresponding carbonyl compounds. [330] As shown below, the similarity in the mechanism of the aminolysis of esters and of the methoxycarbene complexes has been pointed out.

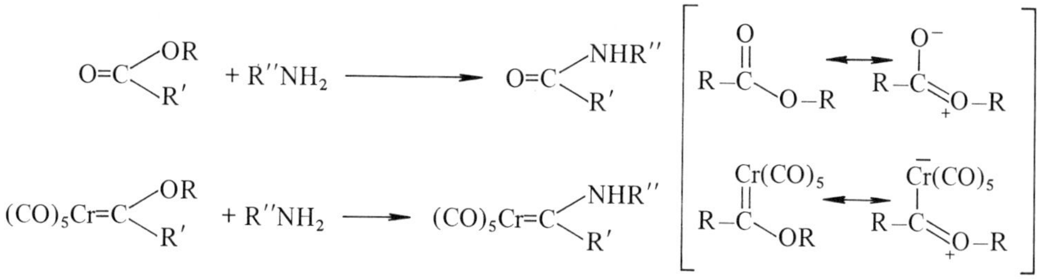

$O=C{\overset{OR}{\underset{R'}{}}}$ + R''NH₂ $\longrightarrow$ $O=C{\overset{NHR''}{\underset{R'}{}}}$

$\left[\quad R-\overset{\overset{O}{\|}}{C}{\underset{O-R}{}} \longleftrightarrow R-\overset{\overset{O^-}{\|}}{C}{\underset{\overset{+}{O}-R}{}} \right.$

$(CO)_5Cr=C{\overset{OR}{\underset{R'}{}}}$ + R''NH₂ $\longrightarrow$ $(CO)_5Cr=C{\overset{NHR''}{\underset{R'}{}}}$

$\left. R-\overset{\overset{Cr(CO)_5}{\|}}{C}{\underset{OR}{}} \longleftrightarrow R-\overset{\overset{\overline{Cr}(CO)_5}{|}}{C}{\underset{\overset{+}{O}-R}{}} \quad\right]$

The reaction of diphenylcyclopropenylidene complex of molybdenum *(303)* with pyridinium ylides gave the stable pyran-2-ylidene complex *(304)*. Similarly 2-pyrone formation took place by the reaction of diphenylcyclopropenone *(305)* with pyridinium ylide. [331] Furthermore, the complex *(304)* was converted into the pyrone *(306)* by oxidation with lead tetraacetate.

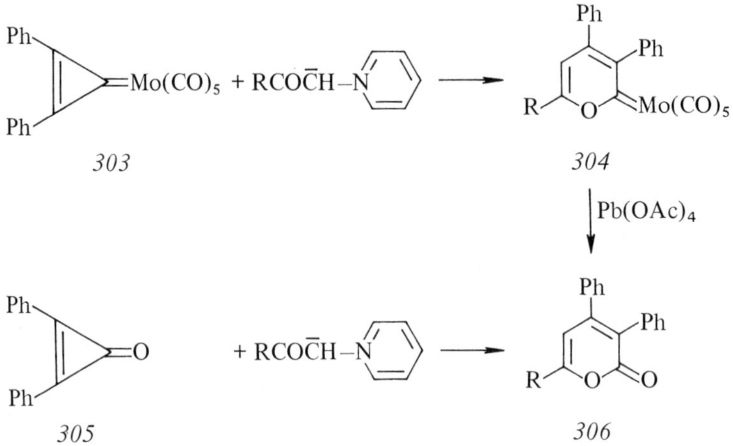

303 + RCOCH—N⟨⟩ → *304*

305 + RCOCH—N⟨⟩ → *306*

From the carbene complexes, several interesting complexes have been prepared. Reaction of phenylmethoxycarbene complex of chromium with 1,4-diazabicyclo[2,2,2]octane *(307)* gave the novel nitrogen ylide complex *(308)*. [332]

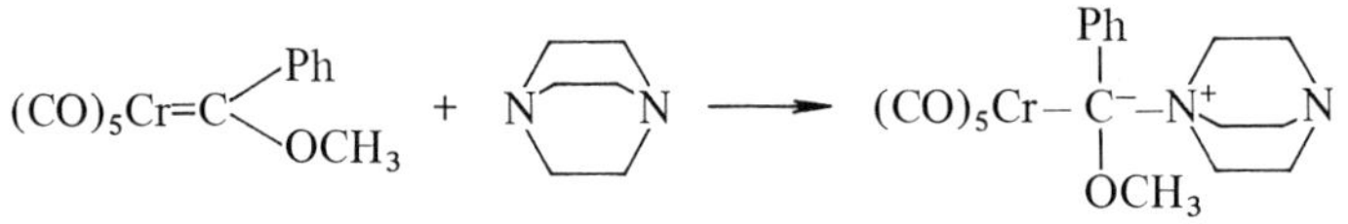

307   *308*

Reaction of boron trihalide on the carbene complex removed the methoxy group to give an interesting carbyne-metal complexes having a transition metal-carbon triple bond *(309)*. [333, 334]

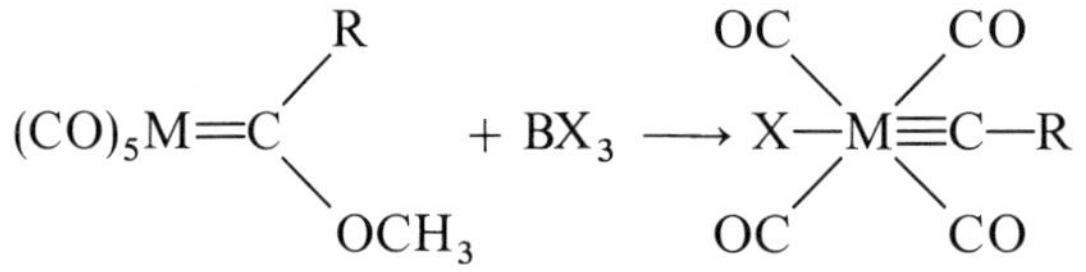

309

IV. Reactivities of σ-Bonds Involving Transition Metals

In the foregoing chapters, various methods of σ-bond formation were surveyed. An important factor to be considered next is the reactivities of the σ-bonds thus formed. In a number of the examples cited in the preceding chapters, the reactivities of various σ-bonds are shown. Further treatment of this problem is given in this chapter.

A carbon atom bonded to electropositive metals such as lithium and magnesium is anionic. In Grignard reactions, the reagents are assumed to have a carbanionic character and their reactions can be understood by the nucleophilic character of the carbon directly bonded to magnesium. On the other hand, the character of the carbon-transition metal and hydrogen-transition metal bonds formed by oxidative addition, insertion reactions and other reactions are much more versatile and cannot be predicted unequivocally. The character depends to a certain degree on the metallic species. For example, the carbon atom bonded to nickel or iron in some complexes does behave like a nucleophile as shown earlier in several examples. The Grignard-like reactions of nickel complexes are typical. However, whether carbon in a transition metal-carbon bond reacts with electrophiles or nucleophiles (or both) depends on many factors in addition to the identity of the metals. Factors such as oxidation state of metals, electronegativity, number and kind of ligands, stability of reduced metal species and reaction medium, contribute to the character of the transition metal-carbon bonds. In addition, the nature of substrates with which the metal complexes react is crucial for determining whether the carbon moiety acts as either a nucleophile or an electrophile.

Different reactivities of various nickel-carbon bonds, classified by Chiusoli, [335] are good examples for understanding the complexity of this problem. The nickel-carbon bonds are cleaved in different ways depending on whether the bonds are part of alkyl, vinyl, ethynyl, aryl, or acyl groups. Thus the nickel-acyl bond, commonly generated in various carbonylation reactions using $Ni(CO)_4$, is cleaved by nucleophilic reagents such as water, alcohols, and various anions.

$$R-\overset{\overset{\displaystyle O}{\|}}{C}-\overset{\overset{\displaystyle |}{}}{Ni}-X + H_2O \longrightarrow RCO_2H + Ni^{2+} + HX$$

Nucleophilic substitution reaction of aromatic halides is also facilitated by the zerovalent nickel complex (p. 40).

On the other hand, alkyl, vinyl, or aryl groups bonded to nickel are usually attacked by electrophilic reagents. The reaction with a carbonyl group is a typical example.

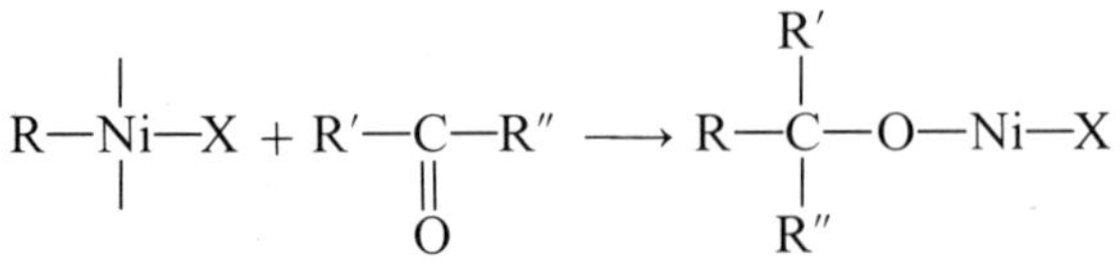

$$R-Ni-X + R'-\underset{\underset{O}{\|}}{C}-R'' \longrightarrow R-\underset{\underset{R''}{|}}{\overset{\overset{R'}{|}}{C}}-O-Ni-X$$

Nickel-allyl bonds are intermediate in nature, and attacked by both electrophilic and nucleophilic reagents depending on the electronegativity of the allyl groups.

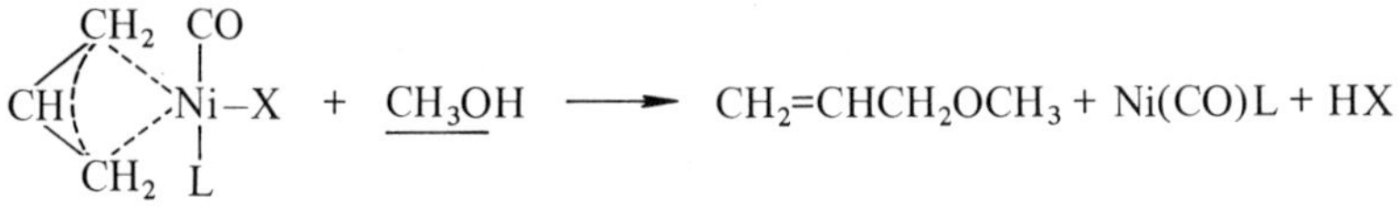

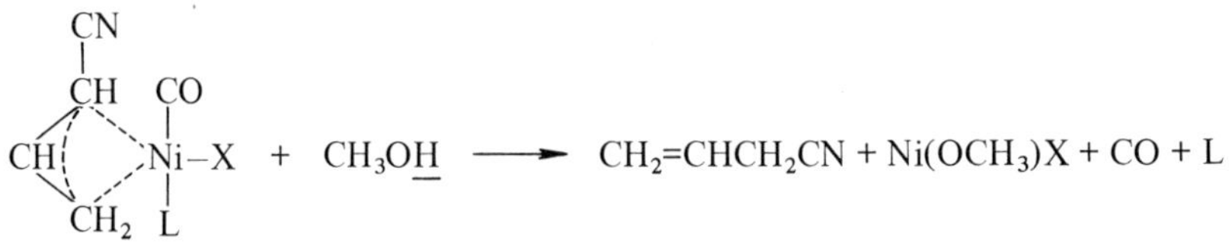

$$\text{allyl-Ni(CO)X} + CH_3OH \longrightarrow CH_2=CHCH_2OCH_3 + Ni(CO)L + HX$$

$$\text{CN-allyl-Ni(CO)X} + CH_3OH \longrightarrow CH_2=CHCH_2CN + Ni(OCH_3)X + CO + L$$

In the first case, the methoxy anion attacks the allyl carbon and zerovalent nickel is formed. In the second reaction, the proton of methanol attacks the cyano-substituted allylic ligand and at the same time bivalent nickel is liberated.

Another example of variation of the character of the carbon bonded to different transition metals is found in the reactions of π-allylic complexes of palladium and nickel. Both π-allylic complexes have similar structures, but are different in their reactivity. For example, π-allylpalladium chloride reacts with nucleophiles, such as enamines and active methylene compounds, where the reduction of bivalent palladium to the zerovalent state is involved. This property is partly due to the high stability of the reduced species. π-Allylpalladium chloride reacted with malonate anion to give allylmalonate. [336, 337]

$$\text{allyl-PdCl} + {}^-CH(CO_2R)_2 \longrightarrow CH_2=CHCH_2-CH(CO_2R)_2 + Pd + Cl^-$$

This reaction was extended to π-allylpalladium complexes formed from olefins and palladium chloride. [114]

On the other hand, the same π-allylnickel complex reacts with a few electrophilic reagents such as carbonyls. [153] For example, π-allylnickel halide reacted with benzaldehyde, cyclopentanone, and acrolein to give alcohols. [338, 339] The reaction is not general; no reaction took place with benzophenone.

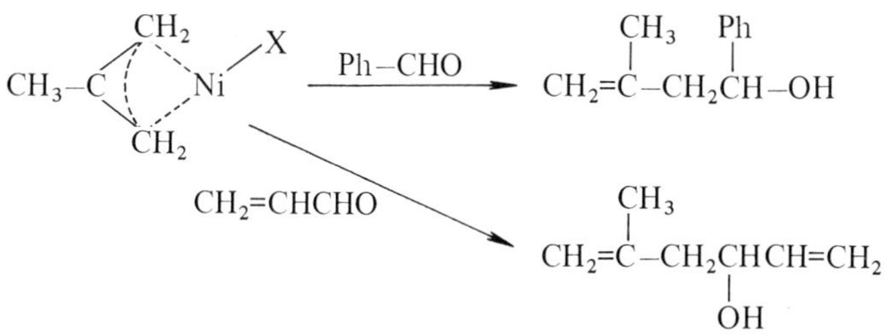

$$CH_3-C\underset{CH_2}{\overset{CH_2}{\langle}}Ni-X \xrightarrow{Ph-CHO} \underset{}{CH_2=\overset{CH_3}{\underset{}{C}}-CH_2\overset{Ph}{\underset{}{CH}}-OH}$$

$$\xrightarrow{CH_2=CHCHO} CH_2=\overset{CH_3}{\underset{}{C}}-CH_2\overset{}{\underset{OH}{CH}}CH=CH_2$$

Styrene oxide was attacked by the allyl group at the α carbon.

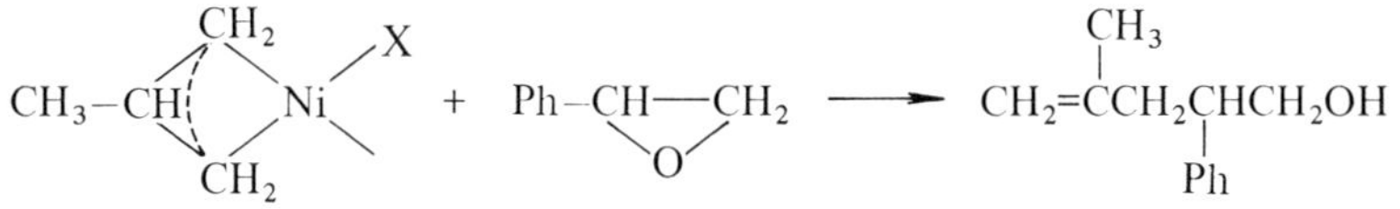

$$CH_3-CH\underset{CH_2}{\overset{CH_2}{\langle}}Ni-X + Ph-\underset{O}{CH-CH_2} \longrightarrow CH_2=\overset{CH_3}{\underset{}{C}}CH_2\overset{}{\underset{Ph}{CH}}CH_2OH$$

In addition to the π-allylnickel halide shown above, bis-π-allylnickel complexes also react with similar carbonyl compounds. A zerovalent nickel complex reacts with butadiene to form a bis-π-allylnickel complex, which in turn reacts with acetone. Notably, two moles of acetone were introduced at the terminal carbons of butadiene. [340]

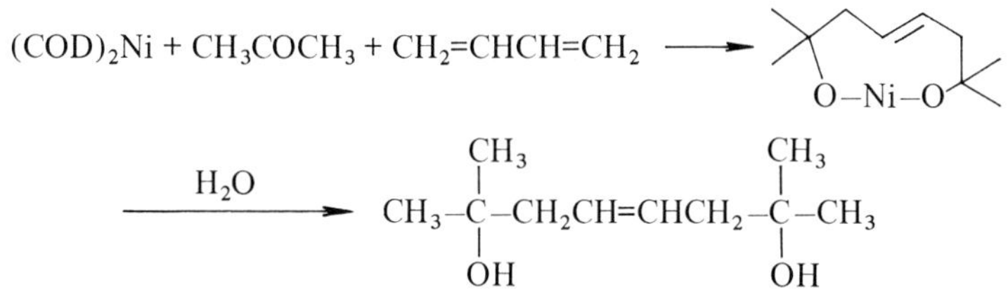

$$(COD)_2Ni + CH_3COCH_3 + CH_2=CHCH=CH_2 \longrightarrow$$

$$\xrightarrow{H_2O} CH_3-\overset{CH_3}{\underset{OH}{C}}-CH_2CH=CHCH_2-\overset{CH_3}{\underset{OH}{C}}-CH_3$$

Similarly the reaction of dodecatrienylnickel *(310)* with one and two moles of acetaldehyde gave the expected alcohols *(311)*, and *(312)* respectively. The ketones *(313)* and *(314)* were obtained by the reaction of acetyl chloride. [341]

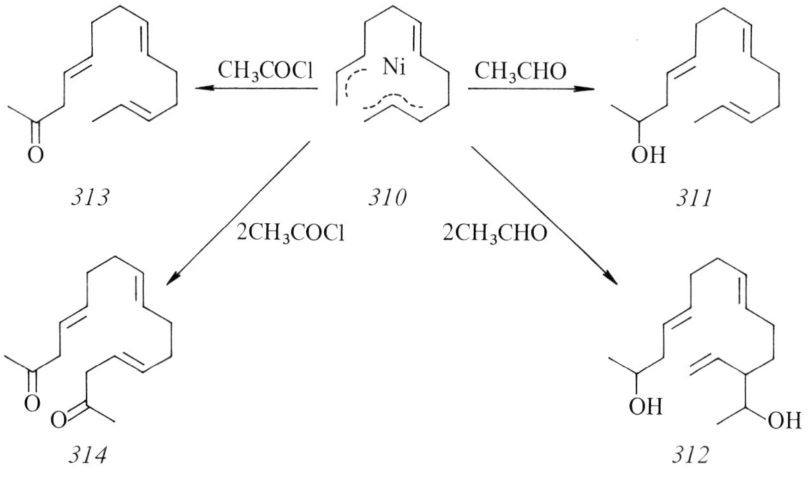

313 *310* *311*

314 *312*

As described above, π-allyl complexes of palladium and nickel behave differently to each other. Another example of different reactivity is observed in the oligomerization of butadiene catalyzed by nickel and palladium complexes. Reactions of butadiene catalyzed by both nickel and palladium complexes proceed *via* formation of π-allylic complexes, but again the products are quite different. The nickel-catalyzed reaction of butadiene, as is well known, affords unsaturated cyclic compounds, mainly 1,5-COD and 1,5,9-CDT. This reaction is discussed in Chapter VII. On the other hand, no cyclization of butadiene takes place in the palladium-catalyzed reactions of butadiene. The main feature of the palladium-catalyzed reaction of butadiene is the incorporation of various nucleophiles to give the 1-substituted 2,7-octadiene *(315)*, accompanied by a small amount of the 3-substituted 1,7-octadiene *(316)*. [150, 342—344] Efficient telomerization of butadiene and nucleophilic reagents takes place in the presence of phosphine complexes of palladium. As nucleophilic reagents, water, alcohols, carboxylic acids, amines, phenol, active methyelene compounds, and enamines take part in the reaction.

$$2CH_2{=}CHCH{=}CH_2 + Y{-}H \longrightarrow \begin{cases} CH_2{=}CHCH_2CH_2CH_2CH{=}CHCH_2{-}Y \\ \qquad\qquad\qquad 315 \\ CH_2{=}CHCH_2CH_2CH_2\underset{\underset{Y}{|}}{C}HCH{=}CH_2 \end{cases}$$

316

This oligomerization reaction can be explained by the attack of these nucleophiles on the π-allylic complex *(317)*, formed from butadiene as an intermediate of the reaction. An experiment with CH_3OD showed that the hydrogen of the nucleophiles(Y—H) migrates to C_6 of the dimeric intermediate chain of the complex *(317)* probably through the oxidative addition reaction of the Y—H bond to the palladium. [345]

$$CH_2{=}CHCH{=}CH_2 \; + \; PdL_n \longrightarrow \overset{1}{\underset{6}{\text{Pd}}} \xrightarrow[CH_3O{-}D]{\text{1,6-addition of}} CH_2{=}CHCHCH_2CH_2CH{=}CHCH_2 $$

317

Although some of these nucleophiles, such as alcohols and amines, can be incorporated into the dimer or monomer of butadiene by using nickel catalysts, the palladium complexes are more efficient and selective catalysts for the telomerization. [346—349] The telomers prepared by this method are useful compounds in that they have a functional group at the end of the carbon chain and a double bond at the other end. For example, n-octanol can be prepared by the reaction of water and butadiene, [350] followed by hydrogenation. Under normal conditions, the reaction of water with butadiene to give 2,7-octadienyl alcohol *(318)* takes place only to a small extent. Efficient telomerization with water was found to occur in the presence of a considerable amount of carbon dioxide in solvents like butyl alcohol, acetone, and acetonitrile.

$$CH_2=CHCH=CH_2 + H_2O \longrightarrow CH_2=CHCH_2CH_2CH_2CH=CHCH_2OH$$
$$318$$

The reaction of acetic acid and butadiene to give a mixture of octadienyl acetates (319, 320), carried out without proper solvents is too slow to be satisfactory from a practical standpoint, whereas the reaction occurs smoothly in proper solvents. A more remarkable effect was observed by the addition of basic compounds. A marked effect of addition of a molar amount of tertiary amines such as 2-(N,N-diethylamino)ethanol has been reported. [351]

$$CH_2=CHCH=CH_2 + CH_3CO_2H$$
$$\longrightarrow CH_2=CHCH_2CH_2CH_2CH=CHCH_2OCOCH_3$$
$$319$$

$$CH_2=CHCH_2CH_2CH_2CHCH=CH_2$$
$$\overset{|}{O}COCH_3$$
$$320$$

Amines produced by the reaction of ammonia and amines with butadiene are useful. The reaction with nitroalkanes and reduction of the resulting unsaturated nitro compounds produce amines bearing long alkyl chains, which have a primary amino group at the center of the carbon chain, [352] and are different from the usual amines which have the amino group at the terminal position. The following products *(321, 322, 323)* were obtained from nitromethane.

$$CH_3NO_2 + CH_2=CHCH=CH_2 \longrightarrow$$

$$\begin{cases} CH_2=CHCH_2CH_2CH_2CH=CHCH_2CH_2NO_2 \\ \qquad\qquad 321 \\ (CH_2=CHCH_2CH_2CH_2CH=CHCH_2)_2CHNO_2 \\ \qquad\qquad 322 \\ (CH_2=CHCH_2CH_2CH_2CH=CHCH_2)_3CNO_2 \\ \qquad\qquad 323 \end{cases}$$

Unexpected products were obtained by the reaction of butadiene with aldehydes catalyzed by palladium complexes. The reaction of aldehydes and butadiene in the presence of palladium acetate and triphenylphosphine gave 1-substituted 2-vinyl-4,6-heptadien-1-ol *(324)* and 2-substituted 3,6-divinyltetrahydropyrans *(325)*. [353—356]

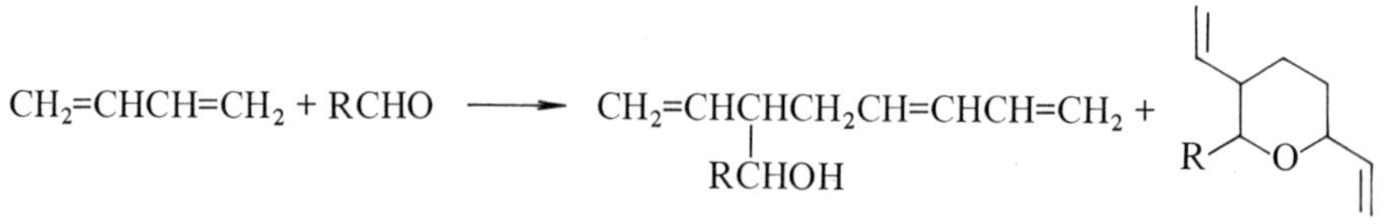
CH_2=CHCH=CH_2 + RCHO
CH_2=CHCHCH$_2$CH=CHCH=CH_2 +
RCHOH
R
O

324 *325*

As previously described (p. 95), π-allylic nickel complexes react with aldehydes stoichiometrically to give alcohols. In view of the fact that aldehydes may undergo oxidative addition to palladium complexes, it is rather peculiar that ketones are not formed *via* oxidative addition followed by insertion of butadiene. The reaction is discussed again in Chapter VII.

Reactions of allene catalyzed by nickel and palladium complexes are also markedly different. Palladium complexes catalyze the dimerization of allene with incorporation of nucleophiles. [357] 3-Methyl-2-methylene-3-butenyl acetate *(326)* was obtained as a main product by the reaction of allene with acetic acid by the following mechanism. Also amines, ammonia, and active methylene compounds such as malonate react similarly with two moles of allene. [358]

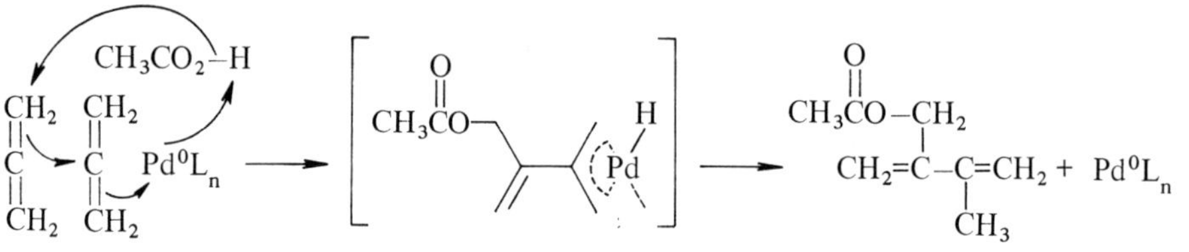
CH_3CO_2–H
CH_2 CH_2
C C Pd^0L_n
CH_2 CH_2
O
CH_3CO
H
Pd
O
CH_3CO–CH_2
CH_2=C–C=CH_2 + Pd^0L_n
CH_3

326

Similar reactions of nucleophiles with allene are also promoted by nickel complexes. [359] The main product obtained by the reaction of allene and amine in the presence of a nickel complex containing phenyldiisopropylphosphine was the oligomer *(326)* formed from three moles of allene and one mole of the amine.

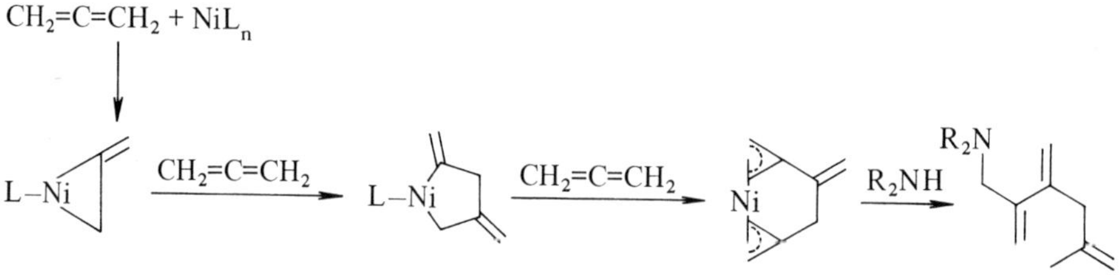
CH_2=C=CH_2 + NiL_n
L–Ni
CH_2=C=CH_2
L–Ni
CH_2=C=CH_2
Ni
R_2NH
R_2N

326

In addition to the reactions shown above which are catalyzed by the palladium and nickel complexes, some iron complexes induce addition reaction of active methylene compounds and nucleophiles. The mode of the reaction, however, is completely different. Dieneiron tetracarbonyl was converted to π-allyliron tetracarbonyl cation *(327)* by protonation with tetrafluoroboric acid. [360] The cationic complex is subjected to react with a wide variety of nucleophiles to give products which arise from 1,4-addition across the diene as shown by the following scheme.

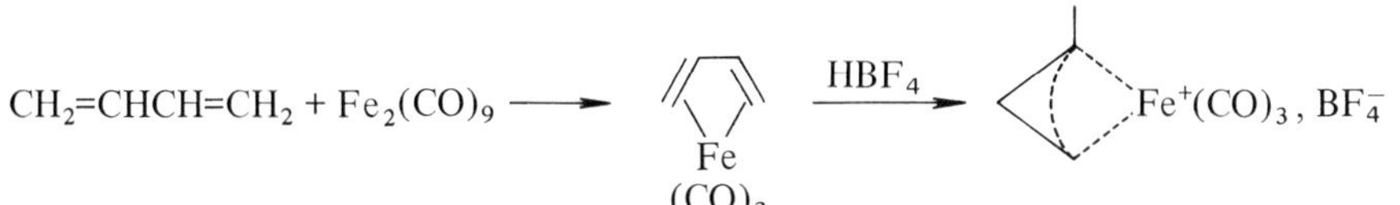

$$CH_2=CHCH=CH_2 + Fe_2(CO)_9 \longrightarrow \underset{\substack{Fe \\ (CO)_3}}{\text{[diene]}} \xrightarrow{HBF_4} \underset{327}{\text{[complex]}} Fe^+(CO)_3, BF_4^-$$

$$\xrightarrow{YH} CH_3CH=CHCH_2-Y$$

The onium salts *(328)* were isolated by the reaction of pyridine or phosphine. [361]

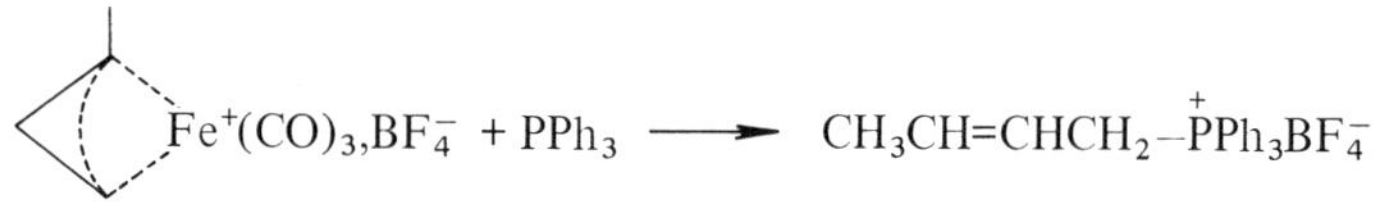

$$\text{[complex]}Fe^+(CO)_3, BF_4^- + PPh_3 \longrightarrow \underset{328}{CH_3CH=CHCH_2-\overset{+}{P}Ph_3BF_4^-}$$

Methyl acetoacetate attacked the π-allyliron complex to give methyl allylacetoacetate *(329)*.

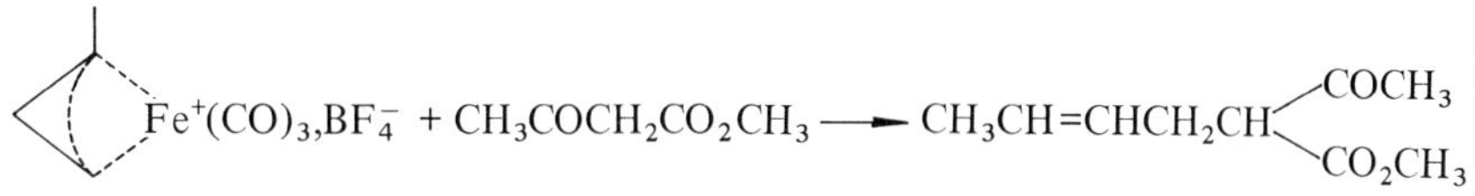

$$\text{[complex]}Fe^+(CO)_3, BF_4^- + CH_3COCH_2CO_2CH_3 \longrightarrow CH_3CH=CHCH_2CH\underset{CO_2CH_3}{\overset{COCH_3}{<}}$$

$$329$$

The products of the Birch reduction of aromatic compounds reacted with iron carbonyl to give (cyclohexa-1,4-diene)iron tricarbonyl as a mixture of isomers. Hydride abstraction from the complexes with triphenylmethyl fluoroborate formed cationic complexes, which are susceptible to various nucleophilic attacks. The complex of 2-methoxycyclohexa-1,3-diene *(330)* formed from anisole was

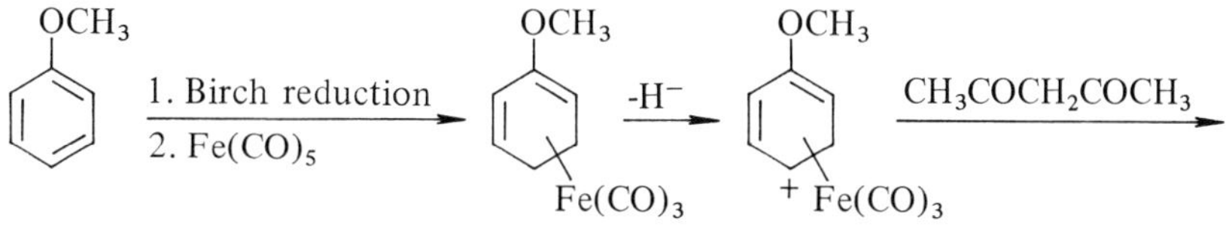

$$330 \qquad 331$$

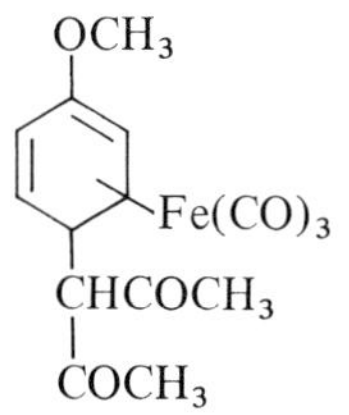

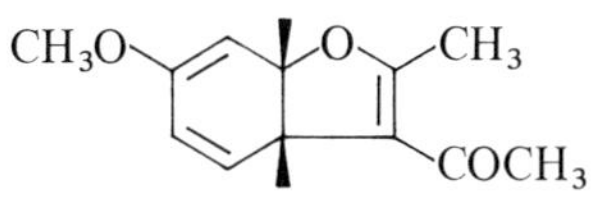

$$332 \qquad 333$$

separated from the isomeric mixture and converted into the cationic complex *(331)* by hydride abstraction. The complex reacted with acetylacetone to give the new coupled complex *(332)*, which underwent oxidative cyclization of the enol form to produce the new dihydrofuranoid ring *(333)*. [362, 363]

The above examples show the importance of a suitable choice of transition metal reagents or catalysts for a specific synthetic purpose.

There are a number of reactions, in which the same substrate takes different reaction courses when different metallic compounds are used as a catalyst. Reactions of highly strained molecules in the presence of various transition metal catalysts are surveyed as typical examples. In the presence of various transition metal compounds, 1,2,2-trimethylbicyclo[1,1,0]butane *(334)* undergoes various ring fission reactions, and many studies have been carried out from theoretical interest. The formation of carbene species by two-bond cleavage was postulated in these reactions. The mode of the bond fission of the bicyclobutane is markedly dependent on the nature of the metals and ligands used as the catalyst.

The reaction of 1,2,2-trimethylbicyclo[1,1,0]butane *(334)* and methyl acrylate in the presence of bis(acrylonitrile)nickel as a catalyst gave the cyclopropane adducts *(335)* in a quantitative yield. [364] The reaction proceeds through C_{1-3}, C_{2-3} cleavage, and the nickel-allylcarbene complex is assumed as an intermediate.

334 $+ CH_2=CHCO_2CH_3$ CO_2CH_3 + CO_2CH_3

335

The reaction of the same bicyclobutane in the presence of $[Rh(CO)_2Cl]_2$ gave the diene *(336)* and the cyclopropane derivative *(337)*. Similar products were obtained by other transition metal complexes such as $RuCl_2(CO)_3$ [365—369]

334   *336*   *337*

On the other hand, the bicyclobutane *(334)* was converted to a mixture of the following different dienes *(338)* and *(339)* via C_{1-3} and C_{1-2} bond cleavage by the action of pentafluorophenylcopper tetramer. Silver tetrafluoroborate showed the similar behavior to give *(338)* and *(339)*.

334   *338*   *339*   *337*

The mode of the fission with palladium complexes is different depending on the ligands. Thus bis(benzonitrile)dichloropalladium gave *(336)* and *(337)*. When π-allylpalladium chloride was used for the rearrangement, the dienes *(338)* and *(339)* were the main products. Only 2% of *(336)* was formed.

These differences in the reaction path were explained in terms of the nature of the resulting allylcarbene-metal complexes *(340)*. [370] The two bond cleavage reaction of the unsymmetrically substituted trimethylbicyclobutane would take place in such a way as to produce a carbenoid with the most preferable resonance stabilization, and hence factors affecting the stability of the carbenic center and the olefin moiety would be the dominating factors.

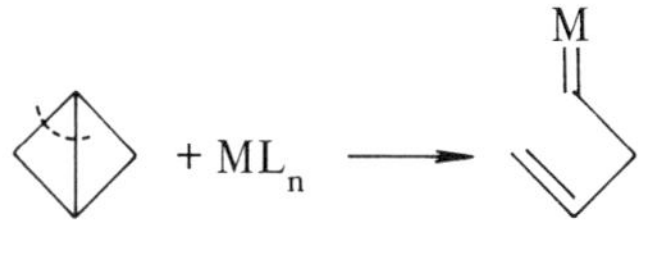

340

Another example of the rearrangement involving the intermediate carbenoid complex is the rearrangement of 1-methyl-2,2-diphenylbicyclo[1,1,0]butane *(341)* catalyzed by $[Rh(CO)_2Cl]_2$ in chloroform at room temperature. [371, 372] Three products *(342, 343, 344)* were obtained as the primary rearranged products, formation of which is reasonably explained on mechanistic grounds if *(341)* is initially converted by the rhodium catalyst to the metal-complexed

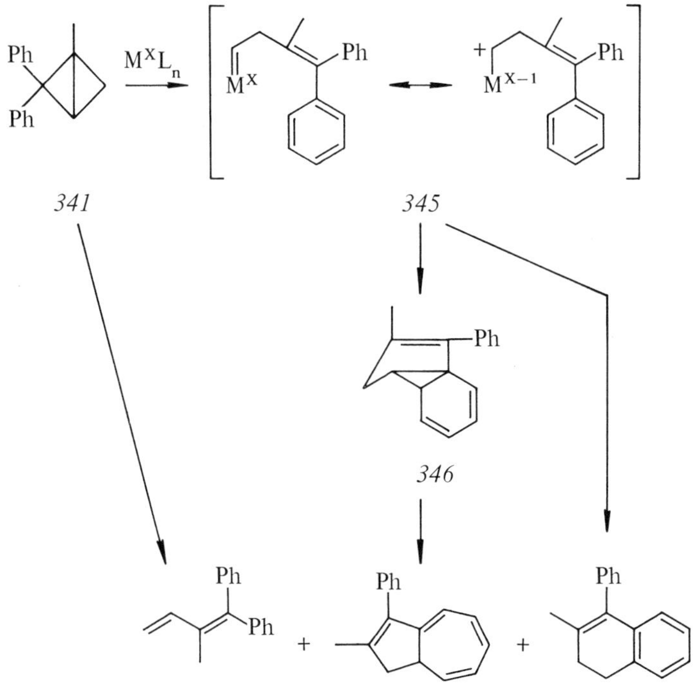

341   *345*

346

342   *343*   *344*

carbene-metal-bonded carbonium ion hybrid *(345)*. Insertion of the carbenoid type intermediate into the benzene ring would produce the norcaradiene derivative *(346)*, which on valence tautomerism would give *(343)*. Direct carbene type insertion to a phenyl C-H bond gives *(344)*.

The carbene complexes of transition metals are also formed from ketenes and isocyanates. The similarity of these carbene complexes with ylide in nature is apparent by the following reactions. When diphenylketene *(347)* was treated with Wilkinson complex *(1)*, tetraphenylethylene *(349)* and tetraphenylallene *(350)* were formed. [373] In the presence of an excess of triphenylphosphine, the latter was the predominant product. These results are understandable by the following mechanism. At first, the strong carbon monoxide abstracting ability of the rhodium complex can convert diphenylketene into the carbene species *(348)* complexed to the rhodium. The coupling of the carbene moieties gives tetraphenylethylene *(349)*. In the presence of triphenylphosphine, the carbene reacts with triphenylphosphine to give the ylide *(351)*, which undergoes the Wittig type of reaction with diphenylketene to give tetraphenylallene *(350)* and triphenylphosphine oxide.

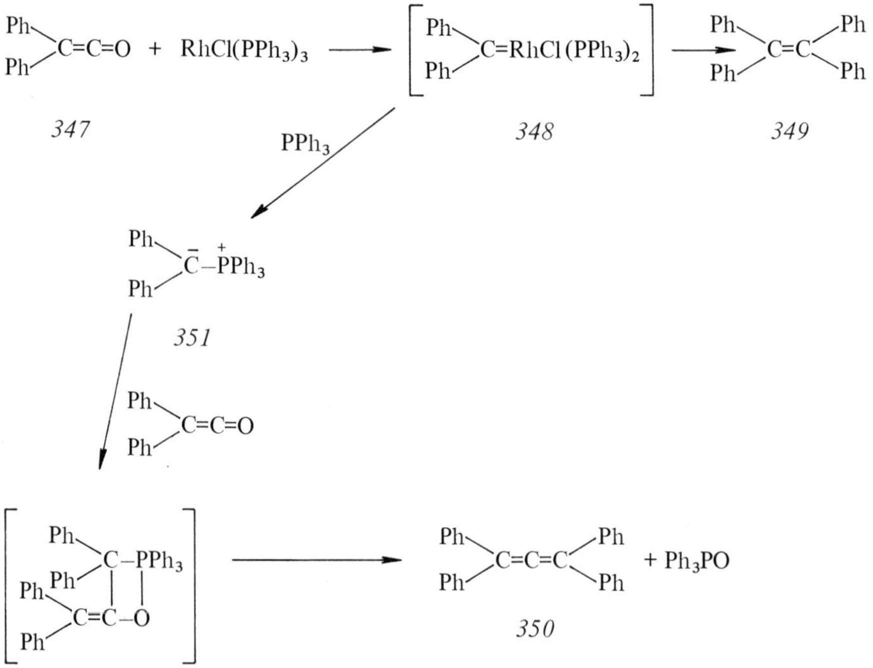

The reaction of diphenylketene to form tetraphenylethylene was achieved in the presence of a catalytic amount of $Co_2(CO)_8$. [373] Here also, the formation of a carbene-cobalt complex, followed by the reaction of the ylide type cobalt complex with the ketene *via* four-membered ring intermediate can account for the formation of tetraphenylethylene.

$$\underset{Ph}{\overset{Ph}{>}}C=C=O \xrightarrow{Co_2(CO)_8} \left[\underset{Ph}{\overset{Ph}{>}}C=Co(CO)_n \right] \xrightarrow{Ph_2C=C=O} \left[\begin{array}{c} \underset{Ph}{\overset{Ph}{>}}C-Co(CO)_n \\ | \quad | \\ Ph-C-C=O \\ | \\ Ph \end{array} \right] \longrightarrow \underset{Ph}{\overset{Ph}{>}}C=C\underset{Ph}{\overset{Ph}{<}}$$

$$349$$

$$Co(CO)_{n+1}$$

Isocyanates were converted into carbodiimides with evolution of carbon dioxide by the action of a catalytic amount of mononuclear metal carbonyls such as $Fe(CO)_5$ and $Mo(CO)_6$. [374] In this reaction, the formation of an iso-nitrile metal complex by decarbonylation is assumed. The complex has a ylide type structure and the reaction of fresh isocyanate with the isonitrile complex gives the carbodiimide *via* the four-membered intermediate.

$$R-N=C=O + Fe(CO)_5$$

$$\longrightarrow \left(\begin{array}{c} (CO)_4Fe-C=O \\ | \quad | \\ R-N=C-O \end{array} \right) \longrightarrow (CO)_4Fe=C=N-R + CO_2$$

$$\Big\downarrow R-N=C=O$$

$$Fe(CO)_5 + R-N=C=N-R \longleftarrow \left(\begin{array}{c} (CO)_4Fe-C=N-R \\ | \quad | \\ O=C-N-R \end{array} \right)$$

It should be pointed out that carbodiimide formation from isocyanates is also catalyzed by phosphine oxide. [375, 376] The similarity of the mechanism of the reaction catalyzed by the metal carbonyls and the phosphine oxide is easily understandable.

$$R-N=C=O + R_3P=O \longrightarrow \left(\begin{array}{c} R_3P-O \\ | \quad | \\ R-N-C=O \end{array} \right) \longrightarrow R_3P=N-R + CO_2$$

$$\Big\downarrow R-N=C=O$$

$$R_3P=O + R-N=C=N-R \longleftarrow \left(\begin{array}{c} R_3P-N-R \\ | \quad | \\ O-C=N-R \end{array} \right)$$

As a related reaction, transition metal oxo complexes reacted with phenyl iso-cyanate to give phenylimido(nitrene) complex with the liberation of carbon dioxide. [377]

$$L_nM=O + Ph-N=C=O \longrightarrow \left(\begin{array}{c} L_nM-O \\ | \quad | \\ R-N-C=O \end{array} \right) \longrightarrow L_nM=N-R + CO_2$$

$L_nM=O : \pi\text{-}C_5H_5TiO_2, (acac)_2VO, PtO_2(PPh_3)_2, \pi\text{-}C_5H_5MoOCl_2, L_2ReOCl_3$

Reaction of aldehyde and phenyl isocyanate in the presence of $Co_2(CO)_8$ gave phenylimines with evolution of carbon dioxide. [378]

$$Ph-N=C=O + R-CHO \xrightarrow{Co_2(CO)_8} R-CH=N-Ph + CO_2$$

Conversion of thiocarbonates of 1,2-diol to olefins is effected by the reaction of $Fe(CO)_5$ and $Ni(COD)_2$. [379, 380] An iron- or nickel-carbene complex was assumed to be an intermediate of the reaction. The same reaction can be carried out by using trimethylphosphine. [381]

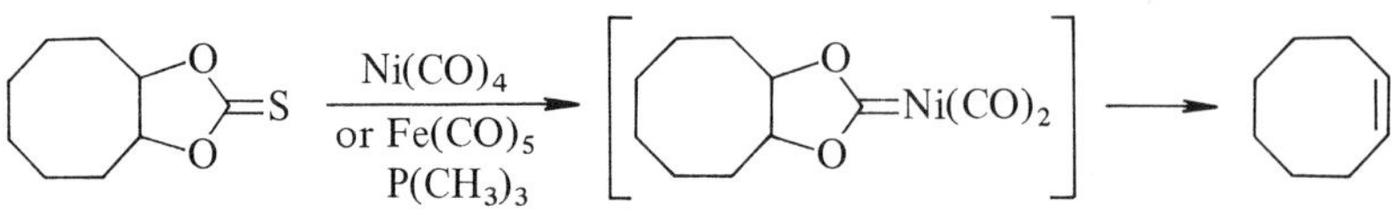

Usually carbon-metal σ-bonds are not stable; they are sensitive to heat, oxygen, water, and acid. [382, 383] This property is very important, because catalysis by transition metal complexes depends on the instability of the metal-carbon bond. However, several complexes are known which have unusually stable carbon-metal bonds. It is generally true that a cobalt-carbon σ-bond is rather unstable and reactive. One exceptionally stable one is alkylcobaloximes (351) which can be formed in several ways, e.g. reductive alkylation, addition of olefins, and substitution reaction of cobaloxime, a model complex of vitamin B_{12}. The surprisingly stable cobalt-carbon bonds thus formed are known to undergo a variety of cleavage reactions, such as β-elimination, coupling, substitution with nucleophiles, and other reactions. The stabilizing effect of this specific ligand on the cobalt-carbon bond is remarkable. [384]

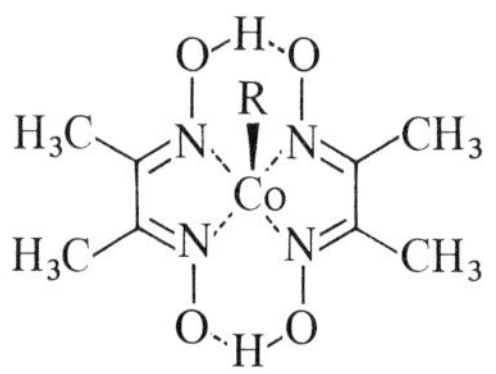

351

Other types of complexes which have stable carbon-metal bonds is metal clusters. [385] Metal clusters are compounds formed from several metal-metal bonds and occur commonly among the carbonyls of iron, cobalt, nickel, and their congeners mostly with the metal in a state of zero valence. They can be regarded as intermediate complexes between metal carbonyls and metallic states. The most typical one is the tricobalt-carbon cluster. A general method of the preparation is the reaction of α,α,α-trihaloalkyl compounds with $Co_2(CO)_8$.

$$5Co_2(CO)_8 + 3CX_3Y \longrightarrow 2Co_3(CO)_9CY + 4CoX_2 + 22CO + [YCX]$$

The complex is also formed from acetylene complexes of cobalt carbonyl.

$$Co_2(CO)_6(RC\equiv CH) \xrightarrow{\text{H}_2\text{SO}_4/\text{CH}_3\text{OH}} RCH_2CCo_3(CO)_9$$

Methinyltricobalt enneacarbonyl, thus prepared, and its derivatives, which are cluster complexes, are interesting in that they have an apical carbon bound to three cobalt atoms and have unusual reactivities due to electronic effects and stereochemical constraints. [386] When methinyltricobalt enneacarbonyl is treated with olefins, the insertion takes place into the carbon-hydrogen bond, rather than the carbon-cobalt bond. The insertion proceeded most easily in the presence of a radical initiator such as azobisisobutyronitrile and hence a free radical mechanism was proposed for this insertion reaction. [387]

$$HCCo_3(CO)_9 \longrightarrow \cdot CCo_3(CO)_9 \xrightarrow{\substack{\diagdown \\ /}C=C\substack{\diagup \\ \diagdown} } \cdot \overset{|}{C}-\overset{|}{C}-CCo_3(CO)_9$$

$$\xrightarrow{HCCo_3(CO)_9} H-\overset{|}{C}-\overset{|}{C}-CCo_3(CO)_6 + \cdot CCo_3(CO)_9$$

Another characteristic property of the cobalt-cluster complexes is the unusual stability of the carbon-cobalt bond to oxidation and thermal degradation. Reaction of the complex with diphenylmercury gave the phenyl substituted derivative. [388] This complex can be subjected to Friedel-Crafts reaction on the benzene ring without decomposition of the cluster structure. [389] Oxidation of the complex with Ce^{4+} ion gave carboxylic acid free from cobalt atoms.

The halo cluster complex is attacked by nucleophiles such as Grignard reagent at the apical carbon atom, yielding the expected alkylated cluster. [390]

$$ClCCo_3(CO)_9 + RMgX \longrightarrow RCCo_3(CO)_9 + MgXCl$$

The cluster complex formed from trichloroacetate and $Co_2(CO)_8$ has a large steric effect as observed for 2,6-substituted benzoate, and can be transformed to the acylium cation only with concentrated sulfuric acid. Addition of HPF_6 to the acid or ester gave the complex (352) as a precipitate, which is a strong acylating agent and reacts easily with amines, ammonia, and alcohols.

$$Cl_3CCO_2R + Co_2(CO)_8 \longrightarrow \underset{\substack{\text{Co} \\ (CO)_3}}{(CO)_3Co\overset{\overset{\displaystyle CO_2R}{\displaystyle |}}{\underset{}{C}}Co(CO)_3} \xrightarrow[\text{2. HPF}_6]{\text{1. H}_2\text{SO}_4} \underset{\substack{\text{Co} \\ (CO)_3}}{(CO)_3Co\overset{\overset{\displaystyle \overset{+}{C}O\ PF_6^-}{\displaystyle |}}{\underset{}{C}}Co(CO)_3}$$

352

$$\xrightarrow{NH_2R} \underset{\substack{\text{Co} \\ (CO)_3}}{(CO)_3Co\overset{\overset{\displaystyle CONHR}{\displaystyle |}}{\underset{}{C}}Co(CO)_3}$$

So far the reactivities of metal-carbon bonds have been surveyed. Another important bond is the metal-hydrogen bond. Hydrogen bound to transition metals exhibits an unusual proton NMR absorption in the τ 15—35 range, higher than any other hydrogens attached to nontransition metals. This characteristic NMR band suggests that the hydrogen is highly shielded by the d electrons of the transition metals; the NMR spectrum of $HCo(CO)_4$ shows a single proton signal at a very high field ($\tau = 20$), and that of $HCo(PF_3)_4$ at 22.5. The acidity and reactivity of such hydrogens vary with the metals. Various hydrides are known to have both H^+ and H^- characters. The property can be determined to a certain degree by the metallic species. However, even with the same metal, the property of the metal-hydrogen bonds changes with the ligands attached to the metals. For example, $H_2Fe(CO)_4$ is more acidic than acetic acid ($K_1 = 3.6 \times 10^{-5}$, $K_2 = 1.1 \times 10^{-14}$). [391, 392] Ligands such as phosphines which are better σ-donors and poorer π-acceptors than carbon monoxide, increase the electron density of the metal, thereby lowering the acidity of the metal hydrides. $HCo(CO)_4$ is as strong an acid as mineral acids, but the electron density on the cobalt is increased by the coordination of phosphine instead of carbon monoxide, and hence the acidity of the hydride becomes lower. The dissociation constant of $HCo(CO)_3(PPh_3)$, 1.09×10^{-7} is compared with that of $HCo(CO)_2[P(OPh)_3]_2$, 1.13×10^{-5}. The acidity of cobalt carbonyl phosphine complexes decreases in the following order. [393, 394]

$$HCo(CO)_4 > HCo(CO)_3(PPh_3) > HCo(CO)_2(PPh_3)_2 > HCo(CO)(PPh_3)_3$$

$$> HCo(dp)_2$$

$$dp = Ph_2PCH_2CH_2PPh_2$$

It was found that the carbonyl stretching wave number of cobalt carbonyl phosphine complexes decreases with an increase in the electron-donating property of the phosphines. [395]

The hydrogen of $HCo(dp)_2$ and $HCo(PBu_3)_4$ can be regarded as H^- rather than H^+. Consequently, the reactivity or catalytic activity of these complexes is somewhat different from that of $HCo(CO)_4$. The difference in iso/normal ratio of aldehydes formed by the cobalt carbonyl-catalyzed oxo reaction of terminal olefins is explained in terms of varying nature of the ligands. [396—400] When a complex coordinated with phosphine such as tributylphosphine is used as a catalyst, the ratio of normal aldehyde increases with respect to that obtained with the corresponding carbonyl complex. Formation of normal aldehydes *via* the olefin insertion is favored with the reduced acidity of the hydride. The increase of the normal aldehydes relative to the branched aldehydes can be explained partly by Markownikoff rule.

$$RCH-CH_3 \longleftarrow RCH-CH_3 \longleftarrow RCH=CH_2$$

$$\begin{array}{ccc} | & | & \\ CHO & Co(CO)_{4-n} & \\ & | & \\ & (PR_3)_n & \end{array}$$

$$H-Co(CO)_{4-n}$$
$$|$$
$$(PR_3)_n$$

$$\longrightarrow RCH_2CH_2 \qquad \longrightarrow RCH_2CH_2CHO$$
$$|$$
$$Co(CO)_{4-n}$$
$$|$$
$$(PR_3)_n$$

In the oxo reaction of 1-octene catalyzed by $Co_2(CO)_8$, n-nonanal was obtained in 43% selectivity. [397] The selectivity became 85% when $Co_2(CO)_3(PBu_3) + 2PBu_3$ was used as the catalyst system. Also tributylphosphine as the ligand favored the reduction of the initially formed aldehydes to the corresponding primary alcohols.

The steric effect of the ligands is another contributing factor for determining the course of the reaction. Since selective formation of the more useful n-butyraldehyde in the oxo reaction of propylene is a serious problem in industry, extensive investigations have been carried out. Many factors such as carbon monoxide pressure, reaction temperature, and ligands of the catalyst, contribute to the change in the ratio of iso/normal aldehydes.

Reaction of benzylideneaniline with alkyl vinyl ether in the presence of $Co_2(CO)_8$ produced the quinoline derivatives *(353, 354)*. [401] Since the reaction also proceeds by the catalysis of boron trifluoride, it seems likely that the strongly acidic $HCo(CO)_4$ plays a similar catalytic role in the reaction. At the same time, the effect of coordination of the reactants to cobalt is a significant contributing factor.

$$\boxed{} -N=CH-Ph + ROCH=CH_2 \longrightarrow$$

353   *354*

Certain metal hydride complexes react with diazomethane to give the corresponding methyl derivatives, indicating the protic nature of the hydrides. [402]

$$HMn(CO)_5 + CH_2N_2 \longrightarrow CH_3Mn(CO)_5 + N_2$$

Generally metal hydrides reduce carbon tetrachloride to chloroform. The reaction of the following platinum dihydride is a typical example. [403]

$$PtH_2(PCy_3)_2 + CCl_4 \longrightarrow PtHCl(PCy_3)_2 + CHCl_3$$

V. Insertion Reactions

1. Introduction

The reactions described in the preceding chapters enable various metal complexes with σ-bonds to be formed. In order to fabricate more elaborate organic compounds from these complexes, it is necessary to carry out further transformations by addition of other molecules. This type of addition reaction is called an "insertion reaction". [7, 404, 405] There are many types of insertion reactions, depending on the σ-bonds and the molecules to be inserted into metal-element σ-bonds. The insertion reaction can be expressed by the following general scheme.

$$-\overset{|}{\underset{|}{M}}-X + A \longrightarrow -\overset{|}{\underset{|}{M}}-A-X$$

Here M is a metal and X and A are monoatomic or polyatomic species. The word "insertion" is used in a rather broad context and carries no mechanistic connotation; it describes only the outcome of the reaction. It merely reflects the overall structural result of cleavage of a metal-element bond(M-X) in complexes by an interposing unsaturated species (A). The major participants in these insertion reactions are shown in Table III.

Table III. Species involved in the insertion reactions

σ-Bond (M—X)	Bonds and molecules to be inserted (A)
M-Hydrogen	$C=C$, $C=C-C=C$, $C\equiv C$
M-Carbon	$>C=O$, $>C=N$, $-N=N-$, $C\equiv N$
M-Halogen	CO, CO_2, SO_2, CS_2
M-Metal	$RN=C$, $RN=C=O$
M-Nitrogen	carbene, benzyne
M-Oxygen	

By the combination of these participants, a large number of the insertion reactions seem to be possible, but not all of them are in fact known. In this chapter, typical insertion reactions are surveyed with examples.

2. Olefin Insertion

The insertion of olefins is by far the most important reaction and the olefin insertions to metal-hydrogen or metal-alkyl bonds actually take place in various reactions of olefins catalyzed by transition metal complexes such as hydrogenation, oligomerization, polymerization, carbonylation, and hydrosilylation. Olefin insertion into a metal-hydrogen bond gives an alkyl complex. In this reaction, a π-olefin complex is formed at first. Then the π-coordinated olefin inserts into the metal bond to form σ-alkyl complex. This process of olefin insertion is called π-σ rearrangement. [406]

$$-\overset{|}{\underset{|}{M}}-H + RCH=CH_2 \longrightarrow \begin{array}{c} RCH=CH_2 \\ \downarrow \\ -\overset{|}{\underset{|}{M}}-H \end{array} \longrightarrow -\overset{|}{\underset{|}{M}}-CH_2CH_2R$$

The reactivity of the metal-hydrogen bond towards unsaturated organic molecules such as olefins and acetylenes is central to many catalytic processes. The most important role of transition metal hydrides in homogeneous catalyses is their ability to add to unsaturated substrates to form catalytic intermediates containing metal-carbon bonds. The most common reaction of this type is the addition of metal hydrides to olefins to form alkyl complexes. The typical reactions involving the olefin insertion are oligomerization and polymerization of olefins, and can be expressed by the following scheme. The olefin insertion to the metal-hydrogen bond gives the alkyl-metal bond, and then continuous olefin insertion proceeds to form polymers.

$$-\overset{|}{\underset{|}{M}}-H + CH_2=CH_2 \longrightarrow -\overset{|}{\underset{|}{M}}-CH_2CH_3$$

$$\xrightarrow{nCH_2=CH_2} -\overset{|}{\underset{|}{M}}(CH_2CH_2)_nCH_2CH_3$$

Conjugated dienes readily react with metal hydride complexes. π-Allyl complexes are formed by the following insertion reaction.

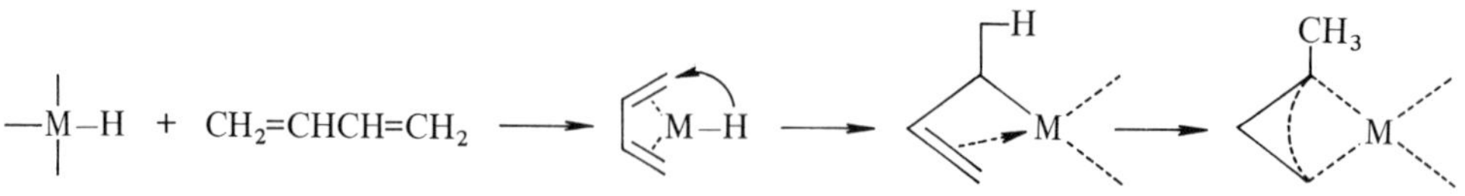

$$-\overset{|}{\underset{|}{M}}-H + CH_2=CHCH=CH_2 \longrightarrow$$

Conjugated dienes, mainly butadiene, undergo a variety of reactions in the presence of transition metal complexes, where the above shown π-allyl complex, formed by insertion, is involved as the essential step. For example, formation of 1,4-hexadiene *(355)* from butadiene and ethylene is catalyzed by iron, cobalt, and nickel complexes of certain phosphines in the presence of alkylaluminum. [407—409] In this reaction, the π-allyl complex is formed by the insertion of butadiene into the metal-hydrogen bond and insertion of ethylene into the π-allyl-metal bond takes place.

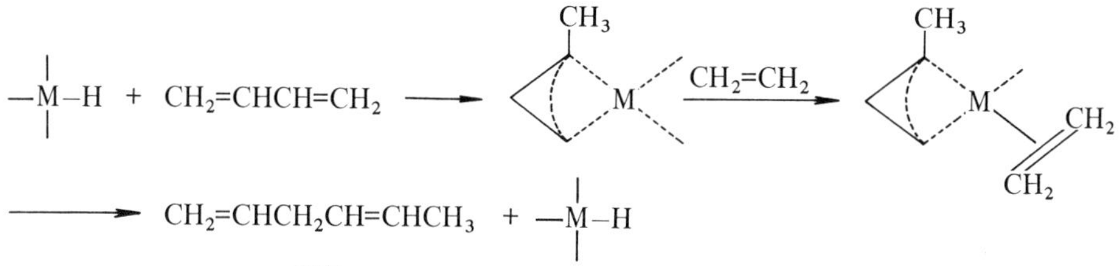

$$-M-H \;+\; CH_2=CHCH=CH_2 \;\longrightarrow\;$$

$$\longrightarrow\; CH_2=CHCH_2CH=CHCH_3 \;+\; -M-H$$

355

As a related reaction, 3-vinylcyclooctene *(356)* was obtained from 1,3-cyclo-octadiene and ethylene. [410] When, as the catalyst for this reaction, π-allyl-nickel chloride, ethylaluminum sesquichloride and an optically active phosphine system was used, the product *(356)* was obtained in an optical yield of up to 70%. The result shows that asymmetric induction takes place at the insertion step.

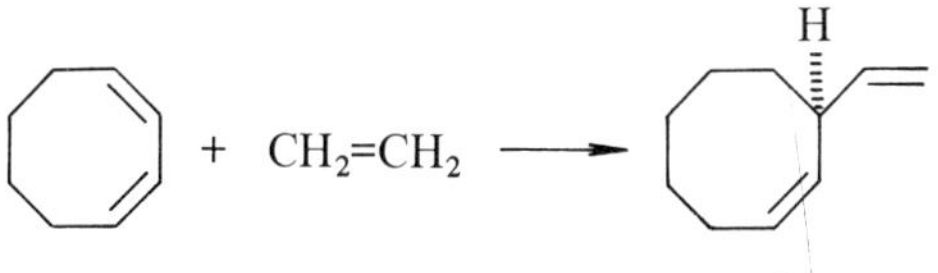

356

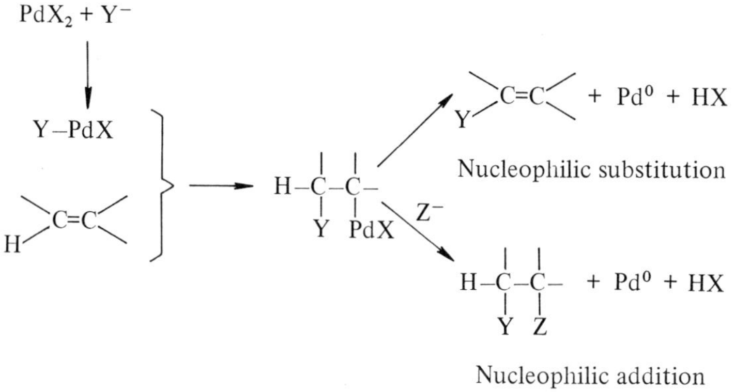

Many useful reactions with bivalent palladium salts involving olefins can be explained by the olefin insertion reactions. [411—416] One of the most characteristic properties of palladium salts is the ability to carry out substitution reactions of vinyl hydrogen and vinyl halogens with nucleophiles. The usefulness of this property is apparent from the consideration that addition to olefins is common in usual organic reactions, but that no substitution reaction of olefinic hydrogen is possible. By coordination to palladium, nucleophilic substitution of olefinic hydrogen becomes feasible. An olefinic bond, when π-coordinated to bivalent palladium salt, reacts with nucleophiles. This substitution reaction involving the olefin insertion, is called "palladation reaction" of olefins. In this type of reaction, the ligand Y on palladium migrates to the π-bonded olefin, which is simultaneously transformed to a σ-bonded alkyl group. Thus the reaction of palladium-olefin complexes with nucleophiles which contribute $^-$OH, $^-$OR, $^-$OCOR, $^-$NR$_2$, carbanions, and others, makes it possible to prepare vinyl compounds by substitution on vinyl carbon with these nucleophiles. In some cases, nucleophilic addition which leads to bifunctional products is possible. There are numerous examples which can be explained by this mechanism.

The reactions of palladium salts which lead to the formation of carbonyl compounds, vinyl acetate, vinyl ether, and acrylate by the reaction of olefins with water, acetate, alcohols, and carbon monoxide are well known.

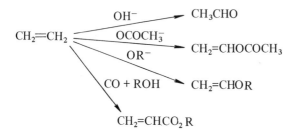

The famous Wacker process is based on the oxidation of olefins by palladium chloride. Industrial production of acetaldehyde from ethylene on a large scale is carried out by using palladium chloride as the catalyst. This process has replaced the method of acetaldehyde production from acetylene catalyzed by mercury salt. The following elementary stoichiometric reactions constitute the Wacker process of acetaldehyde production.

$$CH_2{=}CH_2 + H_2O + PdCl_2 \longrightarrow CH_3CHO + Pd + 2HCl$$
$$Pd + 2CuCl_2 \longrightarrow PdCl_2 + 2CuCl$$
$$2CuCl + \tfrac{1}{2}O_2 + 2HCl \longrightarrow 2CuCl_2 + H_2O$$

$$CH_2{=}CH_2 + \tfrac{1}{2}O_2 \longrightarrow CH_3CHO$$

At first, ethylene is oxidized by palladium chloride to acetaldehyde, and bivalent palladium is reduced to a zerovalent state. This reaction has been known for a long time. The invention of the Wacker process gave a method for facile reoxidation of the reduced palladium to the bivalent state by using cupric salt. In the aqueous reaction medium, zerovalent palladium is reoxidized to the bivalent state by cupric chloride. The cupric chloride is reduced to cuprous state at the same time. The cuprous salt thus produced is easily reoxidized to the cupric state by oxygen. The cycle of these three oxidation and reduction reactions constitutes a catalytic process. The Wacker process can be applied to other olefins, especially to terminal olefins to form methyl ketones. The oxidation reactions of higher olefins are carried out most efficiently in aqueous DMF solution. [417]

$$RCH{=}CH_2 + \tfrac{1}{2}O_2 \longrightarrow \underset{\underset{O}{\|}}{R}CCH_3$$

Many mechanistic studies have been carried out on this reaction. It was found that the reaction performed in D_2O yields acetaldehyde free of deuterium, indicating that all four hydrogens in acetaldehyde must come from ethylene. The question to be answered is how the hydrogen transfer from one carbon to the other takes place by the action of palladium chloride. The reaction involves an intramolecular migration of one hydrogen atom from one carbon of ethylene to the other.

The mechanism shown below was proposed by Smidt and coworkers in order to account for all the experimental findings. [418] At first, π-coordination of ethylene to palladium chloride takes place to form *trans*-monohydroxodichloroethylene π-complex *(357)*. The *trans* complex is then converted to the *cis* complex *(359)* with regard to ethylene and hydroxo ligands *via* the dihydroxo complex *(358)*. In the *cis* complex, essential for the ethylene insertion, ethylene is located favorably for the *cis* insertion into the palladium-OH bond to form σ-hydroxyethyl complex *(360)*. The intramolecular hydrogen migration takes place in the next step. From the σ-hydroxyethyl complex *(360)*, β-hydrogen from OH-bearing carbon is abstracted to give the π-olefin hydride complex *(361)*, in which the double bond of vinyl alcohol forms a π-complex. Then the olefin insertion into the palladium hydrogen bond takes place again to form σ-(α-hydroxyethyl) complex *(362)*. The intramolecular hydrogen shift takes place during the conversion from *(360)* to *(362)*. Finally acetaldehyde is formed by solvolytic decomposition of the σ-complex *(362)* with generation of zerovalent palladium. In connection with the π-complex of vinyl alcohol *(361)*, there is independent evidence to show that nonexisting vinyl alcohol can be isolated as a π-complex. [419—423] The complex was prepared by the coordination of vinyl trimethylsilyl ether to platinum, followed by hydrolysis of the trimethylsilyl group to form vinyl alcohol. Also direct reaction of chloro(acetylacetonato) (π-ethylene) platinum with acetaldehyde in the presence of potassium hydroxide gave potassium chloro(acetylacetonato) (β-hydroxyethyl)platinate.

The palladation is the essential step in the reactions of olefins with palladium salts. The product of the palladation reaction exists only as an active intermediate and in most cases cannot be isolated. However, the product of the palladation

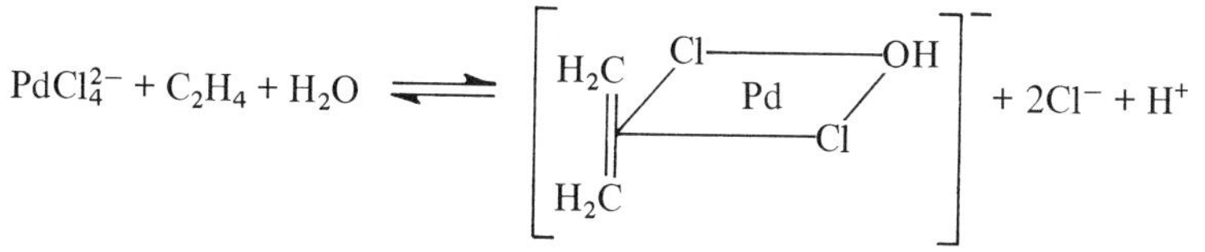

$$PdCl_4^{2-} + C_2H_4 + H_2O \rightleftharpoons \left[\begin{array}{c} H_2C \\ \| \\ H_2C \end{array} \underset{Pd}{\overset{Cl\text{------}OH}{\diagdown \diagup}} \right]^{-} + 2Cl^- + H^+$$

357

$$357 \ + H_2O \rightleftharpoons \left[\begin{array}{c} H_2C \\ \| \\ H_2C \end{array} \underset{Pd}{\overset{Cl\text{------}OH}{\diagdown \diagup}}{}_{OH} \right]^{-} + Cl^- + H^+$$

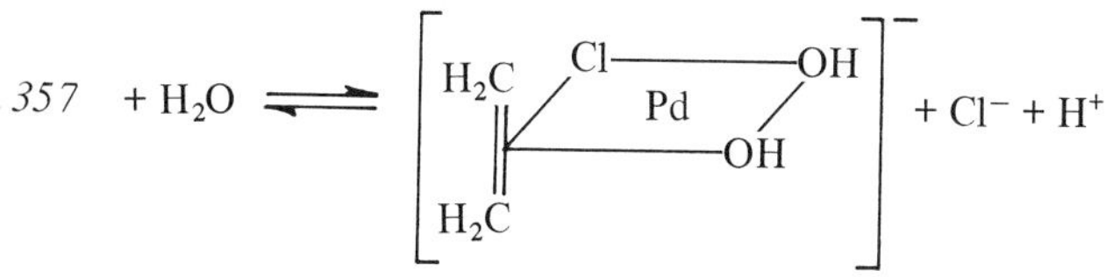

358

$$358 \; + H^+ + Cl^- \; \rightleftharpoons \; \left[\begin{array}{c} H_2C \quad Cl\text{——}Cl \\ \Big\| \quad Pd \\ \quad \quad \text{—OH} \\ H_2C \end{array} \right]^- + H_2O$$

$$359$$

$$\left[\begin{array}{c} H_2C \quad Cl\text{——}Cl \\ \Big\| \quad Pd \\ \quad \quad \text{—OH} \\ H_2C \end{array} \right]^- \; \rightleftharpoons \; \sigma \left[\begin{array}{c} H \\ | \\ HO\text{—CH—CH}_2\text{—PdCl}_2 \end{array} \right]^-$$

$$359 \qquad\qquad\qquad 360$$

$$\sigma \left[\begin{array}{c} H \\ | \\ HO\text{—CH—CH}_2\text{—PdCl}_2 \end{array} \right]^- \; \rightleftharpoons \; \left[\begin{array}{c} H \\ | \\ HO\text{—C} \quad Cl\text{——}Cl \\ \quad \quad Pd \\ \Big\| \quad \quad \text{—H} \\ H_2C \end{array} \right]^-$$

$$360 \qquad\qquad\qquad 361$$

$$\left[\begin{array}{c} H \\ | \\ HO\text{—C} \quad Cl\text{——}Cl \\ \quad \quad Pd \\ \Big\| \quad \quad \text{—H} \\ H_2C \end{array} \right]^- \; \longrightarrow \; \left[\begin{array}{c} OH \\ | \\ CH_3\text{—C—PdCl}_2 \\ | \\ H \end{array} \right]^- \; \xrightarrow{H_2O} \; CH_3\text{—HC} \begin{array}{c} \overset{H}{\overset{+|}{O}}\text{—H} \\ \diagdown \\ OH \end{array}$$

$$361 \qquad\qquad\qquad 362$$

$$CH_3CHO + H_3O^+$$

could be isolated as a stable compound in the reaction of 1,5-COD palladium complex with carbonions such as malonate. [424] The insertion reaction proceeded easily by stirring a heterogeneous mixture of the COD complex, malonate, and sodium carbonate in ether at room temperature forming the new complex *(363)* which has the carbon-palladium σ-bond. The complex *(363)* was converted to the bicyclo[6,1,0]nonene derivative *(364)* by the treatment with a base. Attack of another molecule of malonate also afforded the bicyclo[3,3,0]octane derivative *(365)*. These reactions can be regarded as intra- and intermolecular nucleophilic addition reactions, respectively, to the double bond of COD.

σ-Bonds between aromatic ring and transition metals can be formed by several methods. These bonds are susceptible to insertion reactions. A typical example is observed with palladium complexes. The reaction of arylmercury compounds with palladium salts followed by olefin insertion was described earlier (p. 72). Direct aromatic substitution reaction by olefins in the presence of palladium

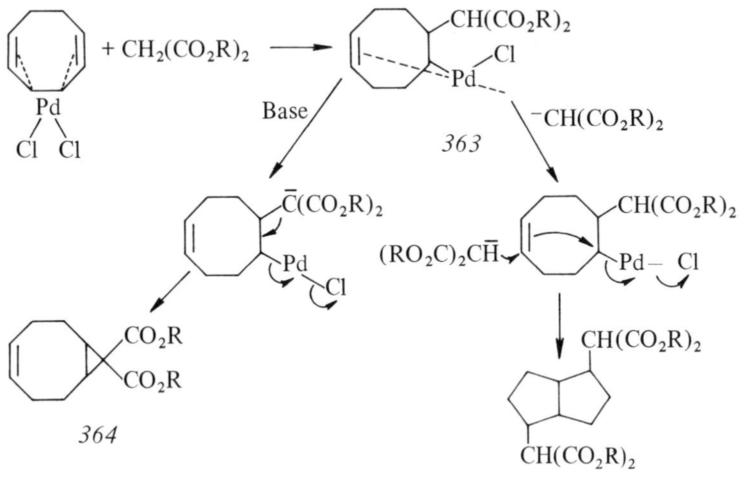

363

364

365

acetate seems to proceed through the olefin insertion. [425] Stilbene was obtained from benzene and styrene.

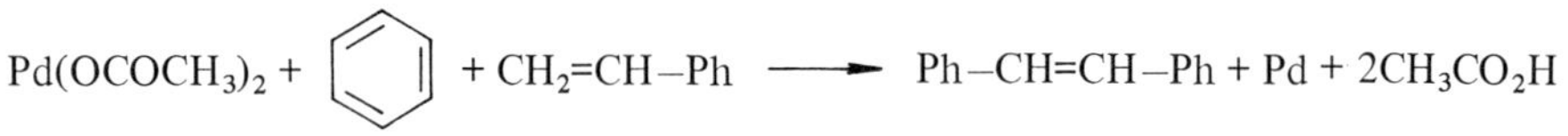

$$Pd(OCOCH_3)_2 + \boxed{} + CH_2{=}CH{-}Ph \longrightarrow Ph{-}CH{=}CH{-}Ph + Pd + 2CH_3CO_2H$$

The following reaction offers evidence for the facile olefin insertion into the benzene palladium σ-bond. The reaction of N,N,-dimethylbenzylamine with palladium chloride gives the complex *(366)* by *ortho*-palladation. [91] Stilbene derivative *(367)* was obtained by the reaction of the complex with styrene at room temperature. [411]

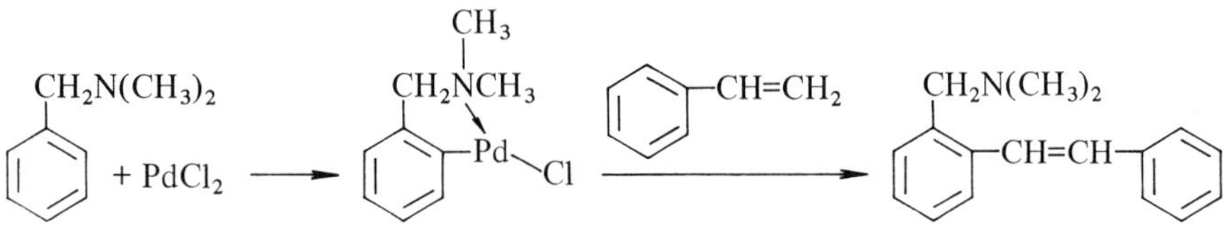

366   *367*

Transition metal catalyzed hydrosilylation of olefins and acetylenes is a well-established reaction. It has a considerable importance in industry. Although, platinum catalyst has monopolyzed the catalysis in hydrosilylation for a long time, many other transition metal complexes, such as rhodium, palladium, cobalt, and nickel complexes were recently found to be active catalysts. The hydrosilylation reaction of olefins can be explained in terms of oxidative addition of silicon-hydrogen bond to the metal, followed by olefin insertion to the hydrogen-metal bond. [426]

$$\begin{array}{ccc}
\overset{\diagdown}{\underset{\diagup}{C}}\underset{\diagup}{\overset{\diagdown}{C}}\!\!\rightarrow\!\text{Pt}- & \xrightarrow{\;-\overset{|}{\underset{|}{Si}}-H\;} & \overset{\diagdown}{\underset{\diagup}{C}}\underset{\diagup}{\overset{\diagdown}{C}}\!\!\rightarrow\!\underset{H}{\text{Pt}}\!\diagdown\!\overset{|}{\underset{|}{Si}}\!\!\diagup
\end{array}$$

Platinum compounds, such as chloroplatinic acid which has high catalytic activity promote hydrosilylation even when no ligand is present. With internal olefins, double-bond migration takes place and the final products are the terminally silylated one. Conversion of internal olefins to terminally functionalized products is very useful. On the other hand, the palladium catalyst is active in the presence of ligands such as phosphines. The palladium phosphine complexes have no ability to migrate internal double bonds to the terminal position and catalyze the hydrosilylation of the internal olefins very slowly.

Asymmetric hydrosilylation of olefins is possible by transition metal complexes coordinated by an optically active phosphine as a ligand. Addition of dichloromethylsilane to α-methylstyrene catalyzed by a nickel complex coordinated by chiral benzylmethylphenylphosphine produced the optically active product (368) in 17.6% optical yield. [427—429]

$$\underset{\text{Ph}}{\overset{CH_3}{\underset{|}{Ph-C=CH_2}}} + HSiCl_2(CH_3) \xrightarrow{NiCl_2L_2^*} \overset{CH_3}{\underset{\underset{H}{|}}{Ph-C^*-CH_2SiCl_2(CH_3)}}$$

$$L = (R)(+)(PhCH_2)(CH_3)PhP \qquad\qquad 368$$

It should be pointed out that each metal catalyst has its own catalytic activity and is used for hydrosilylation with different kinds of silanes. Platinum- and palladium-catalyzed reactions can be carried out most effectively with chlorosilanes, but only slowly with alkylsilanes. On the other hand, the hydrosilylation catalyzed by a rhodium complex is carried out better with alkylsilanes. The chlorosilanes add oxidatively to the rhodium giving stable adducts. Thus, for the hydrosilylation of simple olefins catalyzed by rhodium complexes, alkylsilanes are used. Since alkylsilanes form less stable adducts with the rhodium complexes, they are ready for further reaction.

A conspicuous effect of the kind of silanes on the course of the hydrosilylation reaction was observed in the palladium-catalyzed reaction of butadiene. [430, 431] When trichlorosilane was used for the hydrosilylation of butadiene in the presence of the palladium triphenylphosphine complex as a catalyst, 1-silylated 2-butene (369) was obtained. On the other hand, the hydrosilylation with trimethylsilane proceeded via dimerization and hydrosilylation to give 1-silylated 2,6-octadiene (370).

$$CH_3CH=CHCH_2SiCl_3 \xleftarrow{HSiCl_3} CH_2=CHCH=CH_2$$

369

$$\qquad\qquad\qquad\qquad\qquad\downarrow HSi(CH_3)_3$$

$$CH_3CH=CHCH_2CH_2CH=CHCH_2Si(CH_3)_3$$

370

Hydrosilylation of acetylenes is also catalyzed by transition metal complexes. As one application, the following sequence of reactions to produce the *exo,endo*-cyclic diene *(371)* was reported. [432] Addition of trichlorosilane to propargyl chloride in the presence of chloroplatinic acid gave a mixture of products. 3-Chloro-2-(trichlorosilyl)propene *(372)* was subjected to Diels-Alder reaction with butadiene to give 4-(chloromethyl)-4-(trichlorosilyl)cyclohexene *(373)* in 73% yield. Treatment of this cyclohexene with excess methylmagnesium bromide effected both methylation and elimination of the silyl group to give 4-methylenecyclohexene *(371)* in 43% yield.

$$HC{\equiv}CCH_2Cl + HSiCl_3 \longrightarrow CH_2{=}C\overset{CH_2Cl}{\underset{SiCl_3}{\diagdown}} + \underset{Cl_3Si}{\overset{H}{\diagdown}}C{=}C\overset{CH_2Cl}{\underset{H}{\diagdown}}$$

372

$$\Big| CH_2{=}CHCH{=}CH_2$$

$$\text{(cyclohexene)}\overset{CH_2Cl}{\underset{SiCl_3}{<}} \xrightarrow{CH_3MgBr} \text{(methylenecyclohexene)}$$

373   *371*

3. Carbon Monoxide Insertion

Carbonylation reactions catalyzed by transition metal complexes constitute a very important process used in industry as well as in academic laboratories, including synthetic methods of acid derivatives, aldehydes, and ketones. [405, 433—436] These reactions involve carbon monoxide insertion as an essential step. Both stoichiometric and catalytic reactions are known for carbonylation. As mentioned before, the history of organic synthesis using transition metal complexes actually started with the carbonylation reaction of olefins and acetylenes catalyzed by metal carbonyls. The Oxo and Reppe reactions have been widely applied in industry. In the following, carbon monoxide insertions involved in many catalytic processes are surveyed.

Carbonylation of simple olefins to give saturated carboxylic acids and esters is catalyzed by carbonyls of nickel and cobalt as well as by several noble metal complexes. In the carbonylation reaction, $Ni(CO)_4$ dissociates into nickel tricarbonyl and carbon monoxide; this coordinatively unsaturated species fulfils the role of a real, active catalyst. [437]

$$Ni(CO)_4 \longrightarrow Ni(CO)_3 + CO$$

Hydrogen chloride is often added to the reaction medium. The acid adds to the unsaturated species to give a hydride, which is able to attack olefinic subtrates forming an alkyl complex. Then carbon monoxide insertion takes place to give an acyl complex.

$$Ni(CO)_3 + HCl \longrightarrow H-Ni(CO)_2Cl \xrightarrow{CH_2=CH_2} CH_3CH_2-Ni(CO)_2Cl$$

$$\xrightarrow{CO} CH_3CH_2CO-Ni(CO)_2Cl$$

$$\xrightarrow{H_2O} CH_3CH_2CO_2H + HNi(CO)_2Cl$$

In this carbonylation reaction, no intermediate complex was isolated. $Co_2(CO)_8$ is another active catalyst for the carbonylation of olefins in alcohol to give esters. The mechanism similar to the above one was proposed. [438]

Palladium chloride combined with triphenylphosphine is an active catalyst for various carbonylation reactions. [411, 439] The carbonylation of olefins proceeds smoothly in alcohol in the presence of palladium chloride, triphenylphosphine, and hydrogen chloride to give esters. The reaction is explained by the following mechanism. At first, zerovalent palladium species is formed to which hydrogen chloride adds to form the palladium hydride species (374). Then olefin insertion to give the alkyl complex (375) is followed by carbon monoxide insertion to form the acyl complex (376). The cleavage of the acyl-palladium bond with alcohol gives the ester. At the same time, regeneration of the palladium hydride (374) completes the catalytic cycle. [138]

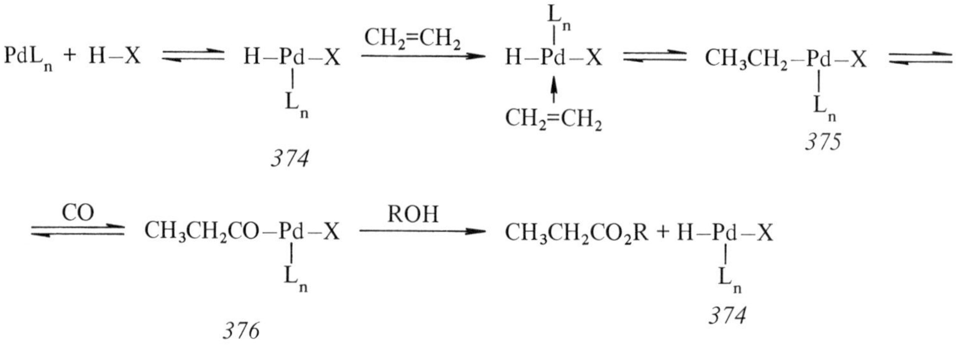

$$PdL_n + H-X \rightleftharpoons \underset{\underset{L_n}{|}}{H-Pd-X} \xrightarrow{CH_2=CH_2} \underset{\underset{CH_2=CH_2}{\uparrow}}{\overset{\overset{L_n}{|}}{H-Pd-X}} \rightleftharpoons \underset{\underset{L_n}{|}}{CH_3CH_2-Pd-X} \rightleftharpoons$$

$$374 \qquad\qquad\qquad\qquad\qquad\qquad\qquad 375$$

$$\xrightarrow{CO} \underset{\underset{L_n}{|}}{CH_3CH_2CO-Pd-X} \xrightarrow{ROH} CH_3CH_2CO_2R + \underset{\underset{L_n}{|}}{H-Pd-X}$$

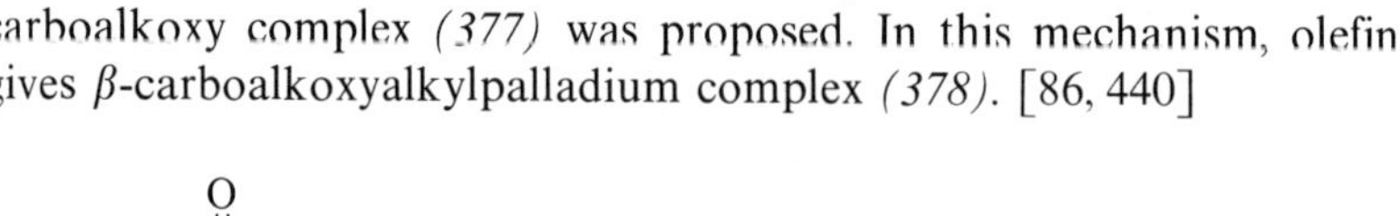

$$376 \qquad\qquad\qquad\qquad\qquad 374$$

Against the mechanism shown above, another mechanism involving the carboalkoxy complex (377) was proposed. In this mechanism, olefin insertion gives β-carboalkoxyalkylpalladium complex (378). [86, 440]

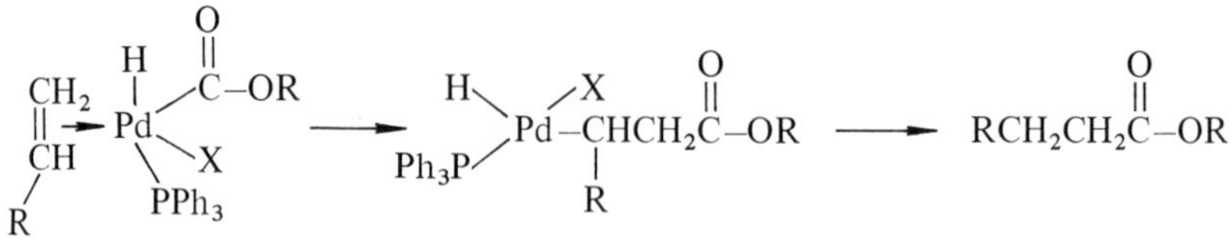

377   378

The palladium-catalyzed carbonylation of COD in alcohol gave mono- or diesters depending on the reaction conditions. [441] But when carbonylation was carried out in THF at 150° and 1000 atm in the presence of $PdI_2(PBu_3)_2$, bicyclo[3,3,1]-2-nonen-9-one *(379)* was obtained in 45% yield. [442] The formation of the bicycloketone can be explained by the intramolecular insertion reaction of the double bond into the acyl-palladium bond. The final step is the β-elimination to give the ketone *(379)* and palladium hydride species.

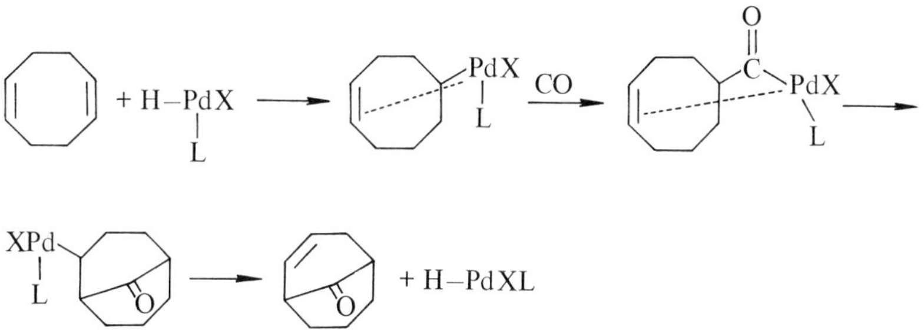

379

Unexpectedly, trialkylimidazoles *(380)* were obtained by the carbonylation of olefins in the presence of ammonia catalyzed by rhodium carbonyl. [443]

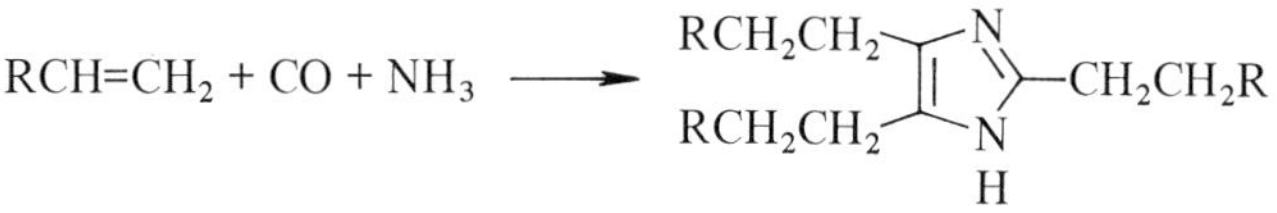

380

By the reaction of ethylene, carbon monoxide, and ammonia, 2,4,5-triethylimidazole was obtained in 52% yield. The ring carbons are all derived from carbon monoxide. In this interesting reaction, rhodium carbonyl catalyzes two types of carbonylation reactions to give diketone *(381)* and carboxamide *(382)*, condensation of which affords the imidazole ring as shown below.

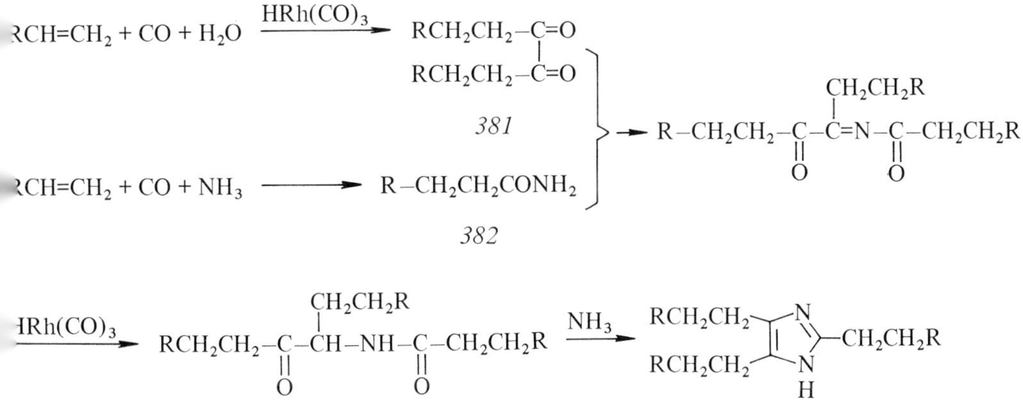

381

382

The oxo reaction was a large scale industrial process used throughout the world. The catalytic species is cobalt carbonyl and the mechanism of the oxo reaction has been elucidated by a number of studies. [444] The balanced oxo reaction can be expressed by the following equation.

$$RCH=CH_2 + CO + H_2 \longrightarrow RCH_2CH_2CHO + \underset{\underset{CH_3}{|}}{R}CHCHO$$

The oxo reaction is carried out in a pressure reactor, and almost any form of cobalt compounds can be used as the catalyst for the reaction. The cobalt compound is reduced by hydrogen to a zerovalent state, whereupon the reaction with carbon monoxide leads to $Co_2(CO)_8$ formation. Under high pressure of hydrogen and at high temperature, $Co_2(CO)_8$ is converted into the true catalyst, $HCo(CO)_4$, by cleavage of the cobalt-cobalt bond.

$$Co_2(CO)_8 + H_2 \longrightarrow 2\,HCo(CO)_4$$

The first step in the reaction is the first order dissociation of the $HCo(CO)_4$ to the coordinatively unsaturated $HCo(CO)_3$ having a vacant coordination site, which facilitates the π-complex formation with olefin. This is followed by an insertion of the π-complexed olefin into the cobalt-hydrogen bond to give alkyl-cobalt complex. The coordinatively unsaturated alkylcobalt tricarbonyl accepts another molecule of carbon monoxide, then the insertion of carbon monoxide into the alkyl-cobalt bond proceeds to give acylcobalt tricarbonyl. Although the reaction is viewed as carbon monoxide insertion, evidence favors an alkyl migration to the coordinated carbon monoxide. [445]

In the presence of carbon monoxide, the unsaturated tricarbonyl is in equilibrium with the tetracarbonyl. The last step is the hydrogenolysis. Interaction of hydrogen with the unsaturated acyl complex is followed by coupling of the acyl group and hydrogen to give aldehyde and $HCo(CO)_3$. The regeneration of $HCo(CO)_3$ completes the catalytic cycle of the reaction.

$$HCo(CO)_4 \rightleftharpoons HCo(CO)_3 + CO$$

$$HCo(CO)_3 + H_2C=CH_2 \rightleftharpoons \underset{\underset{H_2C=CH_2}{\uparrow}}{HCo(CO)_3}$$

$$\underset{\underset{H_2C=CH_2}{\uparrow}}{HCo(CO)_3} \rightleftharpoons CH_3CH_2Co(CO)_3$$

$$CH_3CH_2Co(CO)_3 + CO \rightleftharpoons CH_3CH_2Co(CO)_4$$

$$CH_3CH_2Co(CO)_4 \rightleftharpoons CH_3CH_2COCo(CO)_3$$

$$CH_3CH_2COCo(CO)_3 + CO \rightleftharpoons CH_3CH_2COCo(CO)_4$$

$$CH_3CH_2COCo(CO)_3 + H_2 \longrightarrow CH_3CH_2CHO + HCo(CO)_3$$

Cobalt carbonyl-catalyzed carbonylation of aldehyde and carboxamide offers a new synthesis of N-acylamino acids. [446]

$$RCHO + R'CONH_2 + CO \xrightarrow{Co_2(CO)_8} R\text{—}\underset{\underset{NHCOR'}{|}}{CH}\text{—}CO_2H$$

Since aldehydes are produced from olefins by the oxo reaction catalyzed by $Co_2(CO)_8$, this N-acylamino acid preparation can be carried out from olefin, carboxamide, carbon monoxide, and hydrogen in moderate yields. The following sequence was proposed for the reaction, although the mechanism of the cobalt-carbon σ-bond formation is not clear.

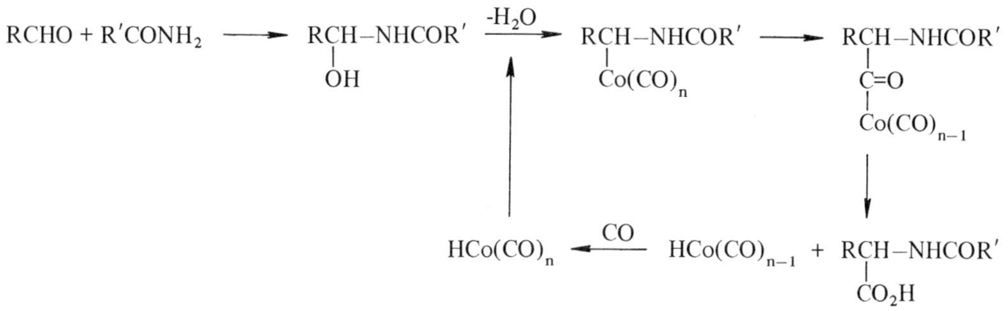

Allyl halides react with carbon monoxide in the presence of $Ni(CO)_4$ or palladium catalyst to give 3-butenoate. [155—157] In this reaction, π-allyl complex is formed at first, and then carbon monoxide insertion into the allyl-metal bond takes place (p. 41). In addition to allyl halides, various allylic compounds such as allylic ethers, esters, and alcohols are carbonylated by the palladium catalyst. In the carbonylation of diallyl ether catalyzed by palladium chloride, stepwise insertion of carbon monoxide at the allylic position occurs to give an allylic ester, which then is converted further to the anhydride of 3-butenoic acid. [156]

$$CH_2{=}CHCH_2OCH_2CH{=}CH_2 \xrightarrow{CO} CH_2{=}CHCH_2CO_2CH_2CH{=}CH_2$$

$$\xrightarrow{CO} (CH_2{=}CHCH_2CO)_2O$$

The products obtained by these carbonylations of allylic compounds are β,γ-unsaturated esters. The new π-allylic complex *(383)* can be prepared from the β,γ-unsaturated esters by reaction with palladium chloride. Any compound having an allylic hydrogen activated by an electron-attracting group, and thus capable of giving a carbanion, undergoes this type of complex formation. [112, 113]

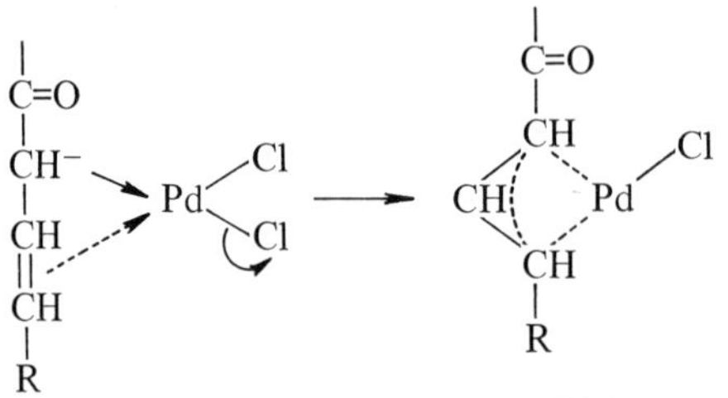

383

Carbonylation of the complex *(384)* prepared from 3-butenoate gave glutaconate, which formed another π-allylic complex *(385)*. Thus the following sequence of complex formation and carbonylation is possible.

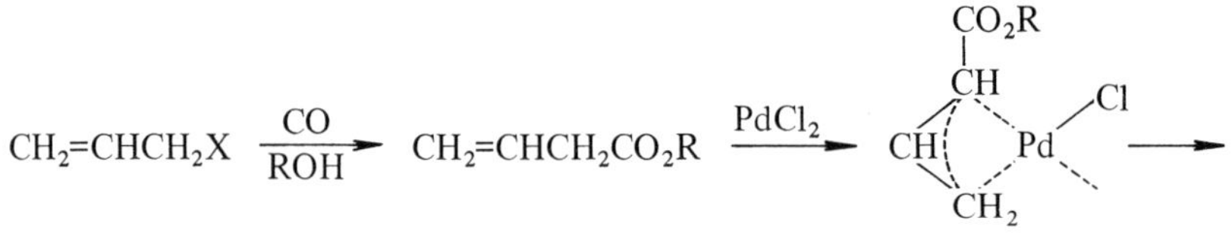

384

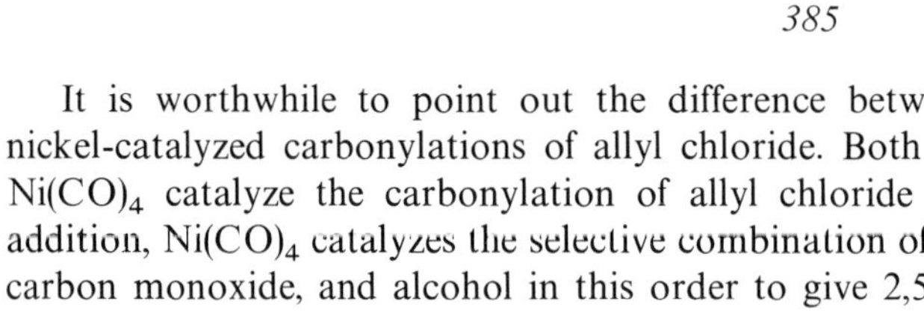

385

It is worthwhile to point out the difference between the palladium- and nickel-catalyzed carbonylations of allyl chloride. Both palladium chloride and $Ni(CO)_4$ catalyze the carbonylation of allyl chloride to give 3-butenoate. In addition, $Ni(CO)_4$ catalyzes the selective combination of allyl chloride, acetylene, carbon monoxide, and alcohol in this order to give 2,5-hexadienoate. [155]

$$CH_2=CHCH_2Cl + HC\equiv CH + CO + ROH$$
$$\longrightarrow CH_2=CHCH_2CH=CHCO_2R$$

On the other hand, palladium catalysts promote reaction of allyl chloride with butadiene (but not with acetylene) and carbon monoxide to form 3,7-octadienoate, although the reaction is not highly selective. [447]

$$CH_2=CHCH_2Cl + CH_2=CHCH=CH_2 + CO + ROH$$
$$\longrightarrow CH_2=CHCH_2CH_2CH=CHCH_2CO_2R$$

Insertion reactions of carbon monoxide and olefins generally occur once or twice in one synthetic reaction. In some cases, the insertion reaction takes place more than once. Successive insertions take place in some reactions, making possible the synthesis of quite a complex molecule. The polymerization reaction of olefins is a typical example. Furthermore, different species also insert successively and orderly. The factors which differentiate oligomerization from polymerization are not clear. Elucidation of the factors which determine whether one step or successive stepwise insertion takes place is an interesting but difficult problem to understand at present. A typical example of the selective and stepwise insertions of different molecules was observed in the above-shown reaction of allyl chloride and acetylene in the presence of $Ni(CO)_4$. Various compounds are formed under varying reaction conditions. [436, 448, 449]

At first, π-allylnickel chloride is formed by the oxidative addition of allyl chloride to $Ni(CO)_4$. Also the reaction is possible by nickel complex formed *in situ* by reduction of nickel chloride with Mn/Fe alloy in the presence of thiourea. The next step is insertion of acetylene and carbon monoxide into the allyl-nickel bond to form *cis*-2,5-hexadienoylnickel complex *(386)*. In methanol, the complex *(386)* is converted into *cis*-2,5-hexadienoate *(387)* in a high yield. In an aprotic solvent, further reaction takes place to give cyclized products. In other words, the intramolecular insertion of the terminal double bond into the acylnickel bond takes place to form (2-oxo-3-cyclopenteno)methylnickel complex *(388)*. Again carbon monoxide insertion takes place to give the acyl complex. If the reaction is terminated at this stage with water present in the reaction medium, the product is 2-oxo-3-cyclopentenylacetic acid *(389)*.

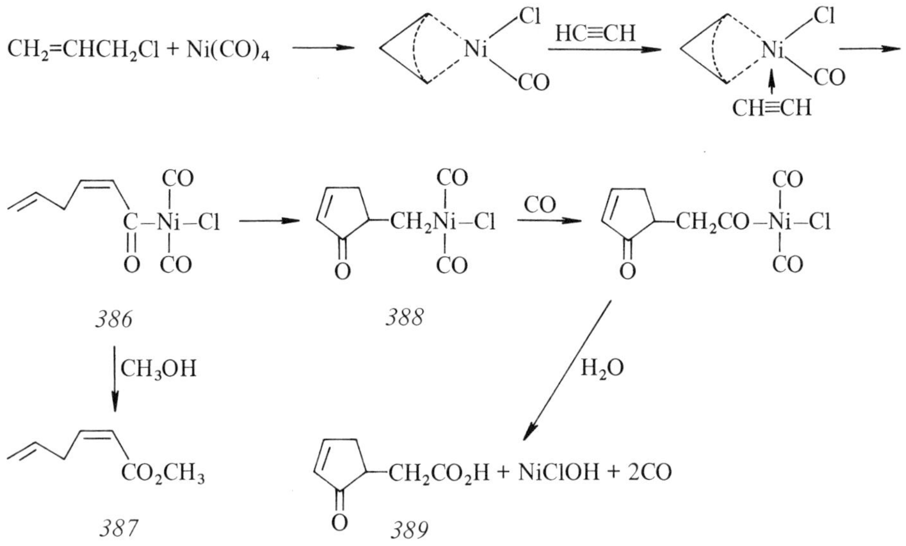

In acetone, further insertion of acetylene and carbon monoxide continues. The last step is the formation of the unsaturated lactone *(390)*. This reaction is understood as an intramolecular insertion of the ketonic carbonyl into the acyl-nickel bond, followed by the elimination of a bivalent nickel compound.

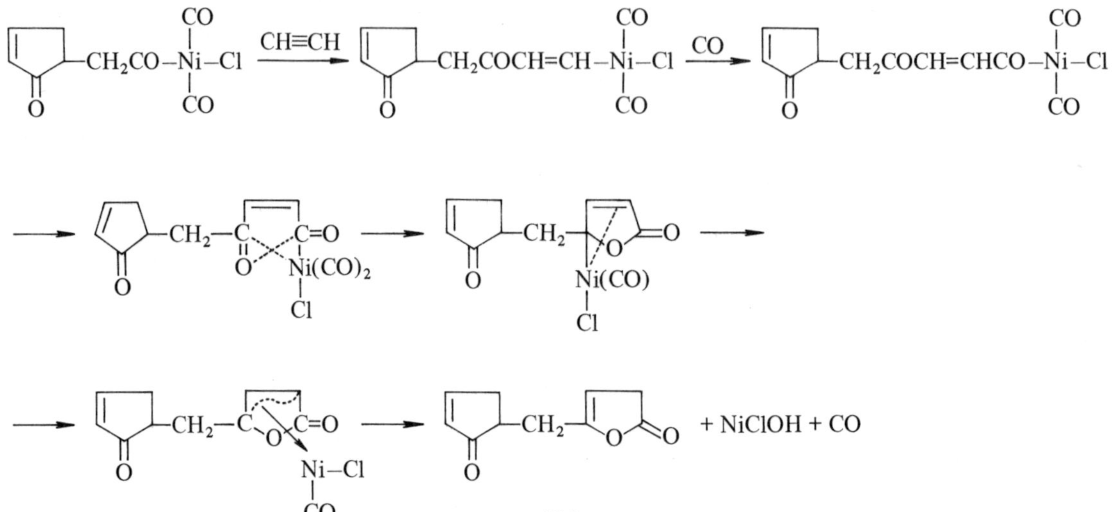

As a competing reaction, acetone itself reacts with the intermediate nickel complex to form the following compound *(391)*. This reaction is similar to the Reformatsky reaction, and involves a nucleophilic attack of the carbon bonded to nickel to the carbonyl group.

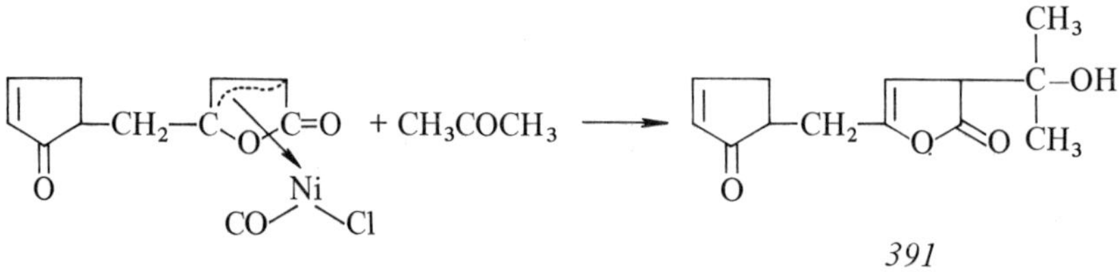

The reaction of methallyl chloride, acetylene, and carbon monoxide produced 5-methyl-2,5-hexadienoic acid *(392)* as a main product and *m*-cresol *(393)* as a minor product. [450] However, *m*-cresol can be synthesized as the main product of the reaction by modifying the reaction conditions. Sodium iodide was added to the reaction medium. In addition, magnesium oxide was used as a neutralizing agent. Iron powder also showed a favorable effect in the cyclization of the intermediate 5-methyl-2,5-hexadienoyl group. In this cyclization reaction the six-membered ketone was formed rather than five-membered ketone. The catalytic reaction was carried out in acetone at room temperature under an atmospheric pressure of carbon monoxide, and *m*-cresol *(393)* was obtained in 74% yield based on methallyl chloride.

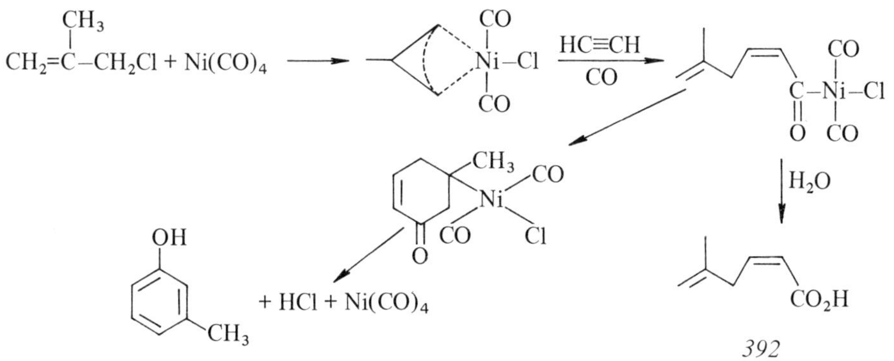

$$CH_2=\overset{\underset{\displaystyle CH_3}{|}}{C}-CH_2Cl + Ni(CO)_4 \longrightarrow$$

393

392

When the carbonylation of allyl chloride was carried out in water in the presence of magnesium oxide, which behaves as an acid catcher, 3-butenylsuccinic acid was obtained in 60% yield. [451, 452] The reaction proceeds *via* the complex *(394)* formed from 3-butenoic acid. The terminal double bond of the coordinated 3-butenoic acid then inserts into the π-allylic bond. Subsequent carbon monoxide insertion and hydrolysis gave 3-butenylsuccinic acid as the final product.

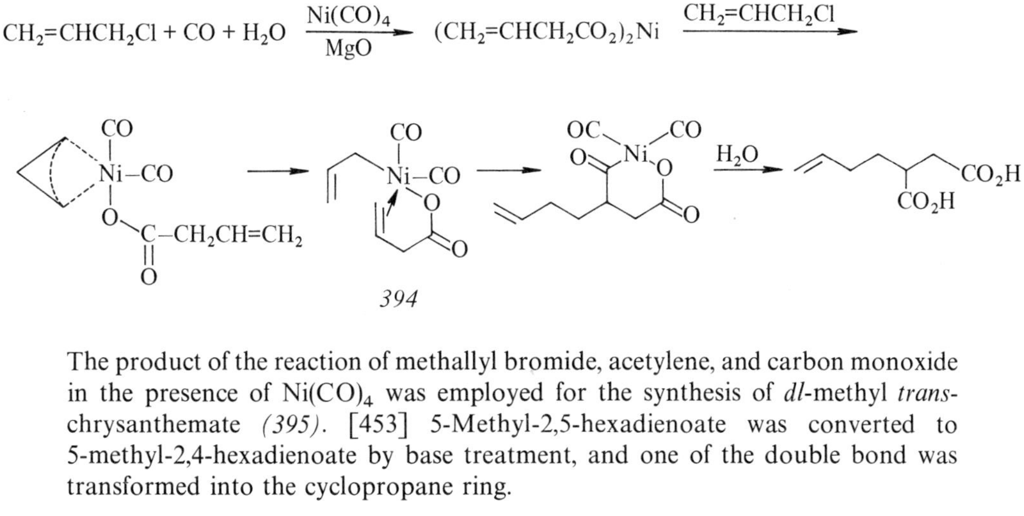

$$CH_2=CHCH_2Cl + CO + H_2O \xrightarrow[\text{MgO}]{\text{Ni(CO)}_4} (CH_2=CHCH_2CO_2)_2Ni \xrightarrow{CH_2=CHCH_2Cl}$$

394

The product of the reaction of methallyl bromide, acetylene, and carbon monoxide in the presence of Ni(CO)$_4$ was employed for the synthesis of *dl*-methyl *trans*-chrysanthemate *(395)*. [453] 5-Methyl-2,5-hexadienoate was converted to 5-methyl-2,4-hexadienoate by base treatment, and one of the double bond was transformed into the cyclopropane ring.

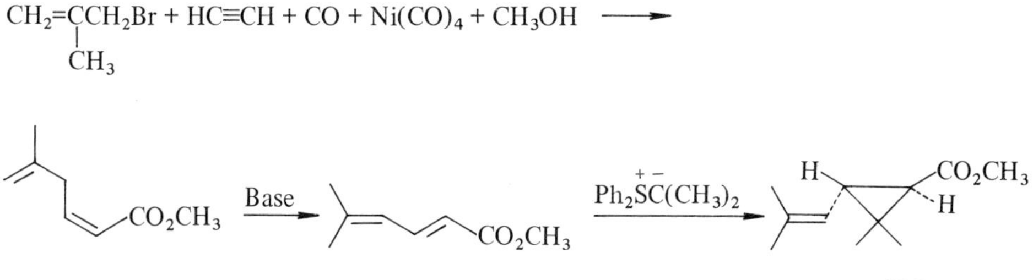

$$CH_2=\overset{\underset{\displaystyle CH_3}{|}}{C}CH_2Br + HC\equiv CH + CO + Ni(CO)_4 + CH_3OH \longrightarrow$$

395

Reaction of acyl halide, acrolein, and acetylene in the presence of Ni(CO)$_4$ gave cyclopentenone derivatives. [454] The reaction starts from the oxidative addition of acyl halide to form an acylnickel complex. The insertion of the carbonyl group of acrolein into the carbon-nickel bond of the acyl complex affords the π-allylic complex (396). Then insertions of acetylene and carbon monoxide take place to give the five-membered ring compounds.

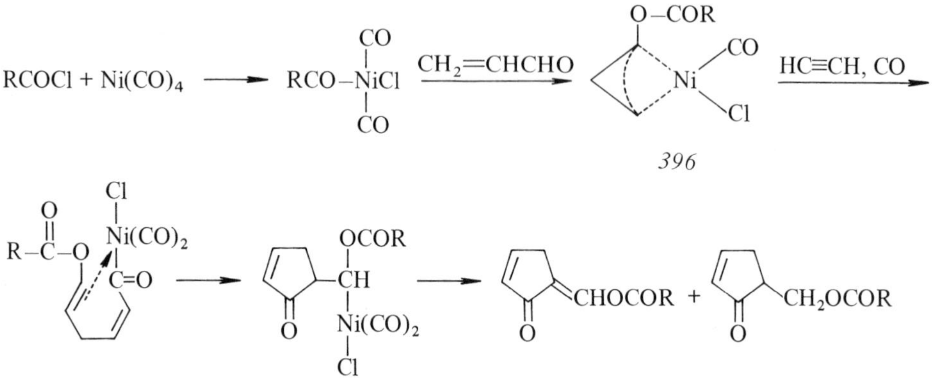

The successive insertion reactions of different molecules shown above are remarkable; somewhat complicated molecules are fabricated from simple molecules such as acetylene and carbon monoxide by a series of orderly insertion reactions.

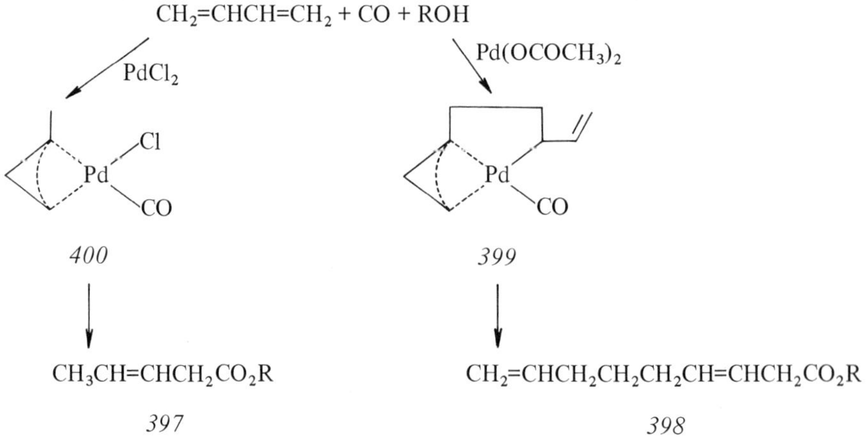

Another carbonylation reaction which proceeds *via* π-allylic complex is the palladium-catalyzed reaction of butadiene. With palladium chloride, in the presence or absence of triphenylphosphine, one mole of butadiene reacted with carbon monoxide in alcohol to give 3-pentenoate (397). [455] On the other hand, two moles of butadiene reacted with carbon monoxide selectively to give 3,8-nonadienoate (398) when palladium acetate was used as the catalyst. [456, 457] The essential factor which differentiates the monomeric from the dimeric carbonylation is the presence or absence of a halide ion coordinated to the palladium. With halide-free palladium complexes, such as palladium acetate combined with triphenylphosphine, 3,8-nonadienoate is obtained almost selectively. The mechanism of these carbonylation reactions of butadiene can be explained by the following scheme. With a halide-free palladium catalyst, two moles of butadiene form the diallylic palladium complex *(399)* and then carbon monoxide insertion takes place to give 3,8-nonadienoate. When chloride ion occupies one coordination site of palladium, the formation of the diallylic complex is not possible, and only the monomeric complex *(400)* which gives 3-pentenoate is formed.

The industrial production of acetic acid has recently been established; it is the rhodium-catalyzed carbonylation of methanol in the presence of certain iodides. [458]

$$CH_3OH + CO \longrightarrow CH_3CO_2H$$

The reaction proceeds under a relatively low pressure of carbon monoxide with very high selectivity. The following mechanism involving oxidative addition of methyl iodide, insertion of carbon monoxide and reductive elimination was proposed for the reaction.

$$CH_3OH + HI \longrightarrow CH_3I + H_2O$$

$$CH_3I + [\text{Rh complex}] \longrightarrow [CH_3\text{—Rh—I complex}] \xrightarrow{CO} [CH_3\overset{\overset{\displaystyle CO}{|}}{Rh}\text{—I complex}]$$

$$[CH_3CO\text{—Rh—I complex}] \xrightarrow{H_2O} CH_3CO_2H + [\text{Rh complex}] + HI$$

Acetylenic compounds are active substrates for transition metal catalyzed carbonylations. The most famous one is the industrial process of the $Ni(CO)_4$-catalyzed acrylic acid production from acetylene. Similar to the carbonylation of olefins, acetylene inserts into the nickel-hydride complex to give a vinylnickel complex, which is subjected to carbon monoxide insertion.

$$CH{\equiv}CH + CO + H_2O \xrightarrow{Ni(CO)_4} CH_2{=}CHCO_2H$$

Palladium is also an active catalyst for the reaction of various acetylenic compounds. When carbonylation of acetylene was carried out in benzene in the presence of palladium chloride under carbon monoxide pressure, chlorides of fumaric, maleic, and muconic acids were obtained. [459] A similar reaction was carried out even at room temperature by passing carbon monoxide and acetylene into a methanolic solution containing palladium chloride and thiourea. The stoichiometry of the process is as follows. [460]

$$CH\equiv CH + 2CO + 2CH_3OH \longrightarrow H_3CO_2CCH=CHCO_2CH_3 + H_2$$

$$2CH\equiv CH + 2CO + 2CH_3OH \longrightarrow H_3CO_2CCH=CHCH=CHCO_2CH_3 + H_2$$

Carbonylation reactions of several acetylenic compounds offer interesting synthetic methods. Carbonylation of propargyl chloride or alcohols is catalyzed by $Ni(CO)_4$ or palladium chloride. [461—465] In the palladium-catalyzed reaction of propargyl alcohol in methanol containing hydrogen chloride, methyl itaconate *(401)* was obtained as a main product. In addition some methyl aconitate *(402)* and methyl 2-(methoxymethyl)acrylate *(403)* were produced. Obviously these products were formed by the reactions of three, two, and one moles of carbon monoxide on the propargyl alcohol.

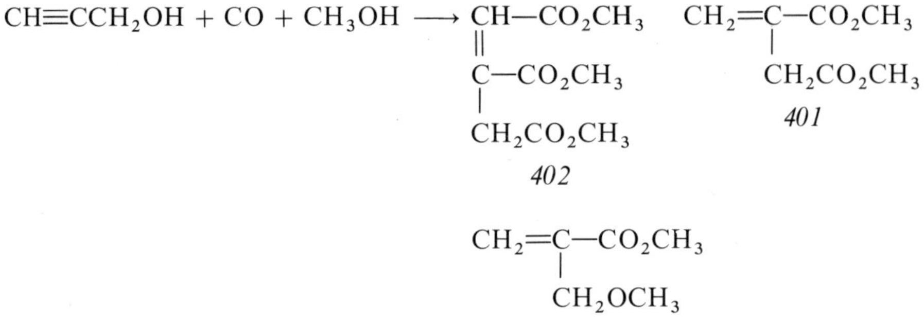

$$CH\equiv CCH_2OH + CO + CH_3OH \longrightarrow$$

CH—CO₂CH₃
‖
C—CO₂CH₃
|
CH₂CO₂CH₃
402

CH₂=C—CO₂CH₃
|
CH₂CO₂CH₃
401

CH₂=C—CO₂CH₃
|
CH₂OCH₃
403

2-Methyl-3-butyn-2-ol *(404)* in methanol was attacked by one or two moles of carbon monoxide in the presence of palladium chloride to produce methyl teraconate and terebate, respectively. In benzene, teraconic anhydride *(405)* was obtained selectively in 42 % yield.

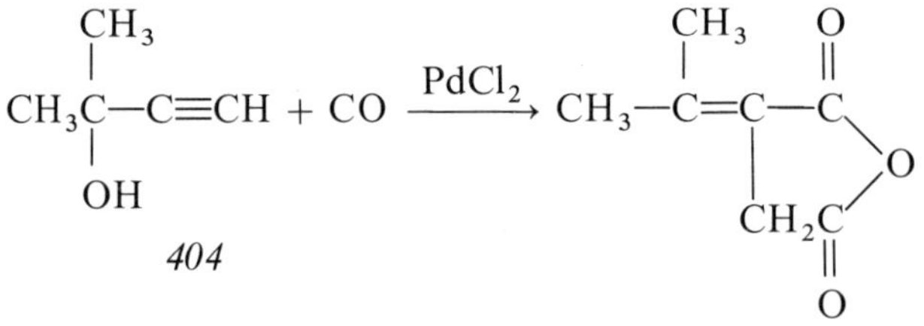

CH₃
|
CH₃C—C≡CH + CO —PdCl₂→
|
OH
404

CH₃ O
| ‖
CH₃—C=C—C
         \
          O
| /
CH₂C
   ‖
   O
405

At this point, it is appropriate to mention the effect of carbon monoxide pressure on carbonylation reactions. [436] For many years carbonylation reactions of olefins and acetylenes have been carried out at high temperature and pressure of carbon monoxide, mainly because the processes leading to coordinative unsaturation and to oxidative addition were not understood. Under high carbon monoxide pressure, it is rather difficult to bring the catalyst complexes to the state of coordinative unsaturation which is necessary for the oxidative addition reaction. Deep understanding of the mechanism of the carbonylation and other reactions catalyzed by transition metal complexes has made it possible to carry out various carbonylation reactions under very mild conditions by appropriate choice of metals and ligands. Carbonylation of allylic compounds in the presence of $Ni(CO)_4$ can be carried out under atmospheric pressure. Acetylene is also converted into muconate in the presence of palladium chloride and thiourea by passing carbon monoxide. Carbonylation reactions under high pressure have not been a laboratory method. But the reactions under mild conditions especially at atmospheric pressure will be a convenient laboratory method if more efficient catalysts are discovered so that special equipment is not necessary.

Cobalt complexes, another type of convenient reagents for carbonylation, effect reactions quite different from those of nickel and palladium. Acylcobalt tetracarbonyl, which is an intermediate of the cobalt-catalyzed oxo reaction, can be prepared by the reaction of sodium cobalt tetracarbonyl with acyl halides or with alkyl halides and carbon monoxide.

$$NaCo(CO)_4 + CH_3COCl \longrightarrow CH_3COCo(CO)_4 + NaCl$$

The acylcobalt complexes thus prepared are very reactive species and suffer insertions of various unsaturated bonds at the cobalt-acyl bond giving interesting products. [466] The cobalt complex reacted with α,β-unsaturated aldehydes or ketones to form 1-acyloxy-π-allylcobalt tricarbonyl *(406)*, which was isolated in the form of the triphenylphosphine complex. [467] The reaction can be explained by the insertion of the carbonyl of the unsaturated ketones into the acyl-cobalt bond followed by π-allylic complex formation.

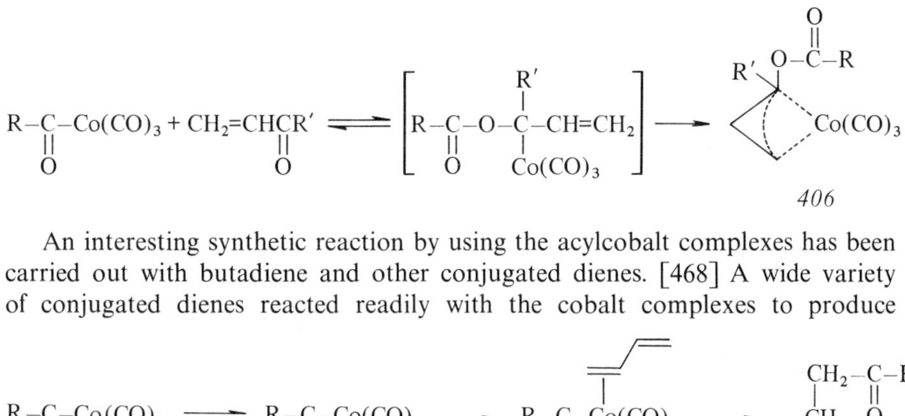

406

An interesting synthetic reaction by using the acylcobalt complexes has been carried out with butadiene and other conjugated dienes. [468] A wide variety of conjugated dienes reacted readily with the cobalt complexes to produce

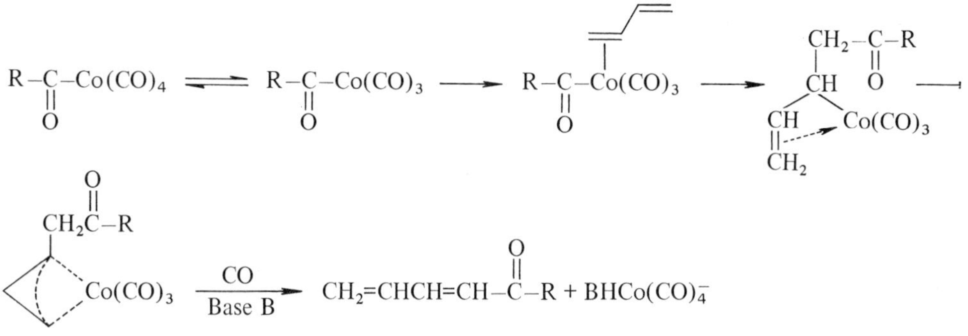

407   *408*

1-acylmethyl-π-allylcobalt tricarbonyl *(407)*. The complex was converted to 1-acyl-1,3-butadiene derivatives *(408)* by treatment with a base.

This reaction affords a general synthetic method of acyldienes, which are not easy to synthesize by conventional methods. The reaction of 1,3-cyclohexadiene with methyl iodide gave 1-acetyl-1,3-cyclohexadiene *(409)*,

$$\text{(1,3-cyclohexadiene)} + CH_3I + CO \xrightarrow{Co(CO)_4^-} \text{(1-acetyl-1,3-cyclohexadiene)}$$

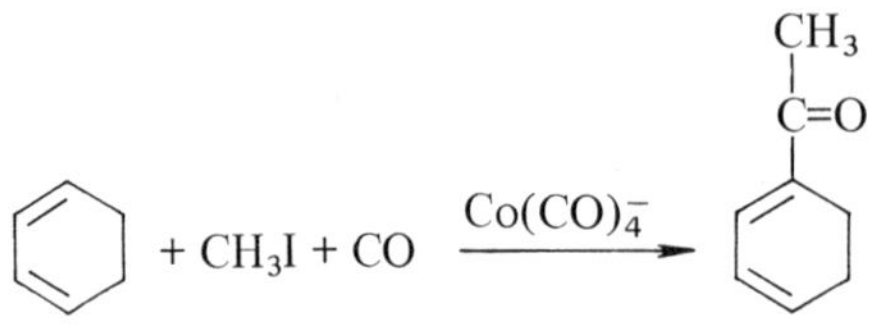

409

As an intramolecular reaction, *trans,trans*-2,4-hexadienoyl(triphenylphosphine)-cobalt tricarbonyl cyclized at 80° to produce 2-methyl-π-cyclopentenone complex of cobalt *(410)*.

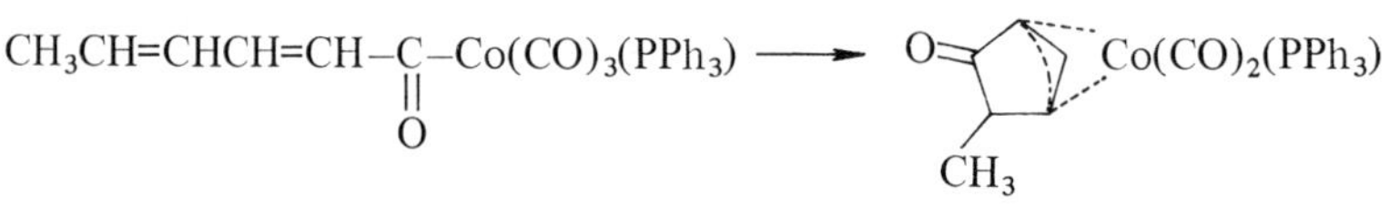

$$CH_3CH=CHCH=CH-\underset{\underset{O}{\|}}{C}-Co(CO)_3(PPh_3) \longrightarrow O=\text{(ring)}\ Co(CO)_2(PPh_3)$$

$$CH_3$$

410

The acyldiene synthesis can be made catalytic if the acyl- or alkylcobalt tetra-carbonyl is prepared from acyl- or alkyl halides and cobalt carbonyl anion in the presence of diene, carbon monoxide, and a base. In this way, cobalt carbonyl anion can be regenerated by the base.

$$RX + CH_2=CHCH=CH_2 + CO \xrightarrow[\text{base}]{Co(CO)_4^-} RCOCH=CHCH=CH_2 + HX$$

Acetylenes also insert into the acyl-cobalt bond. Substituted acetylenes form π-(2,4)-(alkeno-4-lactonyl)cobalt tricarbonyl by the insertion. [469] For example, acetylcobalt tetracarbonyl reacted with 3-hexyne to give 2,3-diethyl-π-(2,4)-(penteno-4-lactonyl)cobalt tricarbonyl *(411)*.

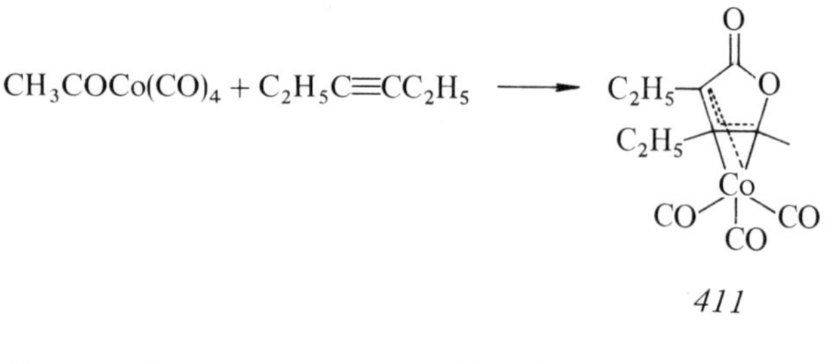

$$CH_3COCo(CO)_4 + C_2H_5C\equiv CC_2H_5 \longrightarrow$$

$$C_2H_5 - \text{(lactone ring with O)}$$
$$C_2H_5$$
$$\underset{\underset{CO}{Co}}{} CO \quad CO$$

411

The reaction can be explained by the successive insertion of the acetylene and carbon monoxide to form β-acylacrylylcobalt tricarbonyl *(412)*. The final step is the intramolecular insertion of the carbonyl bond of the β-acyl group into the cobalt-acyl bond to form unsaturated lactone, which can be stabilized by π-allylic coordination with the cobalt carbonyl.

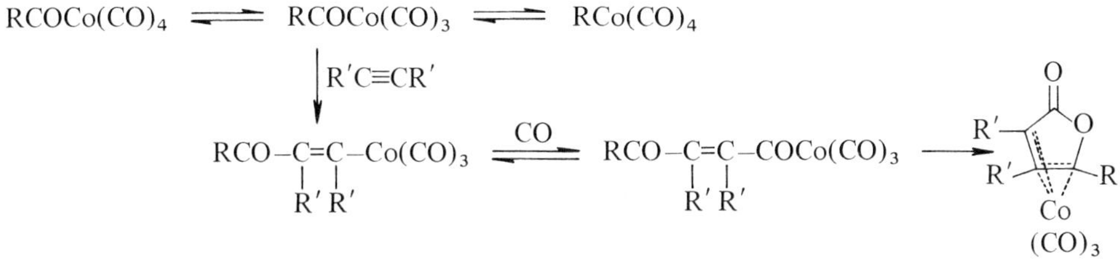

$$RCOCo(CO)_4 \rightleftharpoons RCOCo(CO)_3 \rightleftharpoons RCo(CO)_4$$

$$412 \qquad\qquad 411$$

Carboethoxyacetylcobalt tricarbonyl (413) reacted at 0° with 3-hexyne in the presence of carbon monoxide, forming the lactone complex (414). Since this complex has an active hydrogen at C-5, the base-catalyzed elimination reaction produced 2,4-pentadieno-4-lactone derivatives (415) and hydrocobalt tricarbonyl.

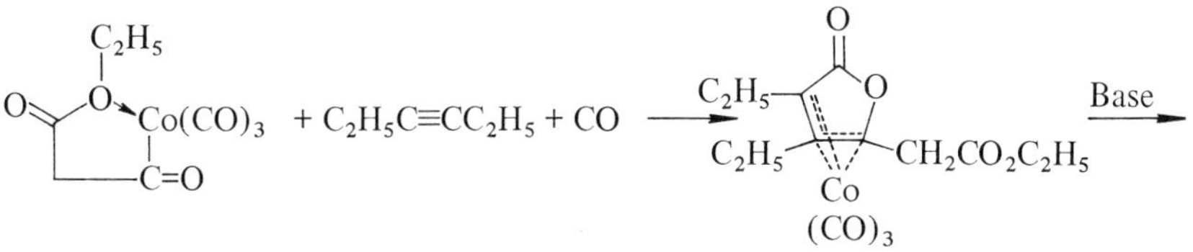

$$413 \qquad\qquad\qquad\qquad 414$$

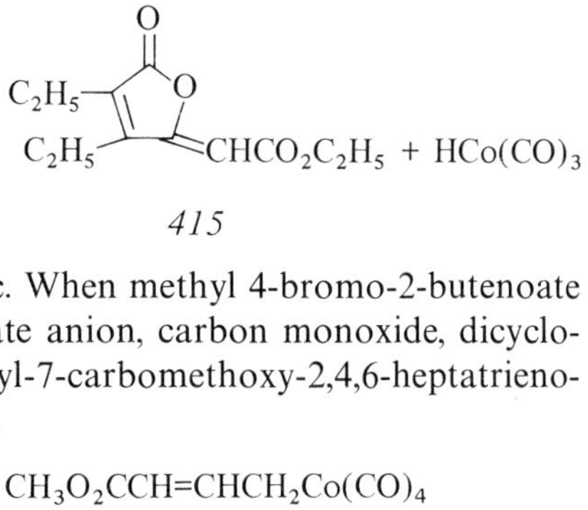

$$415$$

The reaction can be made partially catalytic. When methyl 4-bromo-2-butenoate was allowed to react with cobalt carbonylate anion, carbon monoxide, dicyclo-hexylethylamine and 3-hexyne, 2,3-diethyl-7-carbomethoxy-2,4,6-heptatrieno-4-lactone (416) was obtained in 30% yield.

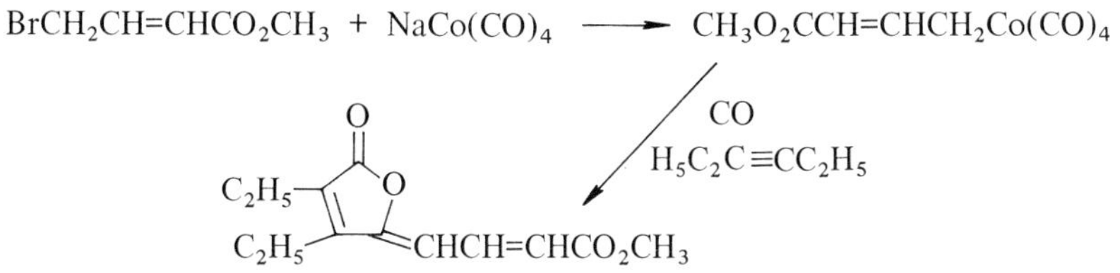

$$BrCH_2CH=CHCO_2CH_3 + NaCo(CO)_4 \longrightarrow CH_3O_2CCH=CHCH_2Co(CO)_4$$

$$416$$

Carbonylation of some aromatic compounds proceeds through carbon monoxide insertion. The following stoichiometric carbon monoxide insertion into an azobenzene palladium complex (64) is a typical example. [470] The complex prepared from palladium chloride and azobenzene (63) by ortho-palladation has a carbon-palladium bond (p. 30). [91] The carbonylation of the complex gave an indazolinone (417) by carbon monoxide insertion into the palladium-benzene bond. Further carbonylation catalyzed by $Co_2(CO)_8$ produced a quinazolinedione (418), hydrolysis of which gave aniline and anthranilic acid (419). The quinazolinedione (418) was also obtained directly from azobenzene by the carbonylation reaction catalyzed by $Co_2(CO)_8$. [471, 472] This method is useful for the synthesis of substituted anthranilic acids.

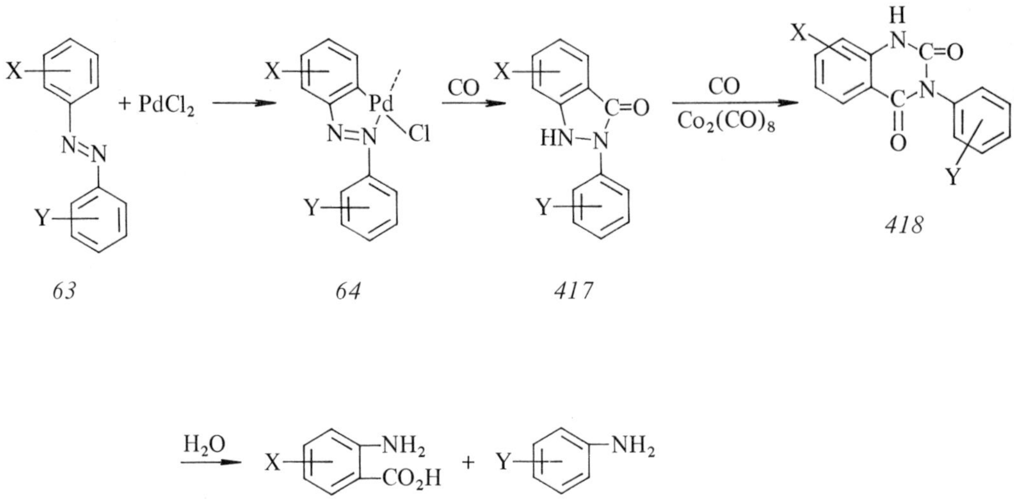

63   64   417   418

419

The *ortho*-metalation reaction of thiobenzophenone (65) with $Fe_2(CO)_9$ is a similar reaction. [93] (p. 30). Removal of the iron from the complex (66) by oxidation with Ce^{4+} ion proceeded through carbon monoxide insertion to give the isobenzothiophene derivative (420).

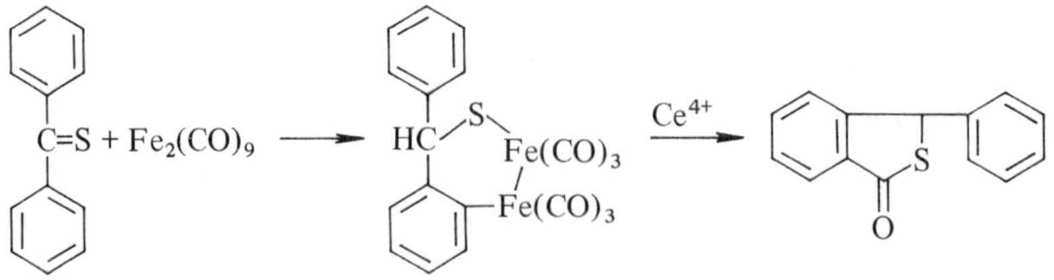

65   66   420

4. Other Insertion Reactions

Insertion of a carbon-nitrogen triple bond is assumed in the hydration reaction of a nitrile to an amide catalyzed by certain noble metal hydroxide complexes. [473] For example, *trans* $Rh(OH)(CO)(PPh_3)_2$ and similar iridium or platinum complexes were found to be good catalysts for hydration of nitriles. The following mechanism involving the insertion of the carbon-nitrogen triple bond into the metal-hydroxide bond was proposed.

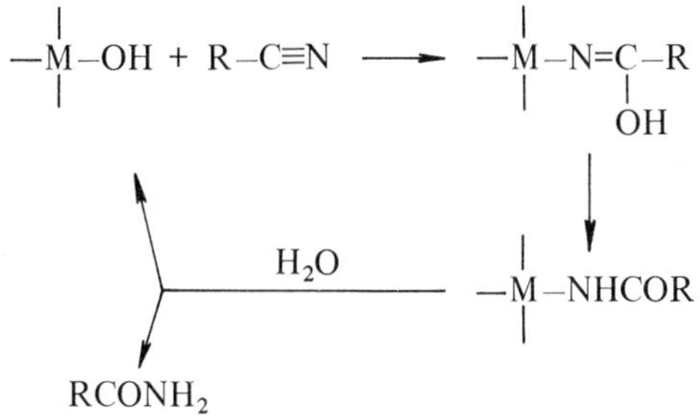

$$-\overset{|}{\underset{|}{M}}-OH + R-C{\equiv}N \longrightarrow -\overset{|}{\underset{|}{M}}-\underset{\underset{OH}{|}}{N{=}C}-R$$

$$\underset{RCONH_2}{\overset{H_2O}{\diagup}} \qquad -\overset{|}{\underset{|}{M}}-NHCOR$$

Insertion of an epoxide into a cobalt-hydrogen bond proceeds with ring opening. [474] Isobutylene oxide reacted with hydrocobalt tetracarbonyl to give the cobalt complex *(421)*, decomposition of which with iodine in methanol gave 3-hydroxy-3-methylbutyrate *(422)* as a main product.

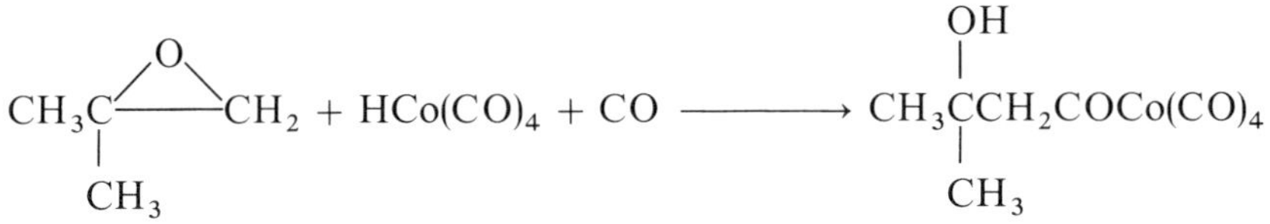

$$CH_3\overset{O}{\overset{\triangle}{C}}CH_2 + HCo(CO)_4 + CO \longrightarrow CH_3\overset{OH}{\underset{CH_3}{\overset{|}{C}}}CH_2COCo(CO)_4$$
$$\underset{CH_3}{}$$

421

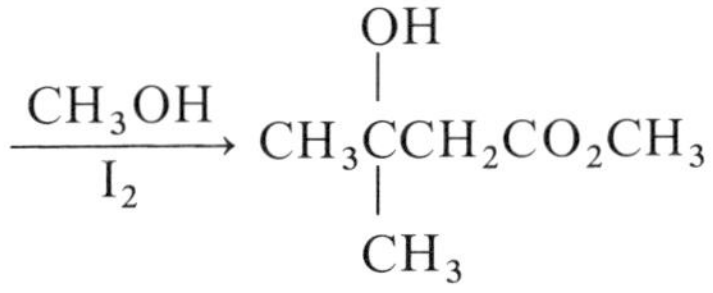

$$\xrightarrow[I_2]{CH_3OH} CH_3\overset{OH}{\underset{CH_3}{\overset{|}{C}}}CH_2CO_2CH_3$$

422

Another example of the insertion with ring opening is given by the reaction of trimethylene oxide *(423)* with hydrocobalt carbonyl. 4-Hydroxybutyrylcobalt tetracarbonyl *(424)*, the intermediate, was converted into γ-butyrolactone by intramolecular attack of the terminal alcohol to the cobalt-acyl bond. The net result is the ring expansion of the trimethylene oxide by carbon monoxide insertion.

$$\square{O} + HCo(CO)_4 + CO \longrightarrow HOCH_2CH_2CH_2COCo(CO)_4 \longrightarrow \square{O}{=}O$$

423   *424*

Isocyanides, like carbon monoxide and phosphines, act as both σ-donor and π-acceptors, and can stabilize low valent states of transition metals. Their π-accepting property as a ligand is stronger than that of phosphines but weaker than that of carbon monoxide. In addition, isocyanides are isoelectronic with carbon monoxide having the following resonance forms.

$$[:\bar{C}\equiv\overset{+}{N}-R \longleftrightarrow :C=\ddot{N}-R \longleftrightarrow :C=\overset{+}{N}=\bar{R}]$$

An insertion of isocyanide, similar to that of carbon monoxide, is also known. [475] Alkyl complexes of platinum and palladium underwent the following insertion. [476]

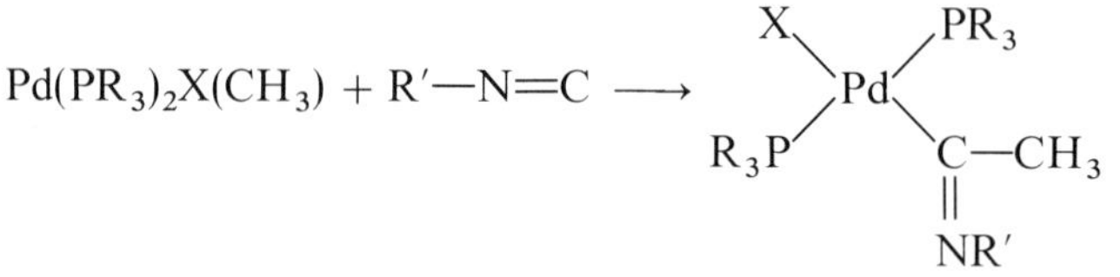

$$Pd(PR_3)_2X(CH_3) + R'-N\equiv C \longrightarrow$$

Reaction of π-allylpalladium chloride with cyclohexyl isocyanide forms the insertion product *(425)* having a new carbon-palladium bond. [477, 478] Alcoholysis of the inserted product gave *N*-cyclohexyl-3-butenimidate *(426)*.

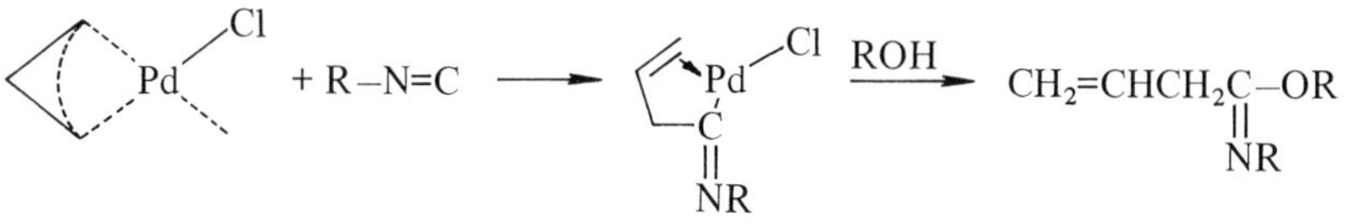

Polymerization of isocyanides is catalyzed by various transition metal complexes. The reaction is explained by the successive stepwise insertion of the isocyanides. As supporting evidence, the tris-imino complexes *(427)* from an iron complex and cyclohexyl isocyanide were isolated and characterized. [479]

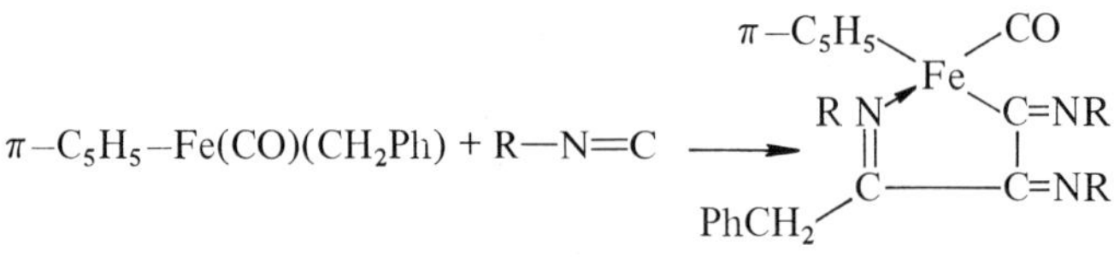

$$\pi-C_5H_5-Fe(CO)(CH_2Ph) + R-N\equiv C \longrightarrow$$

R = C₆H₁₁

A few examples of sulfur dioxide insertion have been reported. [480, 481] NMR and IR analyses showed that the insertion of sulfur dioxide into the alkyl-iron complexes *(428)* proceeds *via* the intermediacy of O-sulfinate, which subsequently rearranges to the thermodynamically stable S-sulfinate *(429)*. [482]

$$(\pi\text{-}C_5H_5)(CO)_2Fe\text{—}R + SO_2 \longrightarrow (\pi\text{-}C_5H_5)(CO)_2Fe\overset{\displaystyle O}{\underset{\displaystyle O}{\overset{\|}{\underset{\|}{\text{—S—R}}}}}$$

$$\underset{428}{} \qquad\qquad\qquad\qquad \underset{429}{}$$

Diastereoisomers of the iron complex *(430)* epimeric at the iron were prepared. The insertion reaction of sulfur dioxide to the metal-carbon bond has been found to proceed with a high degree of stereospecificity with retention of configuration at the iron. [483]

$$\underset{430}{}$$

Formation of *trans*-2-butenyl methyl sulfone *(431)* by the palladium-catalyzed reaction of ethylene and sulfur dioxide was reported. [484] The reaction starts with the insertion of ethylene into a palladium-hydrogen bond, and is followed by sulfur dioxide insertion. [485] Then two molecules of ethylene are inserted into the palladium-sulfur bond. The final step is the β-elimination reaction concomitant with the regeneration of the palladium hydride complex.

$$CH_2{=}CH_2 + H\text{—}Pd\text{—}X \longrightarrow CH_3CH_2\text{—}Pd\text{—}X \xrightarrow{SO_2} CH_3CH_2SO_2\text{—}Pd\text{—}X$$

$$\xrightarrow{2CH_2{=}CH_2} CH_3CH_2SO_2CH_2CH_2CH_2CH_2\text{—}Pd\text{—}X$$

$$\longrightarrow CH_3CH_2SO_2CH_2CH_2CH{=}CH_2 + H\text{—}Pd\text{—}X$$

$$\downarrow$$

$$CH_3CH_2SO_2CH_2CH{=}CHCH_3$$

$$\underset{431}{}$$

Hydrosilylation is a useful reaction not only for the synthesis of silicone compounds from olefins, but also for offering some specific synthetic methods. An unsaturated carbon-oxygen bond of ketones and aldehydes sometimes inserts into a metal σ-bond. Hydrosilylation of carbonyl compounds is possible by using metal catalysts and the reaction proceeds through the insertion of the carbonyl group. The reaction is catalyzed by zinc chloride, reduced nickel, platinum and palladium catalysts, [486, 487] and the Wilkinson complex. [488—490] An application of this reaction is the reduction of ketones and aldehydes by hydrosilylation followed by hydrolysis to give alcohols.

In this reaction, oxidative addition of the silicon-hydrogen bond to the metal takes place to form a silicon-metal bond, to which the carbonyl bond is inserted forming the siloxy complex *(432)*. Finally the silyl ether *(433)* is liberated by reductive elimination.

$$-\!\overset{|}{\underset{|}{M}}\!- \;+\; HSiX_3 \;\longrightarrow\; H\!-\!\overset{|}{\underset{|}{M}}\!-SiX_3 \qquad \overset{R}{\underset{R'}{\diagup}}\!\!\diagdown C\!=\!O$$

$$H\!-\!M \quad SiX_3$$
$$R\!-\!\overset{|}{\underset{\underset{R'}{|}}{C}}\!-\!O \longrightarrow \overset{R}{\underset{R'}{\diagup}}\!C\!\diagdown\!\overset{H}{\underset{OSiX_3}{}} \;+\; -\!\overset{|}{\underset{|}{M}}\!-$$

$$432 \qquad\qquad 433$$

Reduction of simple ketones by the above-shown hydrosilylation has had no particular use in organic synthesis, because the same reduction can be achieved easily with other better reducing agents such as aluminum hydrides or boro-hydrides. Usefulness of the hydrosilylation of carbonyl compounds was found recently in selective reduction. Application of this method to terpene ketones affords a method of stereoselective reduction. Any desired stereoisomeric alcohols can be prepared by the choice of proper hydrosilanes. The use of phenylsilane for the reduction of menthone gave rise to the preferential formation of the less stable alcohol, neomenthol, whereas the reduction by dimethylphenylsilane produced the more stable alcohol, menthol. The reaction was catalyzed by $RhCl(PPh_3)_3$.

The more useful application was found in the reduction of α,β-unsaturated carbonyl compounds catalyzed by the Wilkinson complex. [491] 1,4-Addition to α,β-unsaturated carbonyl compounds gives enol ethers, hydrolysis of which gives saturated carbonyl compounds; 1,2-addition reduces only the carbonyl groups of α,β-unsaturated carbonyl compounds without attacking the double bonds. Both types of the reduction can be carried out. The selectivity of the reduction is greatly influenced by the nature of the hydrosilanes employed. Thus α,β-unsaturated carbonyl compounds are converted into the corresponding saturated carbonyl compounds by the 1,4-addition when monohydrosilanes are used. On the other hand, the same carbonyl compounds are easily reduced to the allylic alcohols by the 1,2-addition when dihydrosilanes are used.

$$\overset{R}{\underset{R'}{\diagup}}\!\!\diagdown C\!=\!CH\underset{\underset{O}{\|}}{C}R'' \quad\xrightarrow{(Rh)}$$

$$\xrightarrow{R_3SiH}\quad \overset{R}{\underset{R'}{\diagup}}\!\!\diagdown CHCH\!=\!\underset{\underset{OSiR_3}{|}}{C}R'' \xrightarrow{H_2O} \overset{R}{\underset{R'}{\diagup}}\!\!\diagdown CHCH_2\underset{\underset{O}{\|}}{C}R''$$

$$\xrightarrow{R_2SiH_2}\quad \overset{R}{\underset{R'}{\diagup}}\!\!\diagdown C\!=\!CH\underset{\underset{OSiHR_2}{|}}{C}HR'' \xrightarrow{H_2O} \overset{R}{\underset{R'}{\diagup}}\!\!\diagdown C\!=\!CH\underset{\underset{OH}{|}}{C}HR''$$

The reaction course is greatly affected by the structure of both hydrosilanes and substrates, as exemplified by the reduction of α-ionone *(434)* and citral *(435)* (Table IV).

Table IV

			ratio of 1,4/1,2	yield %
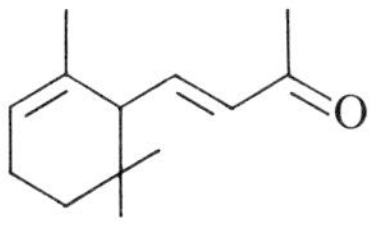 *434*	Et₃SiH	50°, 2 hr	100/0	96
	Ph₂SiH₂	room temp., 30 min	0/100	97
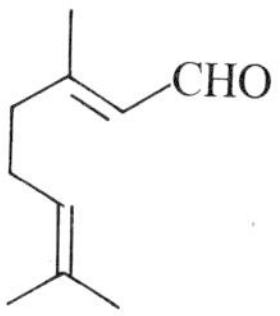 *435*	Et₃SiH	room temp., 1 hr	100/0	97
	Ph₂SiH₂	ice cooled, 30 min	0/100	97

The hydrosilylation of double bonds conjugated to carbonyl groups proceeds much faster than isolated simple double bonds present in the same molecule. Thus the isolated double bonds in α-ionone *(434)* and citral *(435)* were not reduced. This is another useful property.

As in the case of homogeneous hydrogenation of olefins using the rhodium catalysts, asymmetric reduction of carbonyl compounds by hydrosilylation is also possible using rhodium [492—495] and platinum catalysts. [496] The optically active catalysts were prepared by the reaction of two equivalents of the chiral ligand, (−)(S)-benzylmethylphenylphosphine (optical purity 62%) or (+)(R)-benzylmethylphenylphosphine (optical purity 77%) with [Rh(1,5-hexadiene)Cl]₂ or [Rh(COD)Cl]₂.

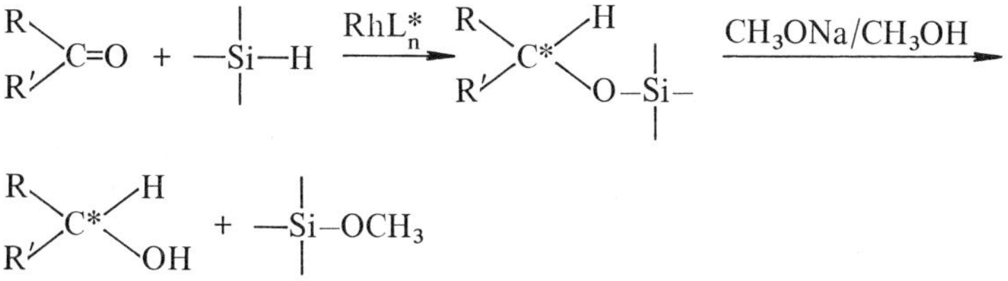

The optical yield depends on the structure of the substrates and the hydrosilanes. The reaction of acetophenone with dimethylphenylsilane catalyzed by the rhodium complex coordinated by (−)-(S)-benzylmethylphenylphosphine produced the optically active silyl ether in 92% yield, [492] methanolysis of which gave

(+)-(R)-1-phenylethanol with an optical yield of 44%. Also dichloro[(R)-(+)-benzylmethylphenylphosphine]di-μ-chlorodiplatinum catalyzes the addition of dichloromethylsilane to a series of alkyl phenyl ketones to give optically active silyl ethers of alkylphenylcarbinols. [496]

Hydrosilylation of imines *(436)* in the presence of the rhodium complex, followed by hydrolysis affords amines. [497]

$$
\begin{array}{c}
\text{R} \\
\diagdown \\
\diagup \\
\text{R}'
\end{array}
\text{C=N—R}'' + \text{Ph}_2\text{SiH}_2 \longrightarrow
\begin{array}{c}
\text{R} \\
\diagdown \\
\diagup \\
\text{R}'
\end{array}
\text{CH—N—R}''
\quad \xrightarrow{\text{H}^+} \quad
\begin{array}{c}
\text{R} \\
\diagdown \\
\diagup \\
\text{R}'
\end{array}
\text{CH—NHR}''
$$

$$\underset{\text{HSiPh}_2}{}$$

436

When the rhodium complex containing the optically active 2,3-O-isopropylidene-2,3-dihydroxy-1,4-bis(diphenylphosphino)butane *(44)* was used as the catalyst in the reaction of the imine formed from acetophenone and benzylamine with diphenylsilane, α-phenylethylbenzylamine was obtained in 50% optical yield. [498]

$$
\begin{array}{c}
\text{Ph} \\
\diagdown \\
\diagup \\
\text{CH}_3
\end{array}
\text{C=NCH}_2\text{Ph} + \text{Ph}_2\text{SiH}_2 \xrightarrow{\text{RhL}_n^*} \xrightarrow{\text{H}^+}
\begin{array}{c}
\text{Ph} \\
\diagdown \\
\diagup \\
\text{CH}_3
\end{array}
\overset{*}{\text{CH}}\text{—NHCH}_2\text{Ph}
$$

As shown by these reactions, the rhodium complexes are very useful catalysts for hydrosilylation reactions.

Palladium chloride and nickel complexes show different activity in the reaction of ketones and hydrosilanes. These catalysts produce silyl vinyl ether and silyl ether, with usually the former predominating. In the presence of palladium chloride and thiophenol, ketones reacted with triethylsilane to give the corresponding silyl vinyl ether *(437)* selectively. [499]

$$
\begin{array}{c}
\diagdown \\
\diagup
\end{array}
\text{CH—}\overset{|}{\text{C}}\text{=O} + (\text{C}_2\text{H}_5)_3\text{SiH} \xrightarrow{\text{PdCl}_2/\text{PhSH}}
\begin{array}{c}
\diagdown \\
\diagup
\end{array}
\text{C=}\overset{|}{\text{C}}\text{—OSi(C}_2\text{H}_5)_3 + \text{H}_2
$$

437

Because carbon dioxide is an abundant and readily available material, its efficient use as an industrial source of carbon is an important target of research. Reactions of carbon dioxide with organic compounds of lithium, magnesium, and other nontransition metals have been studied in some detail. It can be said that carbon dioxide reacts with some nucleophiles. The reaction of carbon dioxide with nontransition metals can be explained as insertion of carbon dioxide into metal-carbon bonds. On the other hand, studies on the reactions of carbon dioxide with transition metal compounds have been initiated only recently. [500] Because of its inert nature, reaction of carbon dioxide with transition metal complexes is sluggish. Formation of carbon dioxide complexes directly from molecular carbon dioxide is known with rhodium, ruthenium, and platinum. [501, 502]

Reaction of the coordinated carbon dioxide has been carried out. A rhodium-carbon dioxide complex is formed by bubbling carbon dioxide into a solution of $RhCl(PPh_3)_3$. [501] This complex consists of two atoms of rhodium and one mole of carbon dioxide. Methyl acetate was formed by treating the complex with methyl iodide. This reaction shows that the carbon dioxide complex has the following partial structure (438); methyl iodide displaces both rhodiums from the complex to give methyl acetate.

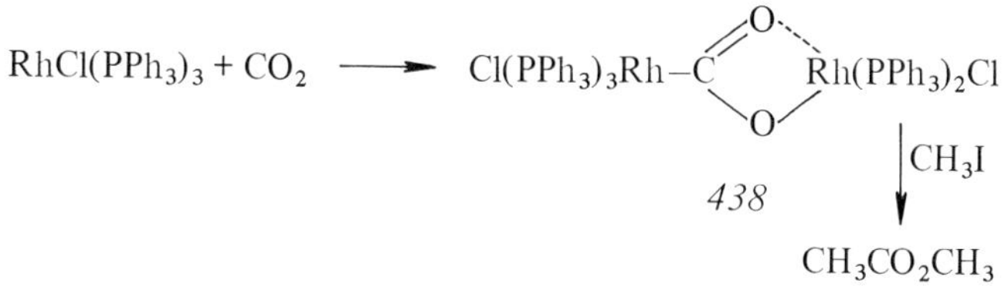

$$RhCl(PPh_3)_3 + CO_2 \longrightarrow Cl(PPh_3)_3Rh-C\diagdown^{O\cdots}_{O}\diagup Rh(PPh_3)_2Cl$$

438

$$\downarrow CH_3I$$

$$CH_3CO_2CH_3$$

A platinum complex of carbon dioxide behaves similarly giving methyl acetate by the reaction of methyl iodide.

A few examples are known for insertion of carbon dioxide into a metal-hydrogen or -carbon bond, the most desirable reaction for organic synthesis. There are two possibilities of the carbon dioxide insertion. One gives a metal-oxygen bond and the other a metal-carbon bond.

$$\overset{+}{M}-\overset{-}{R} + CO_2 \longrightarrow M-O-\overset{\overset{\displaystyle O}{\|}}{C}-R$$

$$\overset{-}{M}-\overset{+}{R} + CO_2 \longrightarrow M-\overset{\overset{\displaystyle O}{\|}}{C}-O-R$$

Formic acid was formed by the insertion of carbon dioxide into a cobalt-hydrogen bond. [503, 504]

$$H(N_2)Co(PPh_3)_3 + CO_2 \longrightarrow H-\overset{\overset{\displaystyle O}{\|}}{C}-OCo(PPh_3)_3 + N_2$$

$$\downarrow CH_3I$$

$$HCO_2CH_3$$

The formation of formamide by the reaction of amine, hydrogen, and carbon dioxide using rhodium, iridium, and cobalt catalysts is assumed to proceed through the insertion of carbon dioxide into a metal-hydrogen bond. [505]

$$L_nMH + CO_2 \longrightarrow L_nM-O-\underset{\underset{\displaystyle O}{\|}}{C}-H \xrightarrow{HN(CH_3)_2} L_nMOH + H-\underset{\underset{\displaystyle O}{\|}}{C}-N(CH_3)_2$$

$$L_nMOH + H_2 \longrightarrow L_nMH + H_2O$$

Reaction of a ruthenium complex with hydrogen and carbon dioxide in methanol produced methyl formate.

$$L_4RuCl_2 + H_2 \longrightarrow L_4RuCl(H) \xrightarrow{CO_2} L_4RuCl(\overset{\overset{\displaystyle O}{\|}}{OC}-H)$$

$$\xrightarrow{CH_3OH} L_4RuCl(OH) + HCO_2CH_3$$

The reaction of dimethyltitanocene *(439)* with carbon dioxide resulted in the formation of titanocene diacetate, which has a metal-oxygen bond. [500]

$$(\pi\text{-}C_5H_5)_2Ti(CH_3)_2 + CO_2 \longrightarrow (\pi\text{-}C_5H_5)_2Ti(\underset{\underset{\displaystyle O}{\|}}{OC}-CH_3)_2$$
$$\textit{439}$$

On the other hand, both modes of the insertion take place in the reaction of ethylcobalt complex with carbon dioxide. Two complexes were obtained, which were converted into methyl propionate and ethyl acetate by treatment with methyl iodide. [506]

$$(PPh_3)_2(CO)Co-C_2H_5 + CO_2 \xrightarrow{CH_3I} \begin{array}{l} C_2H_5CO_2CH_3 \\ CH_3CO_2C_2H_5 \end{array}$$

The formation of acetic acid in a low yield by the insertion of carbon dioxide into methylcobaloxime was reported. [507]

Another example of the carbon dioxide insertion into a metal-carbon bond was observed with diphenyltitanocene *(440)*. In this reaction, carbon dioxide inserted into the *ortho*-position of the benzene ring. The treatment of the resulting complex *(441)* with hydrogen chloride gave benzoic acid, and methylation afforded methyl o-toluate. [500]

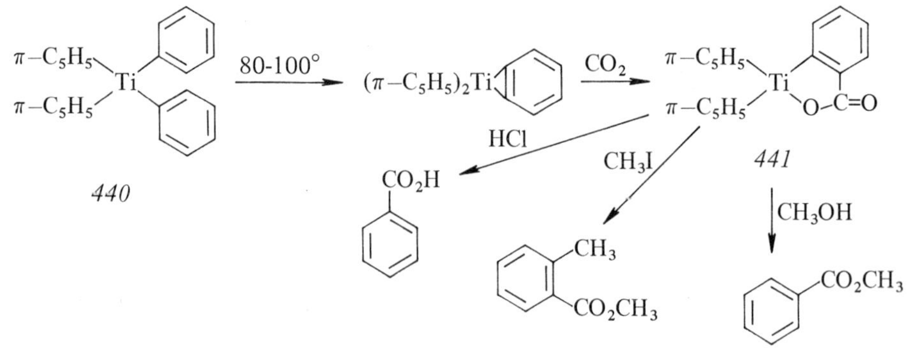

Several zerovalent nickel complexes such as $Ni(PPh_3)_2$ catalyze the formation of alkylene carbonates from certain epoxides and carbon dioxide. [508] When ethylene oxide and carbon dioxide were heated at 100° in benzene containing the nickel complex, ethylene carbonate *(442)* was formed in 95% selectivity. The reaction can be understood as a sequence of oxidative additions of the epoxide to the unsaturated nickel complex, followed by insertion of carbon dioxide and reductive elimination. The yield crucially depends on the structure of the epoxide.

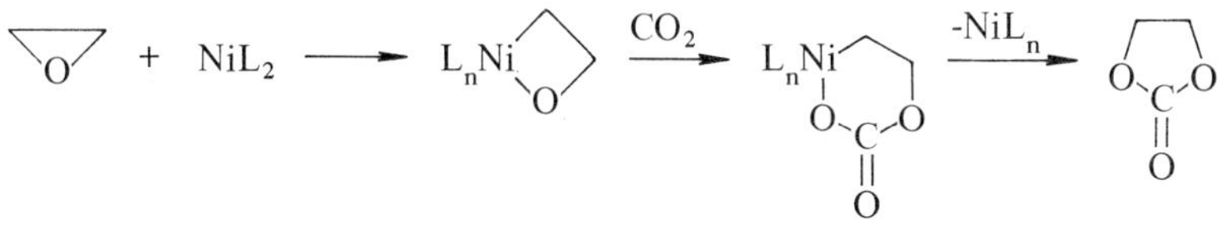

442

L = tricyclohexylphosphine, triphenylphosphine, COD

Reaction of carbon dioxide with an oxygen complex of platinum gave the following cyclic complex *(443)*. [509]

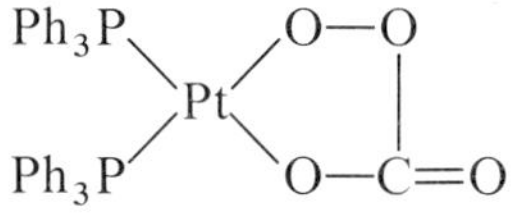

443

Carbon dioxide fixation by insertion into iridium and rhodium hydroxo complexes to form bicarbonato complexes was reported. [510]

$(Ph_3P)_2(CO)M(OH) + CO_2 \longrightarrow (Ph_3P)_2(CO)M(OCO_2H)$

M = Ir, Rh

It should be pointed out that the enzyme carbonic anhydrase catalyzes carbon dioxide hydration involving the following step. [511]

$EnZnOH + CO_2 \longrightarrow EnZnOCO_2H$

Insertions of carbon disulfide into metal-hydrogen and metal-carbon bonds take place as shown by the following examples. [512]

$CH_3RhI_2(PPh_3)_2 + CS_2 \longrightarrow RhI_2(CS_2CH_3)(PPh_3)_2$

$HIr(CO)(PPh_3)_3 + CS_2 \longrightarrow Ir(CO)(CS_2H)(PPh_3)_2$

A dithiocarboxylato structure was proposed for these complexes formed by the insertion of carbon disulfide. [513]

$R-M(CO)_5 + CS_2 \longrightarrow R-\underset{\underset{S}{\|}}{C}-S-M(CO)_4 + CO$

M = Mn, Re

Insertion of a carbon-nitrogen double bond of an isocyanate into a metal hydride is known to form carbamoyl complex (444). [514] The reaction proceeds rapidly in the presence of a catalytic amount of tertiary amine. The reversibility of the reaction was confirmed.

$$(\pi\text{-}C_5H_5)W(CO)_3H + CH_3N{=}C{=}O \underset{}{\overset{Et_3N}{\rightleftharpoons}} (\pi\text{-}C_5H_5)W(CO)_3(CONHCH_3)$$

$$444$$

Carbene is an active species, the stabilization of which by coordination was described (p. 85). The coordinated carbene also undergoes insertion reactions. For example, bis(trifluoromethyl)carbene formed from the diazo compound was inserted at room temperature into a manganese-hydrogen bond to give an alkyl complex. [515]

$$(CF_3)_2CN_2 \longrightarrow (CF_3)_2C: \xrightarrow{HMn(CO)_5} (CF_3)_2CH{-}Mn(CO)_5$$

In the reaction shown below, the carbene generated from diazoacetic acid was inserted into a nickel-carbon bond of π-allylnickel complex to give pentadienoate (445). [516]

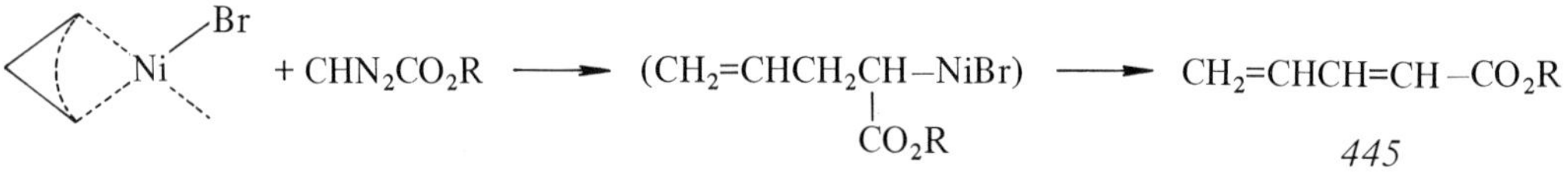

$$(CH_2{=}CHCH_2CH{-}NiBr) \longrightarrow CH_2{=}CHCH{=}CH{-}CO_2R$$
$$\underset{CO_2R}{|}$$

$$445$$

Dichlorocarbene formed by thermolysis of sodium trichloroacetate inserted into a tungsten-hydrogen bond of dicyclopentadienyltungsten dihydride to produce dicyclopentadienyltungsten dichloromethide hydride (446). [517]

$$(\pi\text{-}C_5H_5)_2WH_2 + Cl_3CCO_2Na \longrightarrow (\pi\text{-}C_5H_5)_2WH(CHCl_2) + NaCl + CO_2$$

$$446$$

Insertion of a diazo ketone into a palladium-chlorine bond gave the oxo-π-allyl-palladium complex. [518]

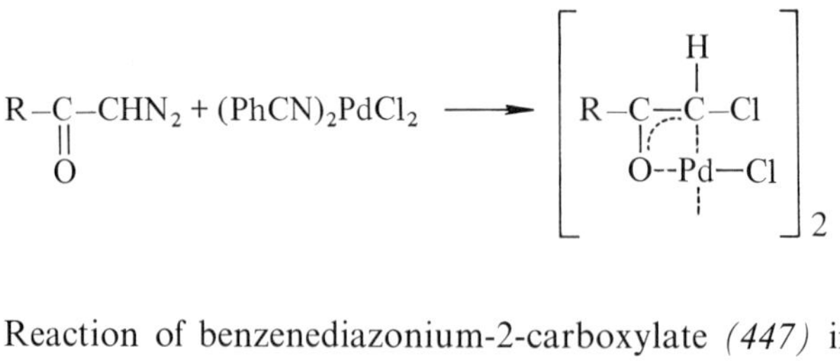

Reaction of benzenediazonium-2-carboxylate (447) in methylene chloride with (phenylethynyl)(trichlorovinyl)bis(triethylphosphine)nickel (448) gave 2-(phenyl-ethynyl)phenyl(trichlorovinyl)bis(triethylphosphine)nickel (449) and 2-phenylethynyl)(trichlorovinyl)benzene *(450)*. [519] The former nickel complex was formed by the insertion of benzyne into the nickel-carbon bond, and the latter by the coupling of the two carbons directly bonded to the nickel by reductive elimination reaction.

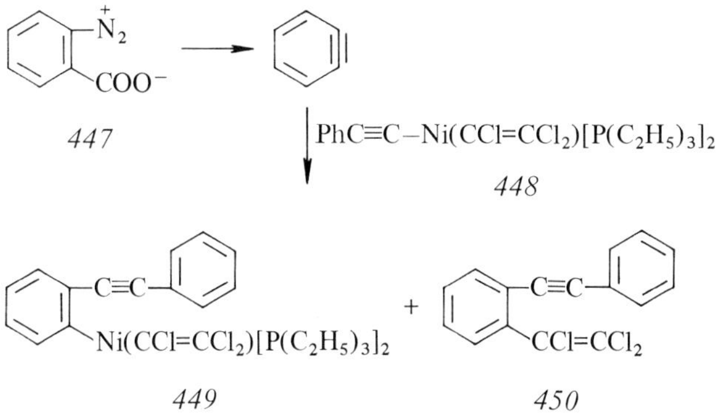

VI. Liberation of Organic Compounds from the σ-Bonded Complexes

The last step in the synthetic reactions *via* transition metal complexes is the liberation of the fabricated organic compounds from the complexes which have metal-carbon σ-bonds. In Grignard reactions, organic products are isolated after hydrolysis of the magnesium-containing compounds. Although some products, as in the Grignard reaction, are isolated after hydrolysis of the complexes produced, there are several ways of liberation characteristic to transition metal complexes. Coordinative lability of the ligands is responsible for this step. The general patterns of cleavage reactions of σ-alkyl bonds to liberate organic products can be summarized in the following way.

$$
\text{R}'\text{—M—CH}_2\text{CH}_2\text{R—} \underset{\text{L}_n}{|}
\begin{cases}
\text{—ML}_n \text{ (low valent)} \begin{cases} \text{R}'\text{CH}_2\text{CH}_2\text{R (reductive elimination)} \\ \text{CH}_3\text{CH}_2\text{R (hydrogen abstraction, hydrogenolysis)} \\ \text{RCH}_2\text{CH}_2\text{CH}_2\text{CH}_2\text{R (coupling)} \\ \text{CH}_2\text{=CHR} + \text{CH}_3\text{CH}_2\text{R (disproportionation)} \end{cases} \\
\\
\text{CH}_2\text{=CHR} + \text{R}'\text{M—H } (\beta\text{-elimination})
\end{cases}
$$

The unstable alkyl transition metal complexes formed initially mostly undergo homolytic fragmentation and elimination reactions leading to alkanes, alkenes, dimeric alkanes plus metal hydride, colloidal or active metal, and low valent metal complexes. [520—522]

One general pattern is reductive elimination with coupling of the coordinated ligands. As stated before, reductive elimination is the reverse process of the oxidative addition. By the addition of neutral ligands such as carbon monoxide, olefin, diolefin, and phosphine, to the metal complexes $\text{L}_n\text{MRR}'$, the coupling of the ligands on the same metal is induced to form R—R' with concomitant reduction of formal oxidation state of the metal by two units. The reaction probably proceeds through an intermediate complex having an expanded co-ordination number. The labilization of the ligands to cause the coupling may occur due to the overfilling of the coordination sphere. There are many examples of the coupling reaction, and the reaction takes place with saturated complexes.

The steric congestion about the metal can be relieved by an irreversible intra-molecular ligand coupling. In other words, an entering ligand or accelerating ligand labilizes the complex to promote the coupling between groups already bonded to the metal. In order for the coupling process to occur, the ligands are assumed to occupy *cis* coordination sites to each other. When the coupling reaction occurs intramolecularly, the outcome is the cyclization (Chapter VII).

Reductive elimination provides a pathway to the cleavage of metal-carbon and of other metal ligand bonds. In some cases, the reductive elimination takes place without entering ligands. For example, the coupling takes place by the thermal decomposition of the following complexes. [523]

$$[(CH_3)_2PhP]_2PtCH_3Cl_3 \longrightarrow [(CH_3)_2PhP]_2PtCl_2 + CH_3Cl$$

$$[(CH_3)_2PhAs]_2PtBr(CH_3)_2(COCH_3)$$

$$\longrightarrow [(CH_3)_2PhAs]_2PtBrCH_3 + CH_3COCH_3$$

Oxidative addition of protic acids followed by reductive elimination provides a pathway for acid cleavage of metal-carbon bonds. [524]

$$[(C_2H_5)_3P]_2PtClCH_3 + HCl \longrightarrow [(C_2H_5)_3P]_2PtCH_3(H)Cl_2$$

$$\longrightarrow [(C_2H_5)_3P]_2PtCl_2 + CH_4$$

π-Allylpalladium acetylacetonate *(452)* undergoes, under carbon monoxide pressure, the coupling reaction of the ligands to form allylacetylacetone *(451)*; the labilizing action of the entering carbon monoxide is notable. [525]

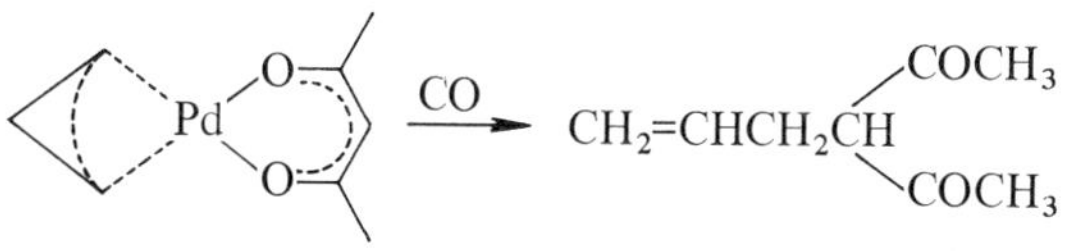

452   451

Similarly the treatment of bis(trityl)nickel *(453)* with carbon monoxide at room temperature generated hexaphenylethane *(78)* and Ni(CO)$_4$ quantitatively by reductive elimination. [116]

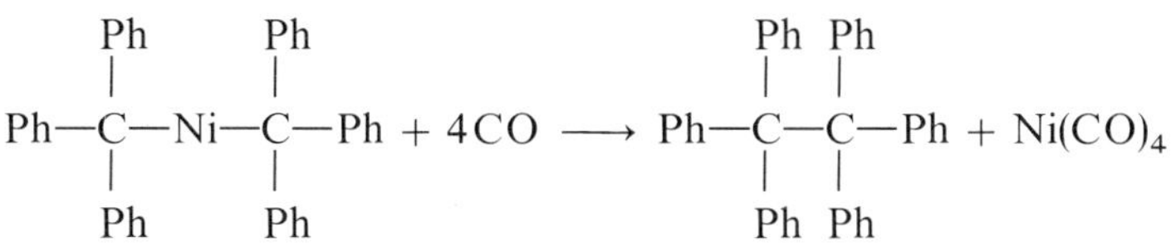

453   78

The accelerating effect of carbon monoxide is also apparent in the coupling reaction between cyclopentadienyl and phenyl groups to give cyclopentadienyl-benzene *(455)* from the nickel complex *(454)*. [526, 527]

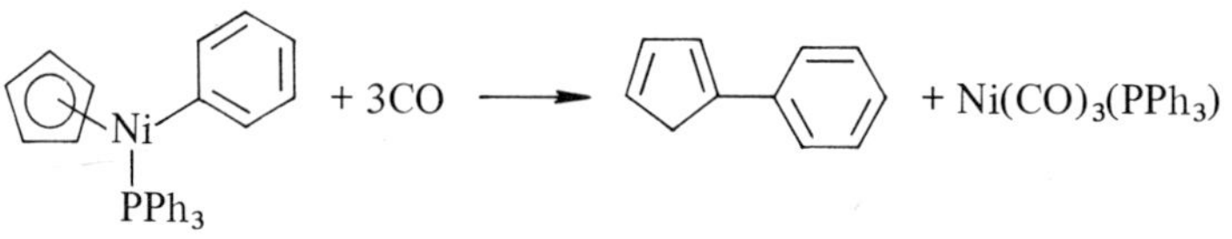

454   455

Also the coupling of a coordinated cyclopentadienyl group in the nickel azobenzene complex *(456)* occurred by the reaction of bromobenzene to form
4-phenylcyclopenta(c)cinnoline *(457)*. [528]

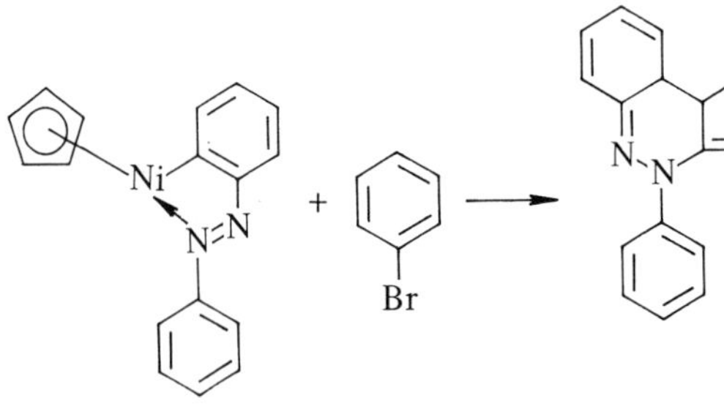

456   457

Another well-established example is the formation of CDT by the accelerating
action of phosphine on the intermediary open-chain complex *(310)* formed from
zerovalent nickel and three moles of butadiene. [529]

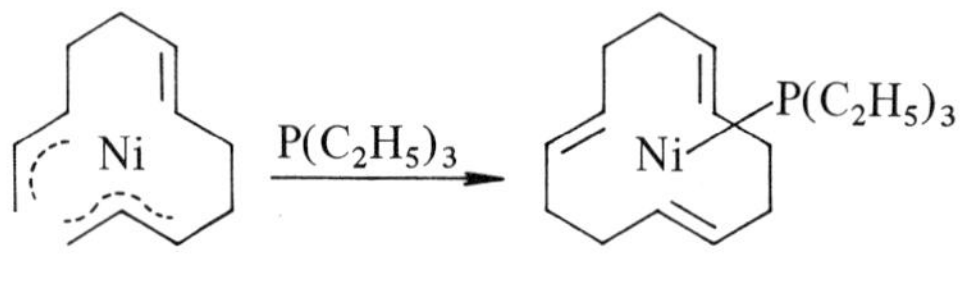

310

When carbon monoxide was added to the above complex *(310)*, the coupling
took place with incorporation of carbon monoxide to form the cyclic ketones,
11-vinyl-3,7-undecadienone *(458)* and 3,7,11-tridecatrienone *(459)*. [530] Isocyanide, which is isoelectronic with carbon monoxide, also effects similar intramolecular cyclization of the complex *(310)*. Hydrolysis of the cyclized imine
derivatives gave the same ketones *(458* and *459)*.

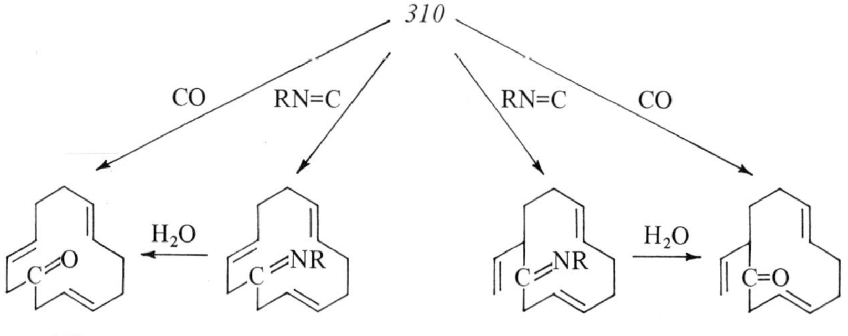

459   458

In the presence of proper ligands, the reversible processes, oxidative addition and reductive elimination, become cyclic, making the whole reaction catalytic. This is the most characteristic property of transition metal complexes. The CDT synthesis can be made catalytic by using a zerovalent nickel complex as a catalyst. The open-chain ligand in the intermediate *(310)* is displaced through cyclization by the entering butadiene with simultaneous formation of the complex *(310)*.

Mechanistic studies on the activation of metal-alkyl bonds by entering olefins were carried out. [531, 532] Diethylbipyridylnickel *(206)* is quite stable up to 100°. However, the formation of butane by the ligand coupling was observed even at a low temperature when the complex *(206)* was treated with certain olefinic compounds such as acrylonitrile. Blue shift in a spectrum was observed when acrylonitrile was added to the complex *(206)*, which is explained by back donation from the nickel to the olefin. It was concluded that the stronger the coordination of the entering olefins, the more the alkyl-metal bond is activated. The diethylbipyridylnickel complex containing acrylonitrile was isolated.

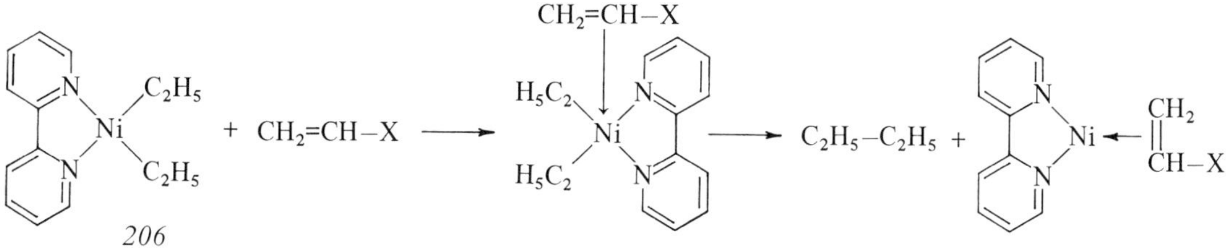

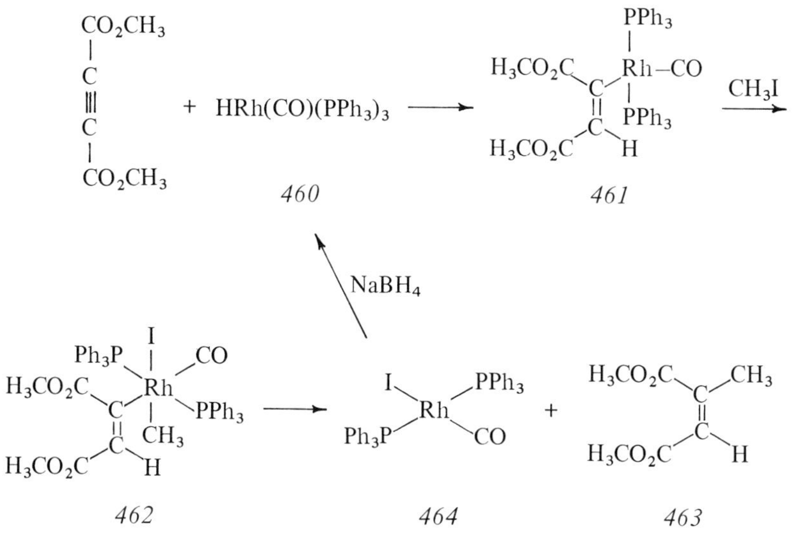

Stereospecific synthesis of trisubstituted olefins is possible from acetylenes by using a rhodium complex. [533] Insertion of acetylenedicarboxylate into hydrido-tris(triphenylphosphine)carbonylrhodium *(460)* gave *trans*-alkenylrhodium complex *(461)*, which has a *cis* double bond. Then this unsaturated complex was allowed to react with methyl iodide at room temperature to give the organo-rhodium complex *(462)*. The reductive elimination of this high valent rhodium complex with the coupling of the organic ligands takes place at 115° to give dimethyl citraconate *(463)*. Treatment of the iodo complex *(464)* thus formed with sodium borohydride produced the original hydrido complex of rhodium *(460)*.

In the reductive elimination reactions, the metal resumes a lower valent state. Therefore a ligand which stabilizes the lower valent state protects the metal from deactivation, as it is by precipitation, and effects the oxidative addition again. In order to stabilize the low valent state, it is necessary to remove the accumulated charge on the metal. For this purpose, ligands which accept charges by back donation, such as triphenylphosphine, are useful. Although a complete theory of the role of the ligands has not been formulated, an approach to explain the role of the ligands in the coupling reaction of π-allyl complexes has been proposed by Traunmuller *et al.* This is based on molecular orbital calculations, taking the ionization potential of the ligands into consideration. [534]

In the foregoing, the formation of organic molecules on transition metal complexes is explained by a stepwise mechanism involving oxidative addition, insertion, and reductive elimination. Another typical example which can be clearly explained by these patterns is provided by the carbonylation and decarbonylation reactions using rhodium and palladium catalysts. [73—75, 535]

One of the characteristic properties of the Wilkinson complex is that it very readily accepts one mole of carbon monoxide to form the very stable carbonyl complex *(5)*.

$$RhCl(PPh_3)_3 + CO \longrightarrow Rh(CO)Cl(PPh_3)_2 + PPh_3$$

1   *5*

Both catalytic and stoichiometric decarbonylations can be effected by utilizing this property. The tendency to form the complex *(5)* is so great that the Wilkinson complex can abstract a carbonyl group from acyl halides under mild conditions to produce olefin and hydrogen halide. If, for structural reasons, olefin cannot be formed, the product is an alkyl halide.

$$R—COX + RhCl(PPh_3)_3 \longrightarrow R—X + Rh(CO)Cl(PPh_3)_2 + PPh_3$$

1   *5*

Similarly, aldehydes are decarbonylated to give the corresponding hydrocarbons.

$$R—CHO + RhCl(PPh_3)_3 \longrightarrow R—H + Rh(CO)Cl(PPh_3)_2 + PPh_3$$

Stoichiometric decarbonylation with the Wilkinson complex *(1)* proceeds in high yields and under extremely mild conditions, sometimes even at room temperature in organic solvents, and is useful in organic synthesis. The introduction of an angular methyl group into the bicyclic enone *(465)* was achieved by Claisen rearrangement and decarbonylation. [536, 537]

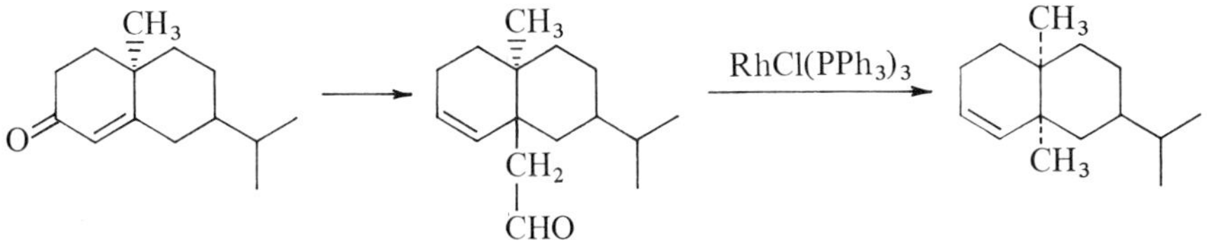

465

Deuterated alkanes can be prepared by the decarbonylation of deuterated aldehydes. [538]

$$\underset{R-C=O}{\overset{\overset{\textstyle D}{|}}{}} + RhCl(PPh_3)_3 \longrightarrow R-D + Rh(CO)Cl(PPh_3)_2 + PPh_3$$

The decarbonylation reaction consumes one mole of the rather expensive rhodium complex *(1)* to form the complex *(5)*. The usefulness of the reaction, however, was greatly enhanced by the discovery of a novel method of reconversion of the complex *(5)* to the Wilkinson complex *(1)*; [539] the oxidative addition of benzyl halide to the complex *(5)* gives the π-benzyl complex *(466)*, which upon treatment with an excess triphenylphosphine in ethanol affords the Wilkinson complex *(1)*.

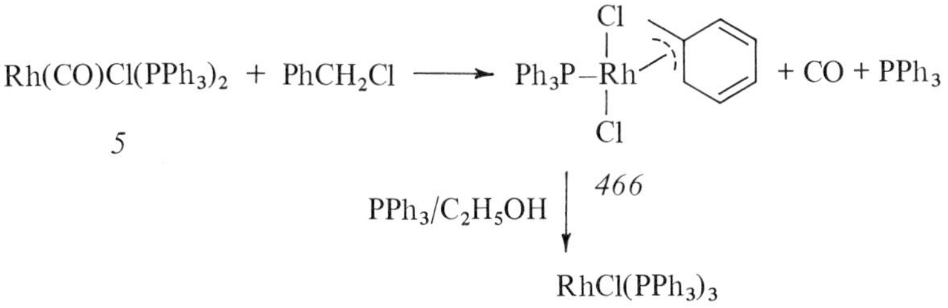

While the above-shown stoichiometric decarbonylation proceeds under mild conditions, catalytic decarbonylation of aldehydes and acyl halides with the Wilkinson complex or the complex *(5)* proceeds smoothly at 200°. [75, 539a] One application of the catalytic decarbonylation is the displacement of an aromatic carboxyl group with a halogen. When halides of aromatic carboxylic acids are heated to 200° with a catalytic amount of the complex *(5)*, carbon monoxide evolves and the halogenated product is obtained in a high yield.

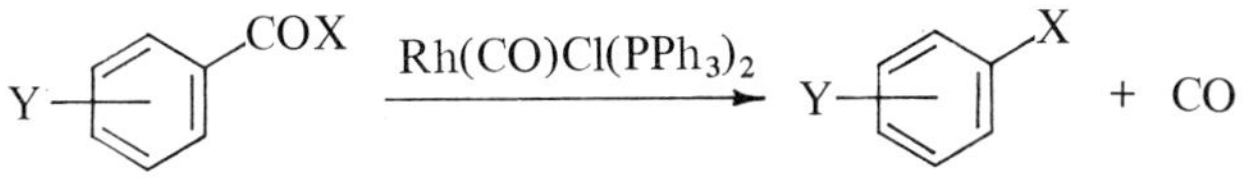

The reverse process of the decarbonylation, namely the carbonylation, is also catalyzed by the complex (5).

$$CH_2\text{=}CHCH_2Cl + CO \xrightarrow{Rh(CO)Cl(PPh_3)_2} CH_2\text{=}CHCH_2COCl$$

These stoichiometric and catalytic decarbonylations as well as the carbonylation can be explained by the following mechanism.

$$Rh(CO)Cl(PPh_3)_2 + RCO\text{—}X \rightleftharpoons RCORh(CO)ClX(PPh_3)_2$$

5 $+CO \upharpoonleft\downharpoonright -CO$ 469

$$RhCl(PPh_3)_3 + RCO\text{—}X \longrightarrow RCORhClX(PPh_3)_2$$

1 $\upharpoonleft\downharpoonright$ 467

$$Rh(CO)Cl(PPh_3)_2 + R\text{—}X \rightleftharpoons RRh(CO)ClX(PPh_3)_2$$

5   (or olefin)   468

The first step of the stoichiometric decarbonylation of acyl halides is the oxidative addition to the Wilkinson complex (1), which is d^8 complex, to form the five-coordinate d^6 acyl complex (467), which is still coordinatively unsaturated. Subsequent acyl-alkyl rearrangement gives the six-coordinate d^6 alkyl complex (468) undergoing carbon-carbon bond cleavage. The final step is the reductive elimination by the coupling of the coordinated ligands or β-elimination to form the complex (5) and olefin or alkyl halide. As supporting evidence of this mechanism, the acyl complex (467) was isolated by the oxidative addition of acyl halide as described earlier (p. 36). Also the alkyl complex (468) was isolated. Heating of these acyl and alkyl complexes gave olefin or alkyl halide.

The catalytic decarbonylation is initiated by the oxidative addition of acyl halide to the complex (5) to form the six-coordinate acyl complex (469). Since the reaction temperature is high, elimination of the coordinated carbon monoxide from the saturated complex (469) takes place to form the unsaturated acyl complex (467), from which the alkyl complex (468) is formed by the acyl-alkyl rearrangement. Finally the reductive elimination forms olefin or alkyl halide. Here, the complex (5) is regenerated, making the whole process a catalytic cycle. The carbonylation reaction can be explained exactly as the reverse process of the decarbonylation; it starts from the oxidative addition of alkyl halides to the rhodium complex (5).

Another general pattern is the formation of hydride-olefin complexes by a β-elimination reaction, in which the hydrogen from a β-carbon is transferred to the metal atom. This is the reverse process of the olefin insertion, as is observed for many alkyl complexes. The driving force of the reaction may be initial coordinate unsaturation. In a catalytic cycle, coordinated olefin formed by the β-elimination is displaced from the complex by the entering olefins. The termination step in oligomerization and polymerization of olefins is an example of this reaction. Fresh olefin is added to the regenerated hydride complex to start the catalytic cycle again. The following scheme for an addition reaction of ethylene to butadiene catalyzed by a rhodium salt illustrates the sequence. [540] 4-Hexenylrhodium complex *(471)* is initially formed by the insertion of ethylene to the allylic complex *(470)*. By the coordination of fresh butadiene, the 4-hexenyl group is liberated as 1,4-hexadiene by the β-elimination to form the rhodium hydride *(472)*. The insertion of butadiene into the rhodium-hydrogen bond regenerates the π-allylic complex *(470)*.

$$CH_3CH=CHCH_2CH=CH_2$$

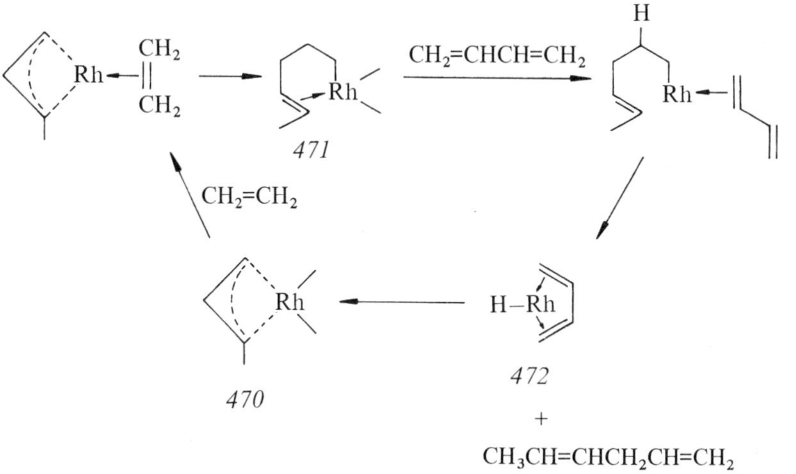

The reversible processes involving the olefin insertion into a metal hydride to form an alkyl complex and the β-hydrogen elimination to regenerate the olefin were demonstrated with the nickel complex, $Ni(C_2H_5)(acac)(PPh_3)_2$ using NMR analysis. Also a disproportionation reaction of the ethyl group to give ethylene and ethane was observed by treatment of the complex with pyridine. [541]

One way of activating metal-carbon bonds is oxidation of the central metal atom. In order to liberate the organic part from the complexes, oxidation of the central metal atom with Ce^{4+} ion is the general method. An example is the liberation of the free cyclobutadiene molecule from its iron carbonyl complex *(170)* (p. 56). Also one electron oxidation of the central metal atom greatly increases the susceptibility of the carbon σ-bonded to the metal to nucleophilic attack. Oxidation of several alkyl-metal carbonyl complexes by Ce^{4+} ion in methanol takes place extremely rapidly to give the methyl ester of the next higher homologous carboxylic acid in a high yield. [542] Oxidation of acyl complexes formed by carbon monoxide insertion makes the complex highly susceptible to nucleophilic attack at the acyl carbon, even with the mildly nucleophilic solvents.

$$RCH_2M(CO)_n \longrightarrow RCH_2COM(CO)_{n-1} \xrightarrow{Ce^{4+}, -e} RCH_2COM(CO)_{n-1}^+$$

$$\downarrow CH_3OH$$

$$RCH_2CO_2CH_3$$

Cupric chloride as an oxidizing agent can be used for a similar purpose. [543] The alkyliron complex *(473)* was converted rapidly into one carbon homologated ester in alcohol at 0° by the treatment with three molar equivalents of cupric chloride.

$$(\pi\text{-}C_5H_5)FeR(CO)_2 + CuCl_2 \xrightarrow[\text{CO}]{\text{R'OH}} RCO_2R' + CuCl + Fe \text{ complex}$$

$$473$$

The synthesis of prostaglandin C series was performed as follows. [544] The key intermediate (474) was treated with $Fe_3(CO)_{12}$ to give an iron carbonyl complex (475) of the conjugated diene via double bond migration. Introduction of the side chain to form (476) was followed by the Collins oxidation (Cr^{6+}) to the ketone. By this oxidation, the complexed iron carbonyl was removed smoothly with regeneration of the conjugated diene system to give prostaglandin C (477). This is an extremely mild method for the oxidative removal of the iron from the complex (476).

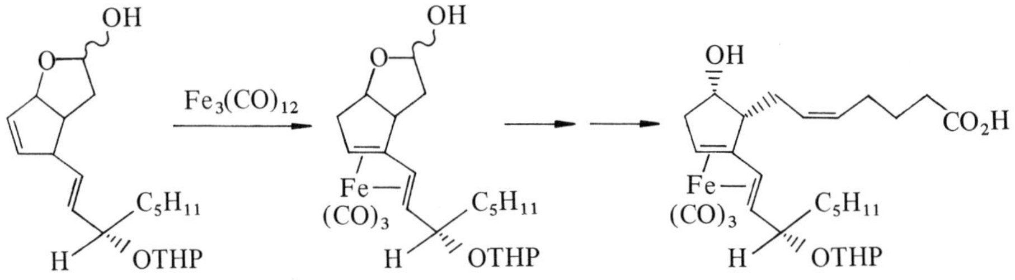

The acetylenic bond was protected by complex formation with $Co_2(CO)_8$ preferentially in the presence of an olefinic bond in the same molecule. The olefin was converted into functional groups such as alcohol by hydroboration. Then the protecting group was removed by oxidation with ferric nitrate. [545]

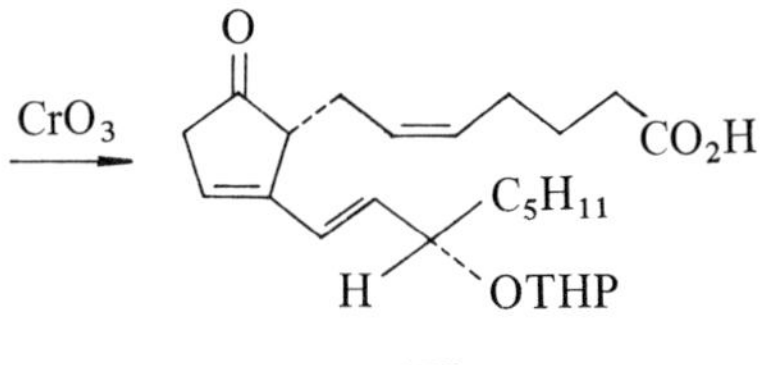

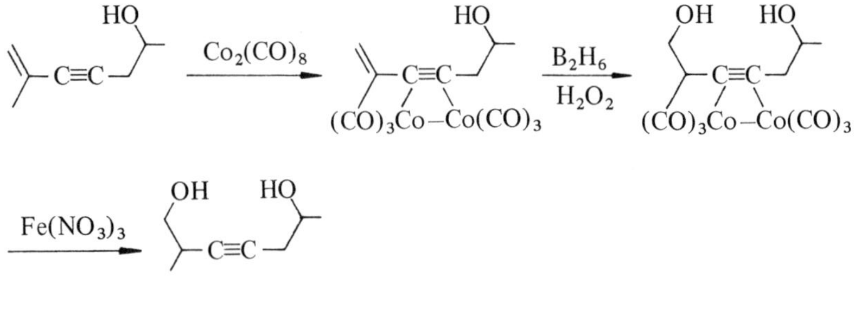

VII. Cyclization Reactions and Related Reactions

One of the most characteristic synthetic reactions using transition metal complexes is the formation of various cyclic compounds from olefinic and some unsaturated compounds. As described before, some cyclization reactions proceed *via* the intramolecular coupling of the coordinated ligands. In the cyclization reactions, formation of a π-complex, as well as σ-complex sometimes plays an important role. In these reactions, the metal exhibits not only an electronic effect, but also a template effect. In the template ligand reactions, the metals facilitate the mutual approach of the reactants and organize the reacting molecules in a suitable orientation. When two ligands coordinate suitably in adjacent positions in a coordination sphere, the possibility of their reaction is greatly enhanced.

The most well-established cyclization reaction is that of butadiene catalyzed by transition metal complexes, especially by nickel complexes. 1,5,9-CDT, 1,5-COD, vinylcyclohexene, 1,2-divinylcyclobutane, and 1-vinyl-2-methylene-cyclopentane are formed by using nickel complexes under different effect of ligands. [150—153] Overfilling of the coordinating sphere is responsible for the cyclization by intramolecular coupling. The entering or accelerating ligands labilize the intermediary open chain complex and the coupling takes place between the two groups bonded to the nickel. By an experiment using an iron complex with deuterated butadiene, it was proved that no hydrogen shift takes place in the cyclization reaction. [94]

In addition to the cyclic oligomers, the transition metal-catalyzed reaction of butadiene affords linear oligomers *via* hydrogen transfer. Whether the cyclization or linear oligomerization occurs depends on the coordination lability. Co-ordinative unsaturation is responsible for the hydrogen transfer by β-elimination reaction to give the linear oligomers and metal hydride.

The above-shown cyclic products of butadiene are formed under different effect of ligands. The reduction of bivalent nickel compounds in the presence of butadiene as the only available ligand produces a catalyst called "naked nickel", which is able to cyclize butadiene to 1,5,9-CDT(all-*trans, trans,trans,cis,* and *trans,cis,cis* isomers). The coupling of three butadiene molecules produces a twelve-membered chain bonded to the nickel atom by two terminal π-allyl groups *(310)*. Interaction of the C_{12} chain with a fresh butadiene molecule leads to the coupling of the terminal allyl groups to form 1,5,9-CDT. [546]

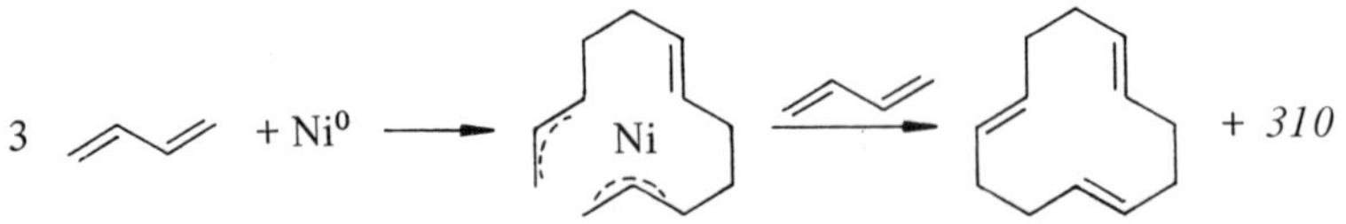

310

When one of the coordination sites of the nickel atom is occupied by a ligand such as phosphine or phosphite, then cyclodimerization of butadiene takes place to give 1,5-COD and vinylcyclohexene *(478)* as the predominant products. [547] 1,2-Divinylcyclobutane *(479)* is formed depending on the reaction conditions. In the presence of tris(2-biphenylyl) phosphite as a ligand of the nickel, divinylcyclobutane was obtained in 40% yield at 85% conversion of butadiene. COD can be formed by either direct cyclization of butadiene or Cope rearrangement of divinylcyclobutane. The ratio of COD to vinylcyclohexene changes depending on the ligand. Phosphites, unlike phosphines, favor the formation of COD over vinylcyclohexene.

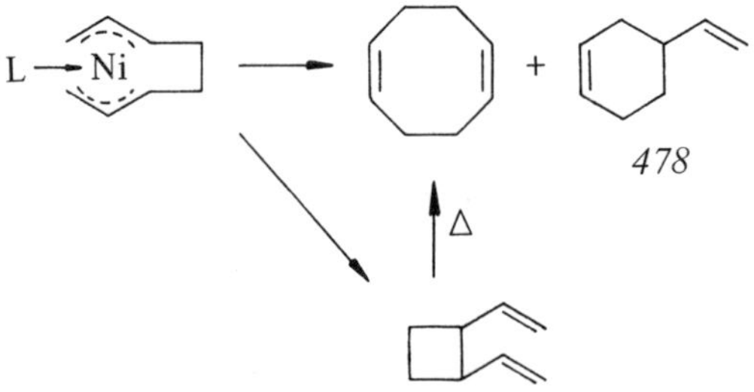

478

479

Selective formation of 1-vinyl-2-methylenecyclopentane *(480)* is possible by the nickel catalyst prepared by the reduction of dichlorobis(tributylphosphine)nickel with sodium borohydride in the presence of alcohol. [548, 549]

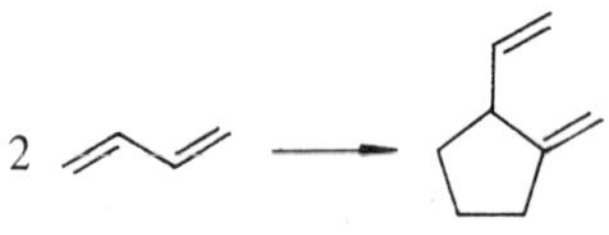

480

The formation of this cyclic dimer should be commented on period. Bis-π-allylic complex is converted into bis-σ-allylic complex, which accepts hydrogen to give σ-allyl-π-olefin nickel complex (481). Then the insertion of the terminal double bond into the carbon-nickel σ-bond causes the cyclization. The final step is the β-elimination to give the cyclic dimer (480) and nickel hydride. This mechanism is supported by the fact that the deuteration of the dimer (480) took place at the 3-carbon atom when the dimerization was carried out in deuterated alcohol. [550]

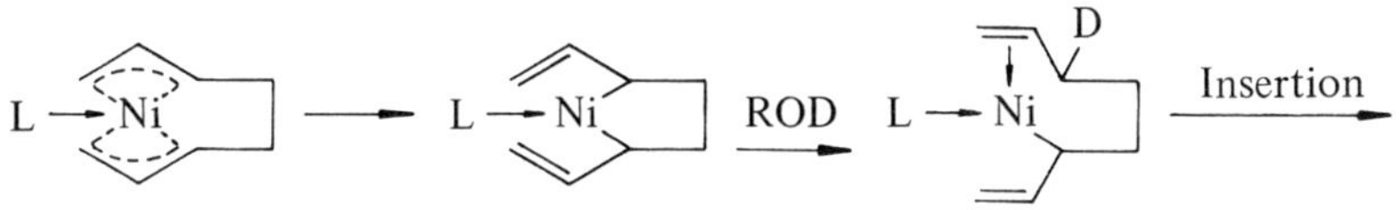

481

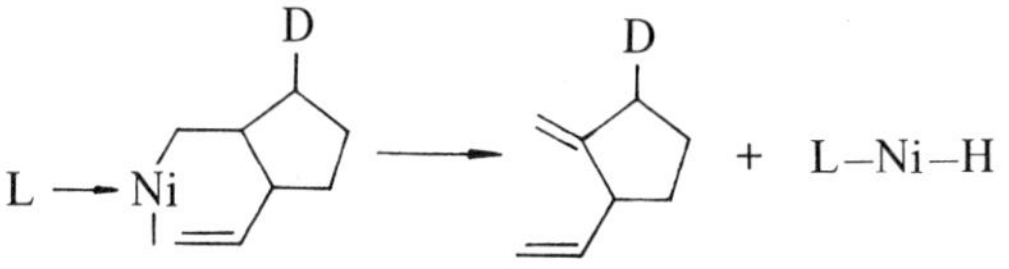

480

Bis(π-allyl)nickel is an active catalyst for butadiene trimerization to give CDT. On the other hand, π-allylnickel chloride catalyzes the polymerization of butadiene. A catalyst formed by the combination of bis(π-allyl)nickel and π-allyl-nickel chloride was found to catalyze the cyclooligomerization of butadiene to give not only COD and CDT, but also cyclic tetramers, hexamers, and large cyclic compounds. [551] One example of the product composition is as follows. $C_8 = 4.8$, $C_{12} = 23.2$, $C_{16} = 4.5$, $C_{20} = 12.2$, $C_{24} = 11.2$, $C_{28} = 5.4$, $C_{32} = 2.7$, and above $C_{32} = 36.1\%$.

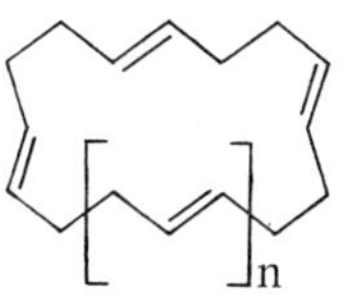

n=1, 2, 3, 4, 5 and above

The cyclic oligomers are useful starting materials for organic synthesis. CDT, produced on an industrial scale, is converted to cyclododecanone, from which nylon 12 is produced.

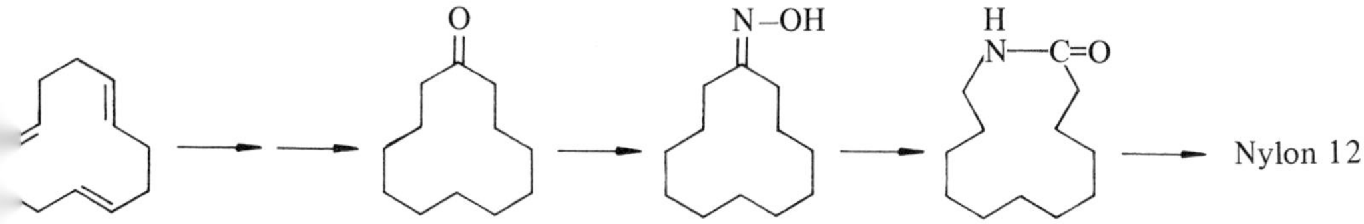

Nylon 12

Two step synthesis of grandisol *(483)*, a key constituent of the male boll weevil pheromone, was achieved by the nickel-catalyzed cyclization of isoprene. The reaction of isoprene catalyzed by a system of Ni(COD)$_2$ and tris(2-biphenylyl) phosphite gave the cyclobutane derivative *(482)* in 12% efficiency based on reacted isoprene. In addition, some 1,5-dimethylcyclooctadiene and vinylcyclo-hexene derivatives were formed. [552] Selective hydroboration reaction of the vinyl group of *(482)* with disiamylborane below room temperature produced grandisol *(483)*.

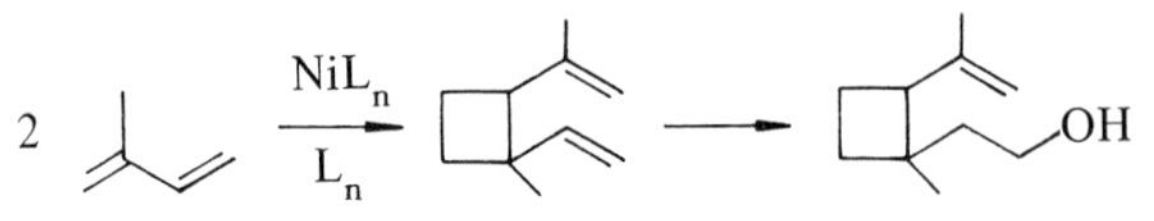

Nickel catalysts are also active for cooligomerization of butadiene with olefins and acetylenes. In the presence of the naked nickel, ethylene reacts with two moles of butadiene to give cyclodecadiene *(484)*. [553] In this reaction, ethylene insertion into the π-allylnickel part takes place to give a nickel complex having alkyl and π-allyl terminals. Coupling of these two terminals gives *(484)*. For this reaction, an acceptor ligand such as triphenyl phosphite is a good ligand.

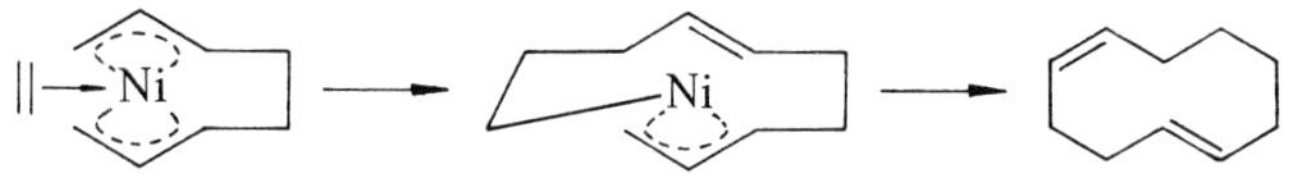

Reaction of butadiene with ethylene in the presence of the Ziegler type titanium catalyst produces vinylcyclobutane *(485)*. [554] 1,4-Hexadiene formation competes with the cyclization. This reaction involves stepwise insertion of butadiene and ethylene into the titanium-hydride bond giving 4-hexenyltitanium complex. The intramolecular insertion of the internal double bond into the titanium-carbon σ-bond gives the cyclobutylethyltitanium complex *(486)*. The final step is the elimination to give vinylcyclobutane *(485)* and titanium hydride.

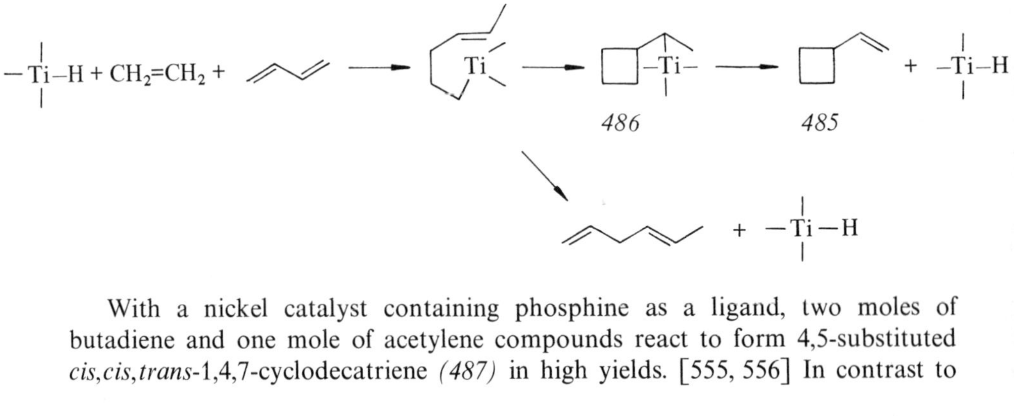

With a nickel catalyst containing phosphine as a ligand, two moles of butadiene and one mole of acetylene compounds react to form 4,5-substituted *cis,cis,trans*-1,4,7-cyclodecatriene *(487)* in high yields. [555, 556] In contrast to the cocyclization of butadiene and ethylene shown above, which is favored by good acceptor ligand such as triphenyl phosphite, the reaction of butadiene and acetylenes proceeds smoothly in the presence of electron-donating ligands such as triphenylphosphine. This difference can be understood by the fact that ethylene is a better donor relative to acetylenes.

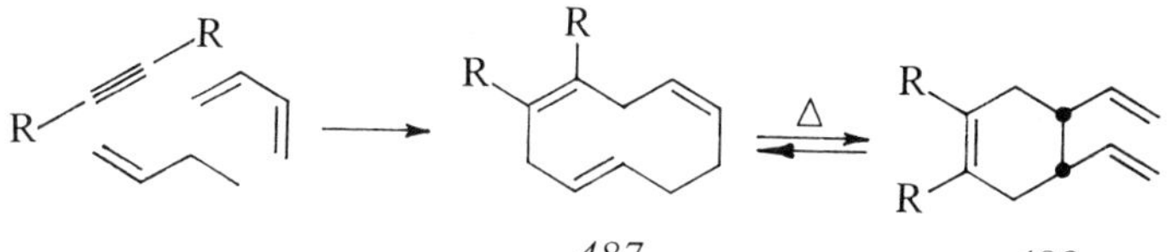

Cyclododecatrienes (487) formed by this reaction are useful compounds for further conversion to other compounds. The trienes (487) undergo thermal Cope rearrangement to form 1,2-disubstituted cis,-4,5-divinylcyclohexenes (488). Since in these trienes the double bond derived from substituted triple bond is more hindered than two double bonds derived from butadiene, the two double bonds derived from butadiene are hydrogenated selectively without attacking the former double bond. This selectivity offers a good method for preparing large ring ketones. When a cyclic acetylene is used for the cocyclization, the corresponding bicyclic compound is formed, selective hydrogenation of which followed by ozonization of the remaining double bond affords the expanded cyclic compound.

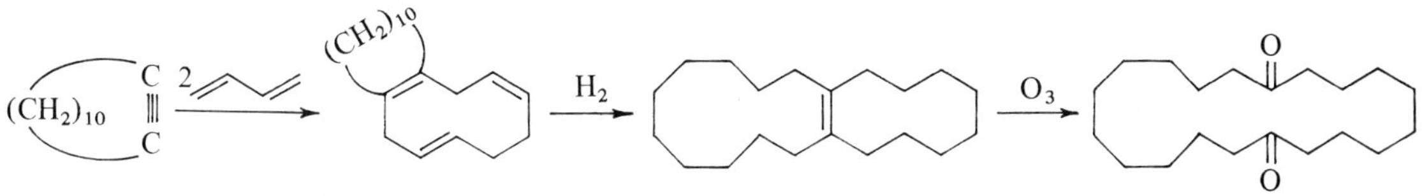

With cyclotetradecadiyne (489), the bicyclic compound (490) (55%) and two tricyclic compounds (491, 492) (40% yield) were obtained. These compounds were converted into 22- and 30-membered ring compounds by the above sequence.

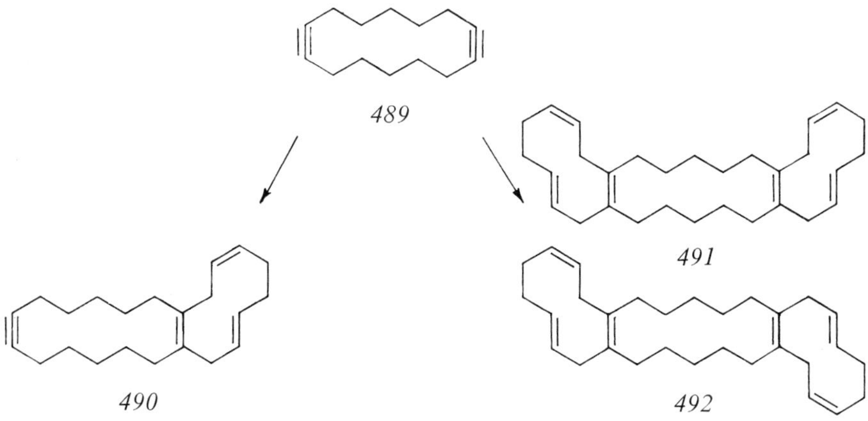

On the other hand, two moles of acetylenecarboxylates react with one mole of butadiene to give 5-vinyl-1,3-cyclohexadiene derivatives. [557] 5-Vinyl-1,3-cyclohexadiene *(493)* was obtained as the main product (60%) by the co-cyclization of two moles of acetylene and one mole of butadiene. This reaction was effected by a nickel complex coordinated by tri-n-alkylphosphines. [558]

$$2 \; CH\equiv CH \;+\; \text{(butadiene)} \;\longrightarrow\; \text{(5-vinyl-1,3-cyclohexadiene)}$$

493

A Diels-Alder type addition of one mole of an acetylene compound and one mole of butadiene is also possible. 1,4-Cyclohexadiene derivatives *(494)* were formed by using an iron complex of cyclooctatetraene (COT). [559]

$$\text{(butadiene)} \;+\; R-C\equiv C-R \;\longrightarrow\; \text{(ring)} \begin{matrix} -R \\ -R \end{matrix}$$

494

The Fe(COT)$_2$ complex is also an active catalyst for the cyclization reaction of two moles of acetylene compounds and one mole of norbornadiene to form a benzene derivative with liberation of cyclopentadiene. [560, 561]

$$\text{(norbornadiene)} \;+\; \begin{matrix} R \\ | \\ C \\ ||| \\ C \\ | \\ R \end{matrix} \;\longrightarrow\; \text{(benzene deriv.)} \;+\; \text{(cyclopentadiene)}$$

Cyclization reaction of allene to give trimer, tetramer, and pentamer is catalyzed by zerovalent nickel complexes. [562—564] The ratio of these cyclic compounds is different depending on the kind of ligands as well as the ratio of nickel and ligand. When triphenylphosphine was used as the ligand, the tetramer *(495)* was obtained. On the other hand, the trimer *(496)* was formed by the coordination of triphenyl phosphite, and the pentamer *(497)* was obtained by using Ni(COD)$_2$. In this reaction of allene, the bis-π-allylic complexes *(498, 499)* as the intermediates, were isolated and characterized.

Rhodium complexes are also active catalysts for cyclization of allene to form various cyclic compounds. [565]

Insertion of allene into the π-allylnickel system of the dodecatrienylnickel complex *(310)* gave the new bis-π-allyl complex *(500)*. Treatment of the latter with carbon monoxide gave the twelve- and fourteen-membered rings and the fifteen-membered cyclic ketone *(501)*, which was converted into muscone *(502)*

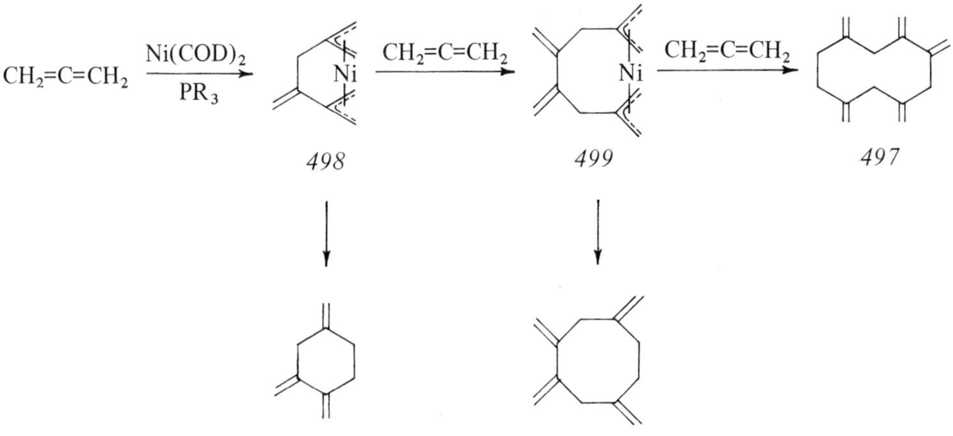

$CH_2=C=CH_2$ $\xrightarrow[PR_3]{Ni(COD)_2}$ [498 structure] Ni $\xrightarrow{CH_2=C=CH_2}$ [499 structure] Ni $\xrightarrow{CH_2=C=CH_2}$ [497 structure]

498   *499*   *497*

[496 structure]   [495 structure]

496   *495*

by hydrogenation. [566] This is a unique synthetic method of muscone, although the yield was only 4 to 5% of the total products.

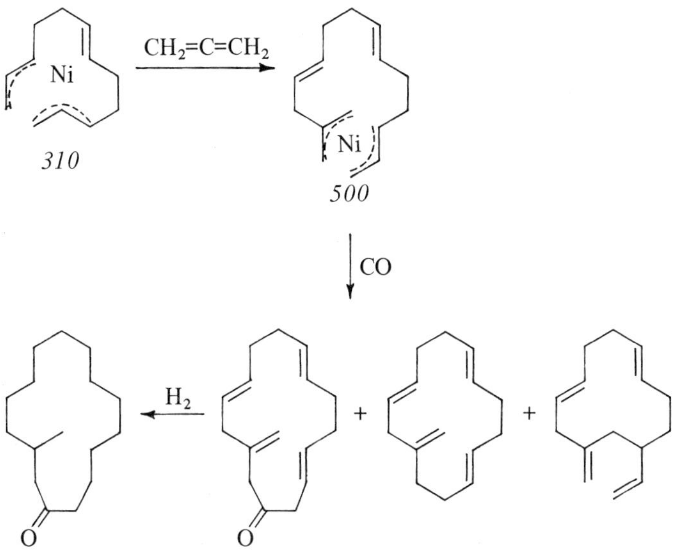

[310 structure] Ni $\xrightarrow{CH_2=C=CH_2}$ [500 structure] Ni

310

500

$\downarrow$ CO

[502 structure] $\xleftarrow{H_2}$ [501 structure] + [structure] + [structure]

O   O

502   *501*

Cyclization of various acetylenes catalyzed by transition metal complexes has been studied extensively. [567, 568] Formation of cyclooctatetraene catalyzed by a nickel complex is an old, unique reaction. Various benzene derivatives are also formed by trimerization. The cyclotrimerization is considered to proceed *via* metallacyclopentadiene *(503)*. [569, 570] Actually palladiacyclopentadiene *(503*, M = Pd) was isolated by the reaction of bis(dibenzylidene)palladium with methyl acetylenedicarboxylate. [571] This complex was converted into hexamethyl mellitate *(504)* by the reaction of dimethyl acetylenedicarboxylate.

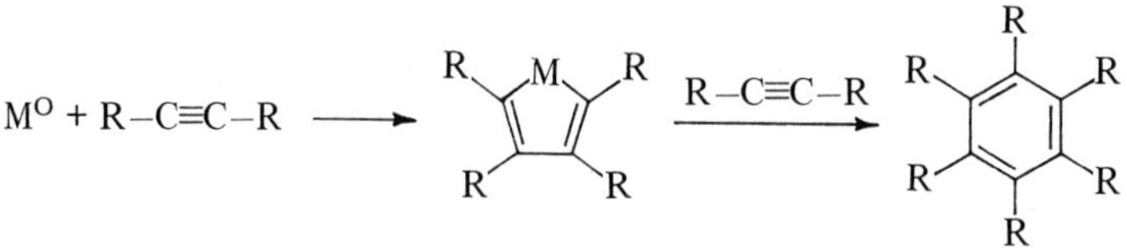

$$M^O + R-C{\equiv}C-R \longrightarrow \text{(503)} \xrightarrow{R-C{\equiv}C-R} \text{(504)}$$

R=CO₂CH₃ → R=CO_2CH_3

R=CO_2CH_3 *503*   *504*

The formation of the metallacyclopentadiene *(503)* was explained by the following mechanism involving a dipolar intermediate.

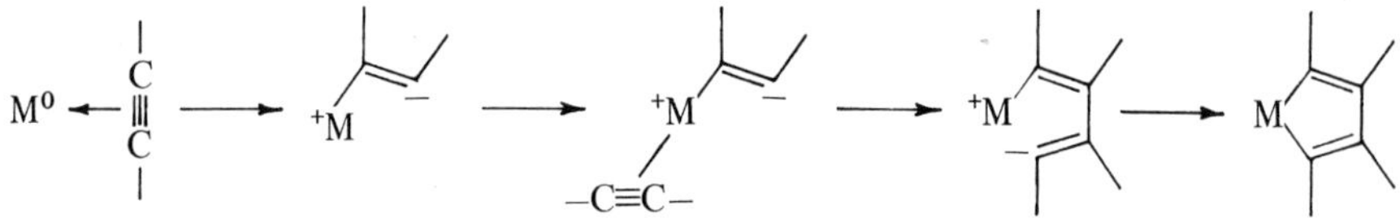

503

Maitlis explained the catalytic cyclotrimerization of acetylenes by the following cycle.

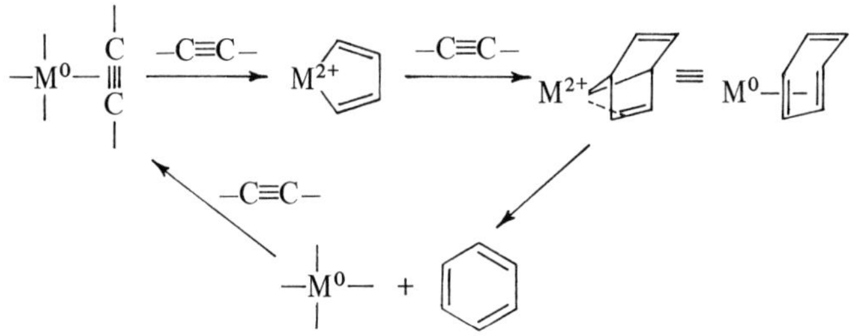

Metallacyclopentadiene formed from acetylene and a transition metal complex is an intermediate of other cyclization reactions. [572] Reaction of cyclopentadienylbis(triphenylphosphine)cobalt with diphenylacetylene gave the cobaltacyclopentadiene complex. Further reaction of this complex with various unsaturated bonds gives heterocyclic compounds. Acetonitrile reacted at the carbon-nitrogen triple bond to give 2-methyl-3,4,5,6-tetraphenylpyridine *(505)* in 33% yield. Also the reaction with carbon disulfide afforded 3,4,5,6-tetraphenyl-1,2-dithiopyrone *(506)* in 50% yield. The treatment with isothiocyanate afforded N-methyltetraphenyl-2-thiopyridone *(507)* in 10% yield. By selecting the proper reaction conditions, these heterocyclic ringformation reactions can be made catalytic with regard to the cobalt complex. [573]

Cyclization involving two moles of an acetylenic compound and one mole of an activated olefin to form cyclohexadiene derivatives is also known. For example, phenylacetylene reacted with N-substituted maleimide in the presence of Ni(CO)₂(PPh₃)₂, to give the adducts *(508 and 509)*. [574]

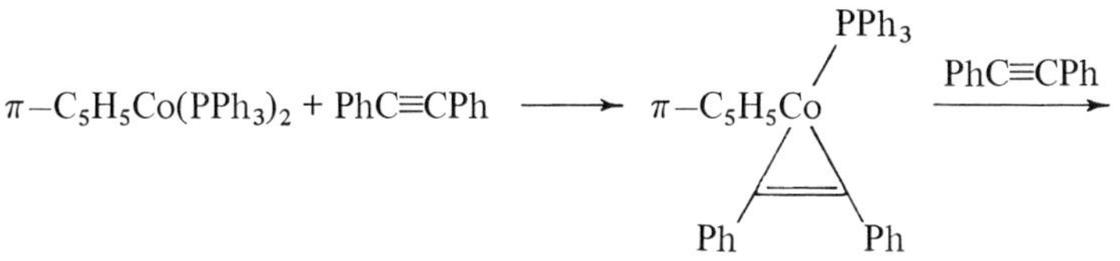

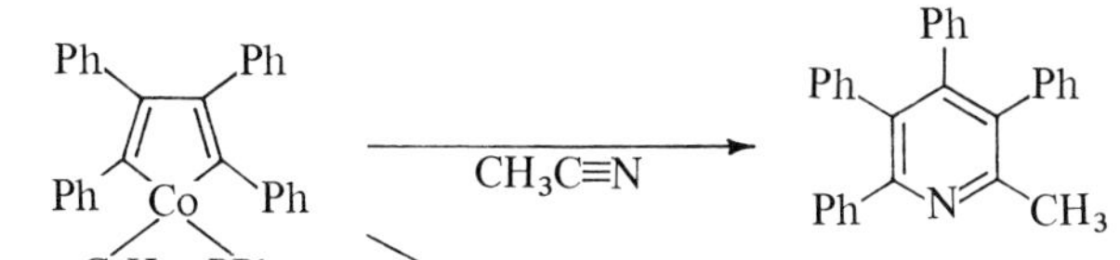

$$\pi - C_5H_5Co(PPh_3)_2 + PhC{\equiv}CPh \longrightarrow \pi - C_5H_5Co \overset{\displaystyle PPh_3}{\underset{Ph \quad\quad Ph}{\bigtriangleup}} \overset{PhC{\equiv}CPh}{\longrightarrow}$$

505

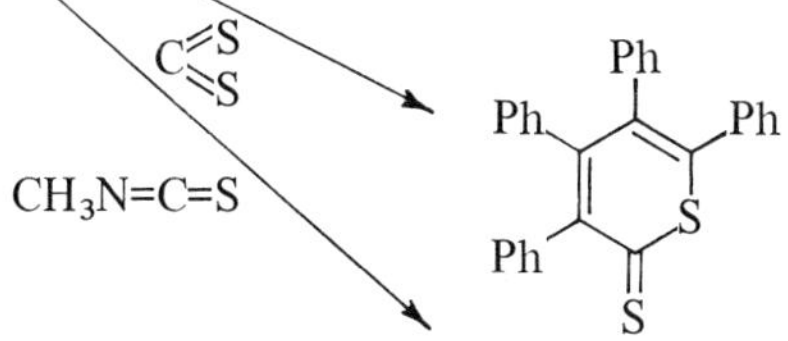

506

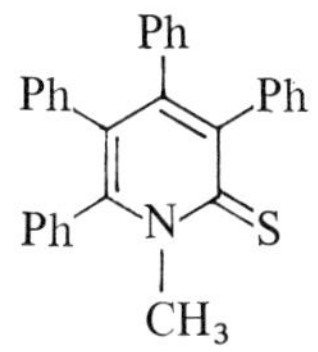

507

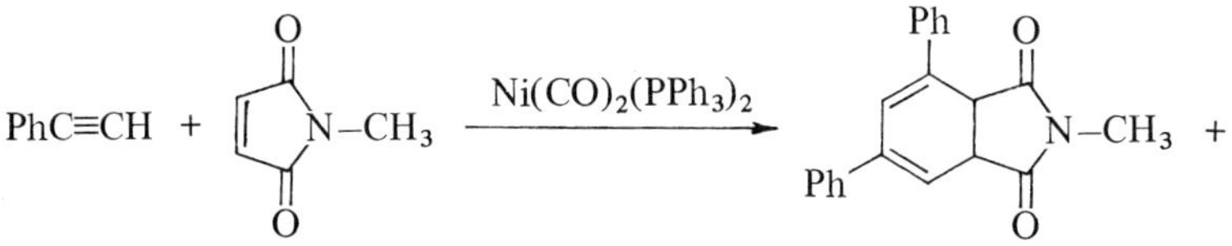

$$PhC{\equiv}CH + \underset{O}{\overset{O}{\|}}\!\!N-CH_3 \xrightarrow{Ni(CO)_2(PPh_3)_2} \text{(structure)} + $$

508   50%

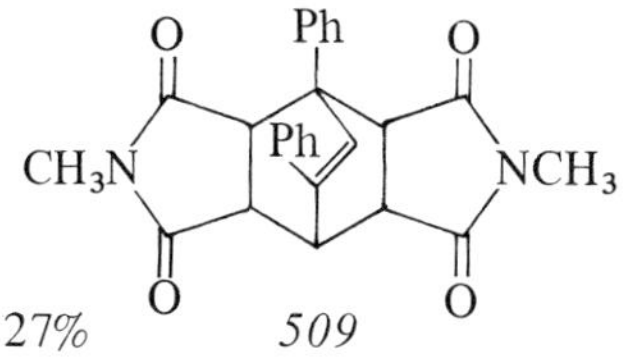

27%   509

But the same nickel complex catalyzed the linear cooligomerization of acetylene and acrylate to form 2,4,6-heptatrienoate rather than cyclization. [575]

$$CH_2{=}CHCO_2R + 2CH{\equiv}CH \longrightarrow CH_2{=}CHCH{=}CHCH{=}CHCO_2R$$

Even non-activated olefins react with two moles of acetylene in the presence of a rhodium catalyst. For example, cyclohexene reacted with 2-butyne in the presence of $[(C_2H_4)_2RhCl]_2$ to give 1,2,3,4-tetramethyl-5,6,7,8-tetrahydronaphthalene by cycloaddition and dehydrogenative aromatization in 20% selectivity, although the main product was hexamethylbenzene. [576]

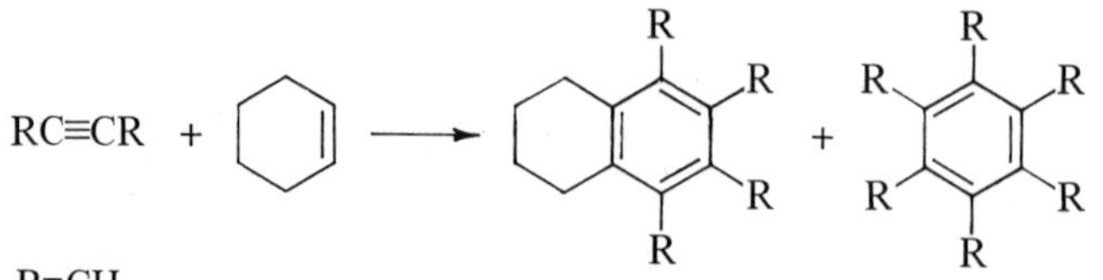

$$RC{\equiv}CR \ + \ \text{cyclohexene} \longrightarrow \text{product} \ + \ \text{product}$$

R=CH$_3$

These cyclization reactions of acetylenes and olefins are explained by the formation of metallacyclopendadiene as an intermediate, which adds to the olefinic bond.

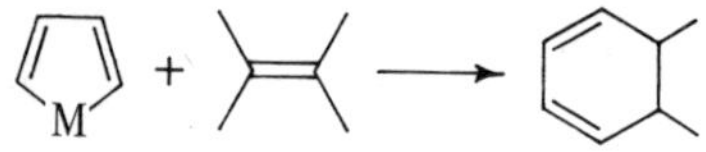

$$\text{metallacycle} + \text{alkene} \longrightarrow \text{cyclohexadiene}$$

Formation of heterocyclic compounds by the reactions of acetylenic compounds with some heterocumulenes via metallocycles in the presence of metal carbonyls has been reported. The reaction of phenylacetylene with phenyl isocyanate in the presence of $Fe(CO)_5$ gave 4-benzylidene-1,3-diphenylhydantoin (510) in 85% yield by the addition reaction and hydrogen shift. [577]

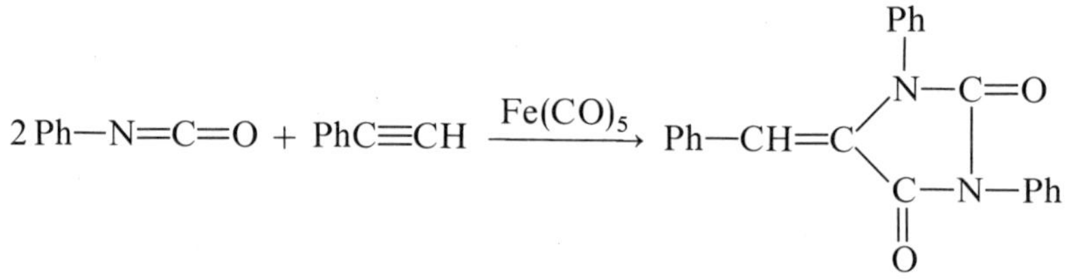

$$2\,Ph{-}N{=}C{=}O + PhC{\equiv}CH \xrightarrow{Fe(CO)_5} Ph{-}CH{=}C\Big\langle\begin{array}{l}N{-}C{=}O\\ C{-}N{-}Ph\end{array}$$

510

The reaction can be explained by the following sequence of oxidative addition and insertion reactions.

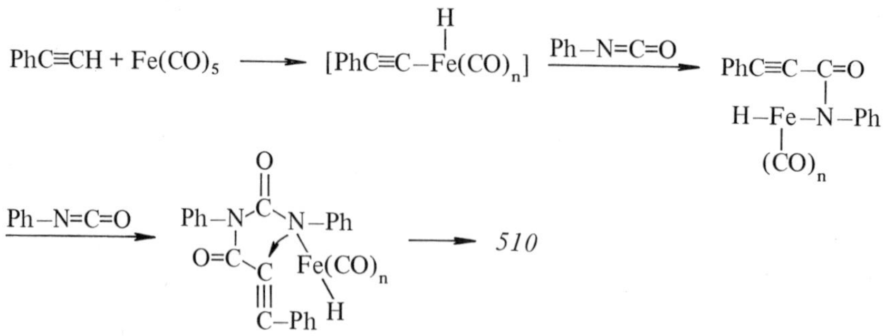

$$PhC{\equiv}CH + Fe(CO)_5 \longrightarrow [PhC{\equiv}C{-}Fe(CO)_n] \xrightarrow{Ph{-}N{=}C{=}O} PhC{\equiv}C{-}C{=}O$$

$$\xrightarrow{Ph{-}N{=}C{=}O} \longrightarrow 510$$

3,4,5,5-Tetraphenyl-3-cyclopenten-1,2-dione *(511)* was obtained as the main product in 62% yield by the reaction of diphenylketene and diphenylacetylene in the presence of Fe(CO)$_5$. [578] On the other hand, the reaction in the presence of Ni(CO)$_4$ yielded 2,2,4,5-tetraphenyl-4-cyclopentene-1,3-dione *(512)* in a low yield.

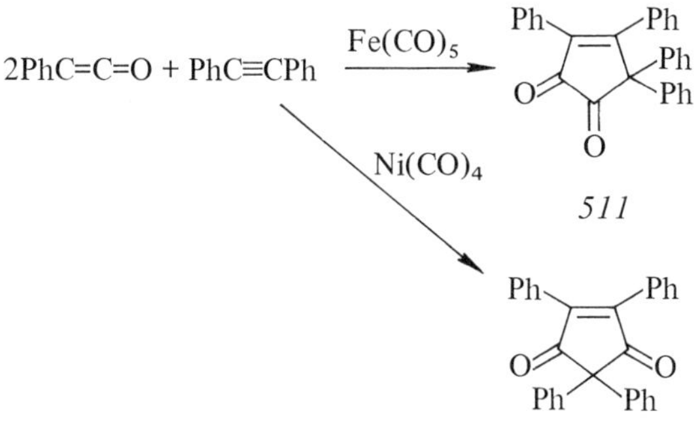

$$2PhC=C=O + PhC\equiv CPh$$

511

512

For these reactions, formation of a "doubly σ-bonded acetylene complex *(513)*" was suggested. Insertions of the ketene and carbon monoxide are followed by reductive elimination to give the five-membered ring.

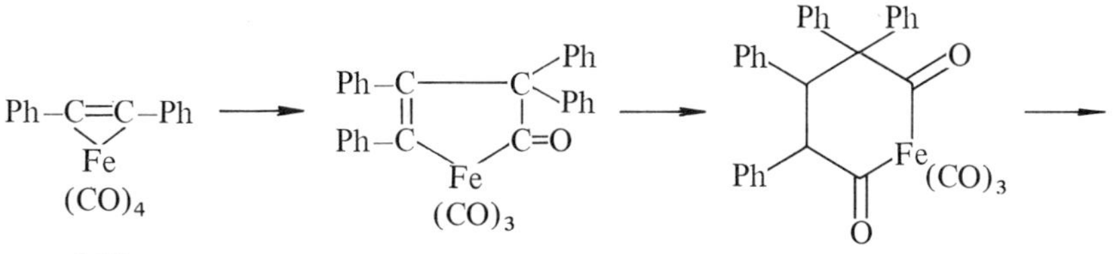

513

The reaction of diphenylbutadiyne with diphenylcarbodiimide in the presence of Fe(CO)$_5$ at 160° produced 1,3-diphenyl-4-phenylethynyl-5-phenylimino-pyrrolin-2-one *(514)*, 1,4-diphenyl-3-phenylethynyl-2,5-bis(phenylimino)pyrroline *(515)* and 1,1',3,3'-tetraphenyl-2,5,5'-tris(phenylimino)-4,4'-bipyrrolin-2'-one *(516)* in yields of 28, 1, and 14%, respectively. The formation of these products was illustrated by the formation of the "doubly σ-bonded acetylene complex", fol-

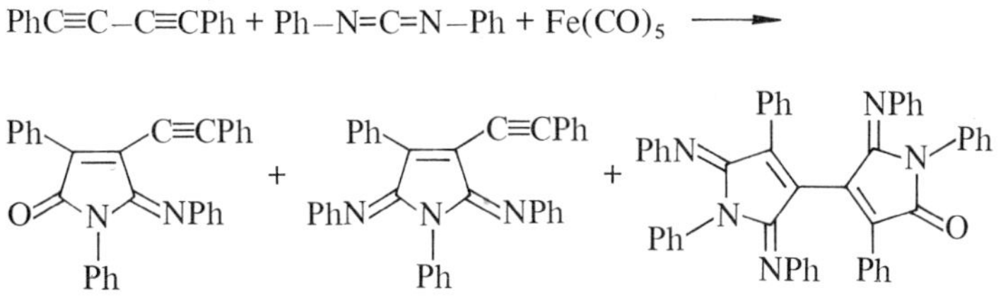

$$PhC\equiv C-C\equiv CPh + Ph-N=C=N-Ph + Fe(CO)_5 \longrightarrow$$

514   *515*   *516*

lowed by the insertion of the carbon-nitrogen double bond of carbodiimide and carbon monoxide or isocyanide. [579] The isocyanide is formed from carbodiimide and $Fe(CO)_5$.

Although cycloaddition reaction of two olefins to form a cyclobutane derivative is formally regarded as a thermally forbidden [2 + 2] process, some specific olefins can undergo the addition reaction in the presence of transition metal complexes. For example, highly selective [2 + 2] cross addition of norbornadiene and the highly strained double bond of methylenecyclopropane to give the adduct *(517)* was achieved with the aid of a zerovalent nickel complex coordinated by triphenylphosphine. [580]

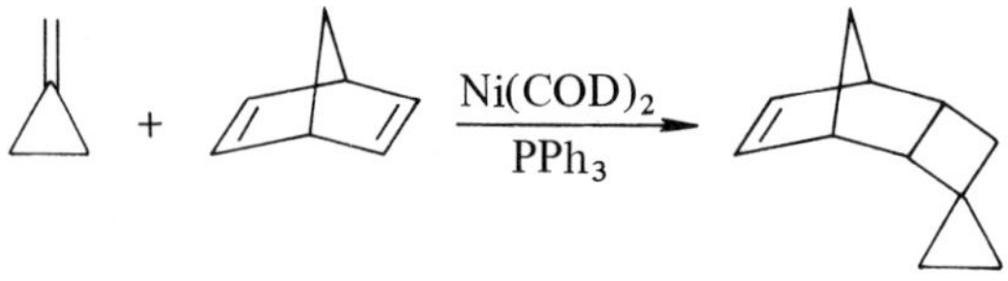

517

When the reaction was carried out in the presence of $Ni(COD)_2$ and an optically active phosphine; $(-)$benzylmethylphenylphosphine, the product was obtained in an optically active form. Addition of methylenecyclopropane to electron deficient olefins under the influence of a nickel catalyst gives cyclopentanes. [581, 582] Thus the reaction of methylenecyclopropane with excess methyl acrylate in the presence of bis(acrylonitrile)nickel gave methyl 3-methylenecyclopentane-carboxylate *(518)* in 82% yield.

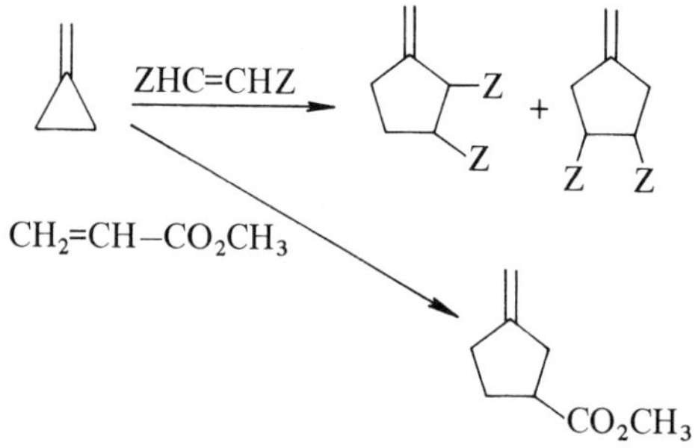

518

Methylenecyclopropane itself dimerizes to give the following cyclobutane, cyclopentane, and cyclohexane derivatives upon treatment with $Ni(COD)_2$. [583, 584]

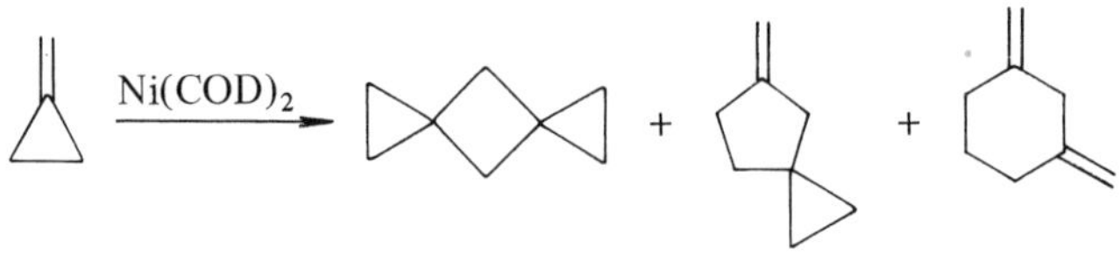

Methylenecyclopropane was converted into 1-vinyl-2-methylenecyclopentane *(480)* in 91% yield by treatment with the nickel catalyst prepared from bis(tri-n-butylphosphine)nickel dibromide and n-butyllithium. [585] The product obtained from the labeled methylenecyclopropane revealed that the conversion proceeded through the intermediacy of butadiene. The formation of butadiene was explained by addition of nickel hydride to the double bond, conformational change, cyclopropylcarbinyl-allylcarbinyl rearrangement and elimination of the nickel-hydride catalyst. The formation of vinylmethylenecyclopentane from butadiene is a well-known reaction as described earlier, (p. 154).

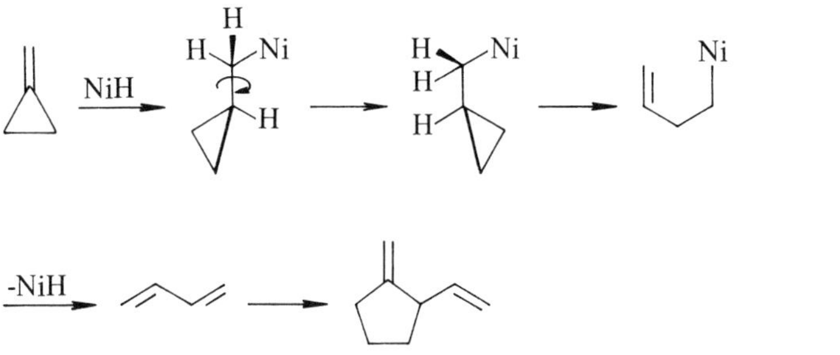

480

The reaction of 5,5-dideuterated bicyclo[2,1,0]pentane with methyl acrylate in the presence of bis(acrylonitrile)nickel gave the following three 1:1 adducts. [586]

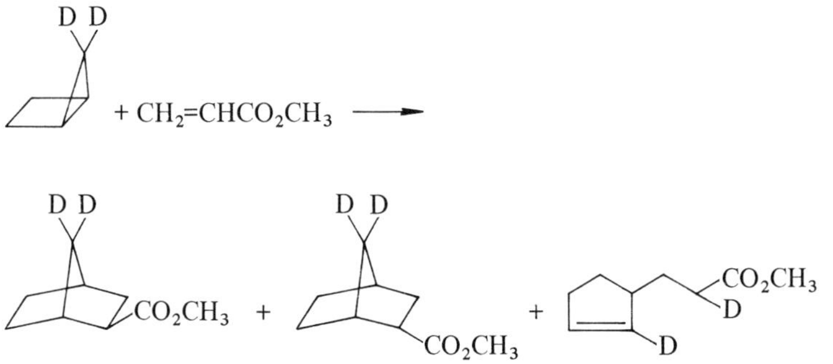

The following mechanism involving the oxidative addition, olefin insertion, β-elimination, and reductive elimination, was proposed for the formation of these adducts.

A case in which π-coordination of olefins to metallic complexes plays a decisive role is the remarkable reaction called olefin metathesis. [587] Although the reaction is not a cyclization reaction in an exact sense, it involves some kinds of cyclized intermediates, and hence it is worthwhile to survey it here because of its uniqueness and utility in organic synthesis. Calderon *et al.* reported the reaction in which two olefinic bonds are broken and two new olefinic bonds

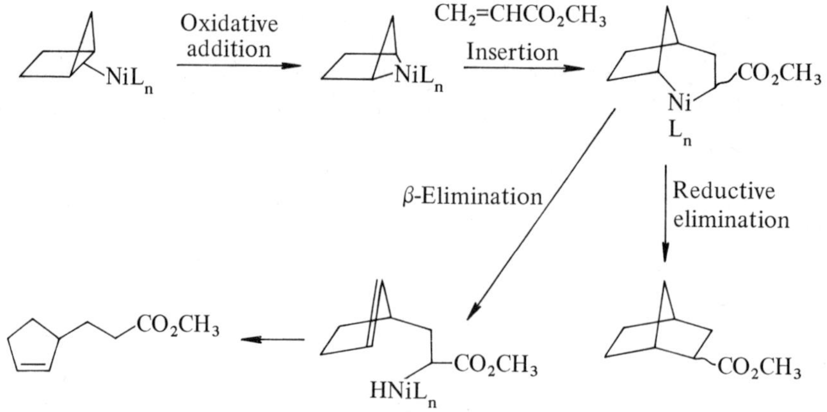

Oxidative addition → Insertion ($CH_2=CHCO_2CH_3$) → (CO_2CH_3, $Ni^. L_n$)

β-Elimination

Reductive elimination

(CO_2CH_3), $HNiL_n$

(CO_2CH_3)

are formed *via* a four-membered ring, e.g. 2-pentene was converted into 3-hexene and 2-butene. [588]

$$C—C=C—C—C \longrightarrow C—C=C—C + C—C—C=C—C—C$$

The reaction can be generalized by the following scheme to form an equilibrium mixture of olefins.

$$2R—CH=CH—R' \rightleftharpoons R—CH=CH—R + R'CH=CH—R'$$
$$50\% 25\% 25\%$$

The catalysts used in the metathesis reaction are Ziegler type catalysts formed from WCl_6 or $MoCl_5$ and ethylaluminum dichloride or butyllithium. For the mechanism of the reaction, formation of some kind of a four-membered ring was proposed on the basis of the experiments using deuterated olefins. Thus in the metathesis reaction of 2-butene and 2-butene-d_8, the newly formed 2-butene has a mass number of *60*; and no butene of mass numbers of *61* or *59* was formed. [589]

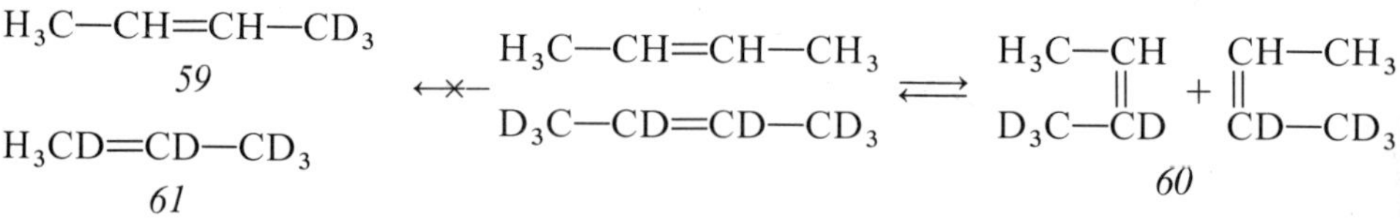

$$H_3C—CH=CH—CD_3$$
$$59$$
$$H_3CD=CD—CD_3$$
$$61$$

$\xleftarrow{\;\times\;}$

$$H_3C—CH=CH—CH_3$$
$$D_3C—CD=CD—CD_3$$

$\rightleftharpoons$

$$H_3C—CH \quad CH—CH_3$$
$$\| + \|$$
$$D_3C—CD \quad CD—CD_3$$
$$60$$

Since the reaction proceeds rapidly under moderate conditions, its synthetic utility is obvious. Conversion of internal olefins into terminal olefins would be very useful. However, it is thermodynamically "uphill" and can occur only if a second reagent is consumed concurrently, as observed for hydroboration. The conversion can be done by metathesis in the form of an "ethenolysis" reaction. In this reaction, internal olefins are treated with ethylene to form two moles of terminal olefins. The reaction of ethylene and 2-heptene gave a mixture of olefins consisting of propylene (6%), 2-butene (3%), 1-hexene (19%), 2-heptene (66%), and 5-decene (7%).

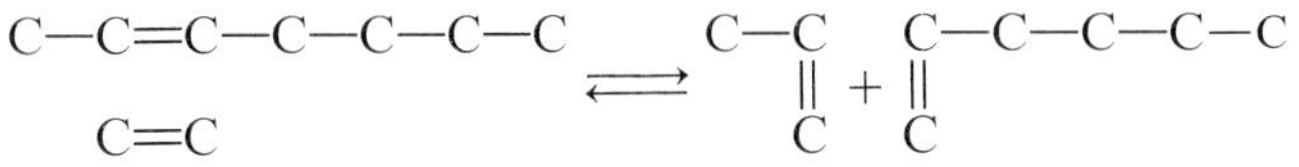

$$C-C=C-C-C-C-C \atop C=C \quad \rightleftharpoons \quad {C-C \atop \underset{C}{\overset{\|}{}}} + {C-C-C-C-C \atop \underset{C}{\overset{\|}{}}}$$

The reaction is not selective due to the following competitive reactions

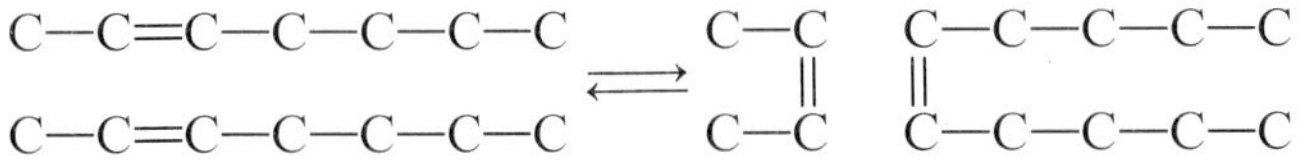

$$\begin{matrix} C-C=C-C-C-C-C \\ C-C=C-C-C-C-C \end{matrix} \rightleftharpoons \begin{matrix} C-C \\ \| \\ C-C \end{matrix} \quad \begin{matrix} C-C-C-C-C \\ \| \\ C-C-C-C-C \end{matrix}$$

to form 2-butene and 5-decene. The metathesis reaction has industrial importance and is called the "triolefin process". For example, propylene can be formed from ethylene and 2-butene. In industrial applications, the reaction is most conveniently carried out as a gas-phase reaction using solid catalysts containing molybdenum or tungsten oxides. Actually the metathesis reaction in the gas phase has been discovered before the homogeneous reaction. [590] It should also be pointed out that the metathesis reaction had been carried out without notice of the true nature of the reaction in the polymerization reaction of cyclic olefins such as cyclopentene and cyclobutene. [591—592] The application of the metathesis reaction to cyclic olefins affords large-membered cyclic olefins as shown below.

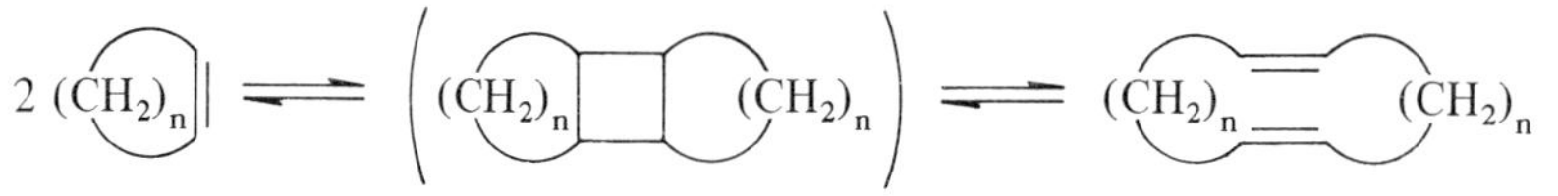

Of course the reaction with cyclic olefins does not necessarily stop at the first step, but proceeds further to form trimer, tetramer, and finally polymer by the sequential metathesis reactions. [593—595] Wideman carried out the metathesis of cyclooctene and obtained 1,9-cyclohexadecadiene in 20% yield, from which sixteen-membered mono- and diketones were synthesized. [596]

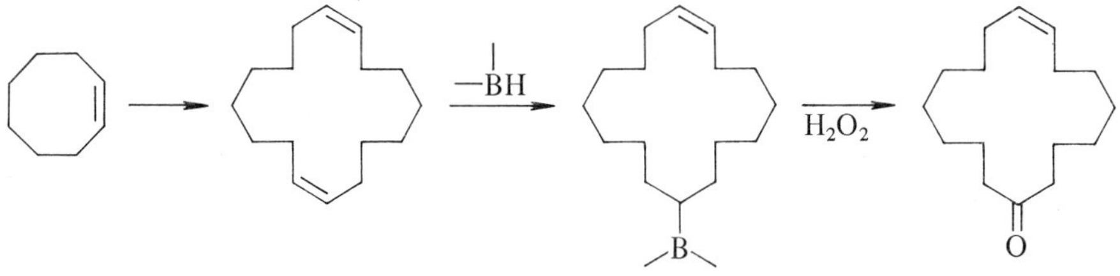

Furthermore, interlocked ring systems were obtained by the metathesis reaction of cyclododecene. Vapor phase chromatography of the oligomers showed the presence of C_{24}, C_{36}, C_{48}, C_{60}, C_{72} and higher oligomers, and mass spectrometric studies revealed that the oligomers have interlocked structures, called "catenanes" *(519)*. [597]

519

The metathesis reactions described so far are concerned only with simple olefins. The nature of the catalysts, especially aluminum or lithium compounds can not tolerate the presence of functional groups on the olefins. The usefulness of the metathesis reaction was enhanced by the finding that the catalyst system formed from WCl_6 and tetramethyltin can be used for the metathesis reaction of unsaturated esters. Bifunctional compounds are synthesized by this reaction. For example, methyl elaidate was converted into 9-octadecene and dimethyl 9-octadecenedioate, although the conversion was not high. [598]

$$R-CH=CH(CH_2)_nCO_2R \longrightarrow \begin{matrix} R-CH \\ \| \\ R-CH \end{matrix} + \begin{matrix} CH(CH_2)_nCO_2R \\ \| \\ CH(CH_2)_nCO_2R \end{matrix}$$

Regarding the mechanism of the metathesis reaction, at first the formation of an electronically excited four-membered ring coordinated to the metal was postulated.

On the basis of the concept of the conservation of orbital symmetry, Mango and Schachtschneider demonstrated that a reaction pathway does exist for orbital symmetry conservation if the two olefinic ligands are coordinated to the transition metal in a four-membered ring. [599—601] The formation of a cyclobutane by the cycloaddition of two olefinic bonds is forbidden in the absence of the transition-metal catalyst. The role of the catalyst is the switching of symmetry-forbidden to symmetry-allowed, through an appropriate manipulation of bonding and non-bonding electrons. The catalyst behaves as an electron relay switch and need not undergo significant charge generation or electron excitation. This consideration gave some insight into the mysterious role of transition metal catalysts in the metathesis reaction.

Pettit critisized the formation of the cyclobutane intermediate and proposed that the metathesis reaction of olefins proceeds *via* reversible transformation of the two coordinated olefinic bonds into the multi-three-centered species, called "tetramethylene complex" *(520)*, in which four CH_2 units of sp^3-hybridized carbon atoms are involved. [602]

$$\|\text{[·M·]}\| \rightleftharpoons \underset{520}{\overset{\text{H}_2\text{C}\diagdown\diagup\text{CH}_2}{\underset{\text{H}_2\text{C}\diagup\diagdown\text{CH}_2}{\diagup\text{M}\diagdown}}} \rightleftharpoons \diagup\text{M}\diagdown$$

Rooney and coworkers supported this intermediate by the fact that the site on the molybdate catalysts which disproportionate olefins also selectively convert adsorbed methylene formed from diazomethane into ethylene. [603] The carbon-metal σ-bonded metallocyclic species (521) was proposed as another intermediate based on the experimental fact that the reaction of WCl_6 with 1,4-dilithiobutane gave a metallocyclic compound which decomposed to give ethylene. [604] A mechanism involving, i) rearrangement of the complexed olefin to the metallocyclic intermediate (521a), ii) rearrangement to another metallocyclic intermediate (521b), and iii) cycloreversion to the complexed olefin, was proposed.

$$\underset{}{\overset{R\diagdown\quad\diagup R}{\|\text{[·M·]}\|}} \rightleftharpoons \underset{521a}{\overset{R\diagdown\quad\diagup R}{\diagdown\text{M}\diagup}} \rightleftharpoons \underset{521b}{\overset{\diagup R}{\text{M}\diagdown R}} \rightleftharpoons \|\text{[·M·]}\|\diagup{\overset{R}{\diagdown R}}$$

The triphenylphosphine complex of palladium catalyzes a unique cocyclization reaction of butadiene with aldehydes to form tetrahydropyrans. [353—356] In the presence of palladium acetate and triphenylphosphine, two moles of butadiene and one mole of an aldehyde reacted to give 2-substituted 3,6-divinyl-tetrahydropyran (325) and 1-substituted 2-vinyl-4,6-heptadien-1-ol (324). The ratio of PPh_3/Pd in the catalyst system showed a decisive effect on the course of the reaction. The latter was formed predominantly when the ratio was one, whereas the pyran was formed as the main product when the ratio was larger than three.

$$2 \diagup\diagdown\diagup + \text{RCHO} \longrightarrow \underset{325}{\overset{}{R\diagup\overset{}{\underset{\text{O}}{\bigcirc}}}} + \underset{324}{\overset{}{\underset{\underset{\text{H}}{|}}{R-\text{C}-\text{OH}}}}$$

The following mechanism was proposed for this reaction.

At first, a π-allylic palladium complex is formed and the insertion of the carbonyl group of aldehyde takes place. Depending on the number of triphenylphosphine coordinated to the palladium, the unsaturated alcohol (324) is formed by a hydrogen shift, and the pyran (325) is formed by coupling of the carbon and oxygen σ-bonds.

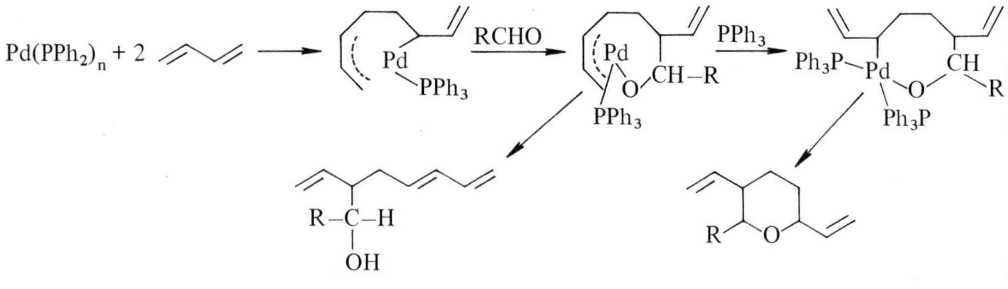

324   *325*

A similar cocyclization reaction of butadiene is also possible with isocyanates to give 1-substituted 3,6-divinylpiperidone *(522)*. Isoprene and phenyl isocyanate were converted to *trans-* and *cis-*3,6-diisopropenyl-1-phenyl-2-piperidone *(522)* in the presence of palladium acetate and triphenylphosphine. [605]

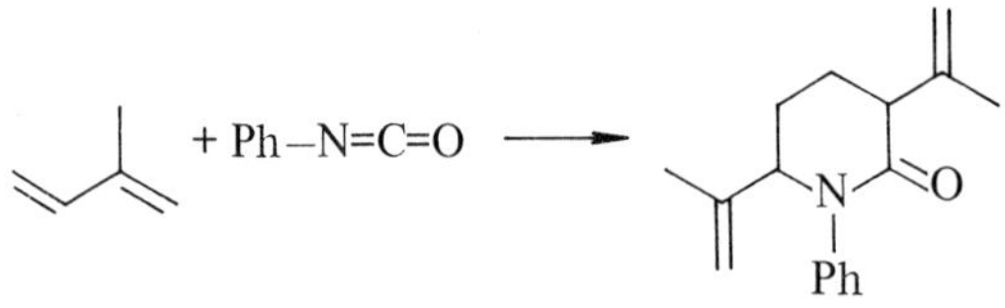

522

Carbonylation reactions of unsaturated compounds sometimes give cyclized products such as lactones, lactams, and other heterocyclic compounds. In the carbonylation reactions catalyzed by transition metal complexes, a metal-acyl bond is formed, which is cleaved by nucleophiles such as water, alcohol, and amines to give carbonyl compounds as the products. The cyclization product can be produced when the cleavage takes place intramolecularly. Several ex-

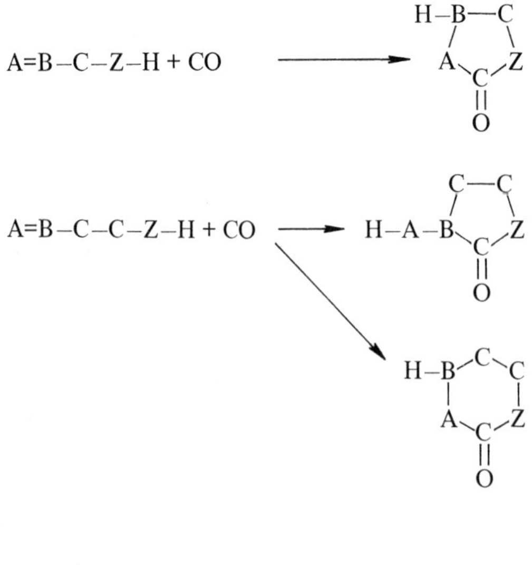

amples of the cyclization involving carbon monoxide have already been described in the preceding chapters. When an unsaturated bond and an active
hydrogen group are present in the same molecule at the position suitable for
five- and six-membered rings, then the cyclization reaction proceeds. Falbe
expressed the cyclization reactions involving carbon monoxide by the following
general schemes. [606,607]

Not only carbon-carbon double bonds, but also unsaturated carbon-nitrogen,
and nitrogen-nitrogen bonds take part in the reactions. [608] Possible *ortho*-
metalation is also a significant driving force for the cyclization. In this reaction,
an aromatic *ortho*-hydrogen behaves as an active hydrogen. The cyclization
reactions involving carbon monoxide are catalyzed most effectively by $Co_2(CO)_8$,
and the reaction is explained by the following mechanism, which involves the
insertion of the unsaturated bond into the cobalt-hydrogen bond.

$$A=B-C-Z-H + HCo(CO)_4 \longrightarrow \overset{\overset{H}{|}}{A-B}-C-Z-H \underset{\underset{Co(CO)_4}{|}}{} \xrightarrow{CO}$$

$$\underset{\underset{Co(CO)_4}{|}}{\overset{\overset{H}{|}}{A-B}-C-Z-H} \overset{|}{\underset{|}{C=O}} \longrightarrow \underset{\underset{O}{||}}{\overset{H-B-C}{A\diagdown C \diagup Z}} + HCo(CO)_4$$

Carbonylation of allylic alcohols catalyzed by $Co_2(CO)_8$ gives lactones, but
the yield is not high. Extensive isomerization of the double bond to give carbonyl
compounds proceeds. [609]

$$RCH=CH-CH_2OH + CO \longrightarrow RCH_2CH_2CHO +$$

Allylic amines react with carbon monoxide to give lactams in a satisfactory
yield. N-methylallylamine formed N-methylpyrrolidone *(523)* in 78% yield.
[610, 611]

$$CH_2=CHCH_2NHCH_3 + CO \longrightarrow$$

523

Metal carbonyls other than $Co_2(CO)_8$ are not satisfactory. With $Fe(CO)_5$, a
different reaction took place. [611] When allylamine was treated with $Fe(CO)_5$
at 160°, deamination took place to give the intermediate unsaturated amine *(524)*.
This intermediate was converted into 3,5-dimethyl-2-ethylpyridine *(525)* by the
reaction at 250°.

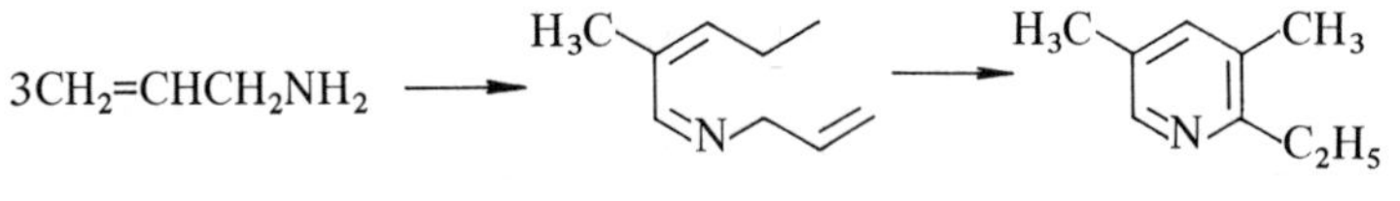

$$3CH_2=CHCH_2NH_2 \longrightarrow$$

524   525

N-monosubstituted acrylamide was converted into N-substituted succinimide *(526)* in a high yield. [612]

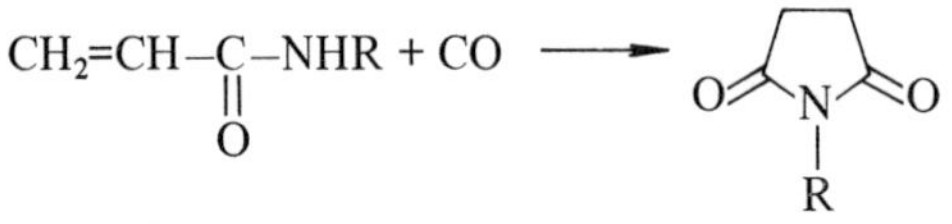

$$CH_2=CH-\underset{\underset{O}{\|}}{C}-NHR + CO \longrightarrow$$

R

526

Preparation of heterocyclic compounds by the carbonylation of unsaturated carbon-nitrogen or nitrogen-nitrogen bonds conjugated to aromatic ring is a very useful method. These compounds have the possibility of *ortho*-metalation, and carbon monoxide is introduced directly at the *ortho*-position of the aromatic ring. Schiff bases of aromatic aldehydes or ketones are converted to phthalimidines in high yields. [613, 614] Benzylideneaniline *(527)* reacted with carbon monoxide under 100—200 atm pressure at 200° in the presence of $Co_2(CO)_8$ in benzene to give N-phenylphthalimidine *(528)* in 72% yield.

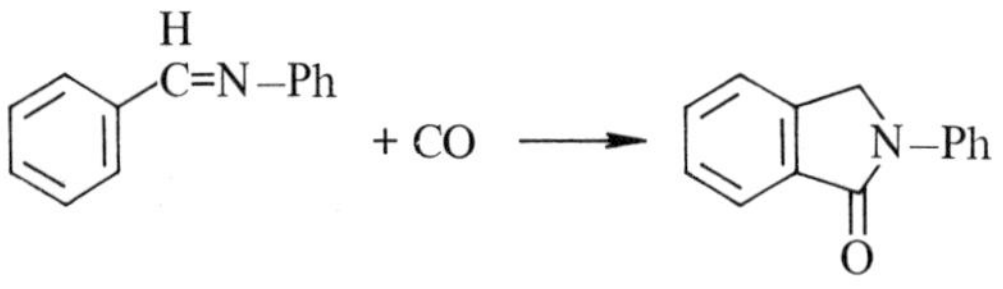

527   528

The cyclization seems to be catalyzed by $HCo(CO)_4$ formed from $Co_2(CO)_8$ and hydrogen present in carbon monoxide. The first step is the insertion of the carbon-nitrogen bond to the cobalt-hydrogen bond; then carbon monoxide insertion takes place. *Ortho*-metalation affords the cobalt-aromatic carbon bond. It is reasonable to assume that ring closure takes place by the coupling of the acyl group and the aromatic carbon.

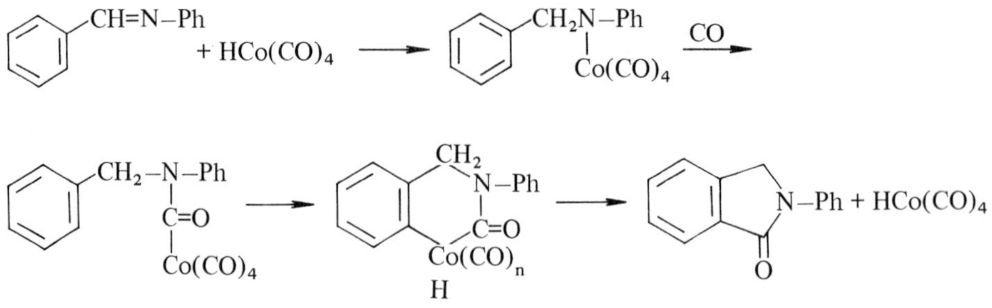

Azobenzene *(63)* is another active compound for carbonylation catalyzed by $Co_2(CO)_8$. [471, 472] Azobenzene reacted with one mole of carbon monoxide at a pressure of 150 atm and at 190° to form 2-phenylindazolinone *(417)* in 55% yield. When the reaction was carried out at 230°, two moles of carbon monoxide took part in the reaction, and 3-phenylquinazolinedione *(418)* was obtained. The mechanistic interpretation that the former is the intermediate of the latter was confirmed by the experimental result that the quinazolinedione *(418)* was formed by the carbonylation of the indazolinone *(417)* at 230°. The mechanism of the carbon monoxide insertion into the nitrogen-nitrogen bond in *(417)* to form *(418)* is not clear.

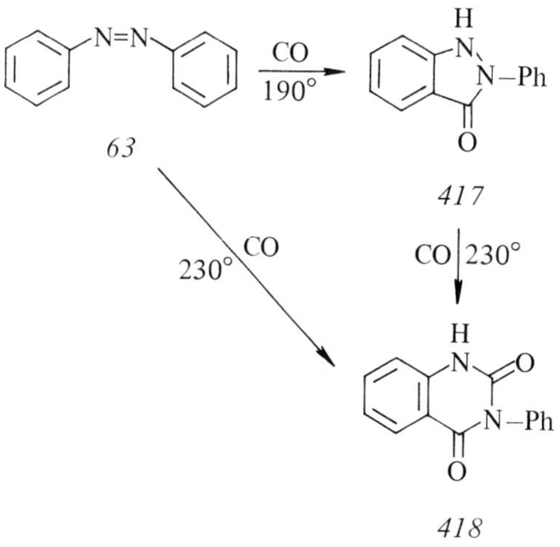

VIII. Concluding Remarks

The general patterns of synthetic reactions involving transition metal complexes have been surveyed and elaborated with typical examples. It should be emphasized again that much has to be done before we can clearly understand what actually happens in the process of forming organic compounds in the coordination spheres of transition metals. In this sense, we can still anticipate the discovery of unexpected and exciting new reactions in the course of further studies of reactions promoted by transition metal complexes.

For additional information on this field, several books listed in the References are recommended. [615—618]

References

1. Lehmkuhl, H., Reinehr, D.: J. Organometal. Chem. **25**, C47 (1960).
2. Lehmkuhl, H., Reinehr, D.: J. Organometal. Chem. **34**, 1 (1972).
3. Lehmkuhl, H., Reinehr, D., Brandt, H., Schroth, G.: J. Organometal. Chem. **57**, 39 (1973).
4. Parshall, G. W., Mrowca, J.: Advan. Organometal. Chem. **7**, 157 (1968).
5. Collman, J. P.: Accounts Chem. Res. **1**, 136 (1968).
6. Collman, J. P., Roper, W. R.: Advan. Organometal. Chem. **7**, 54 (1968).
7. Heck, R. F.: Accounts Chem. Res. **2**, 10 (1969).
8. Halpern, J.: Accounts Chem. Res. **3**, 386 (1970).
9. Tolman, C. A.: Chem. Soc. Rev. **1**, 337 (1972).
10. Shriver, D. F.: Accounts Chem. Res. **3**, 231 (1970).
11. Osborn, J. A., Jardine, F. H., Young, J. F., Wilkinson, G.: J. Chem. Soc. A, 1711 (1966).
12. Arai, H., Halpern, J.: Chem. Commun., 1571 (1971).
13. Koerner von Gustorf, F., Grevels, F. W., Hogan, J. C.: Angew. Chem. **81**, 918 (1969).
14. Balzani, V., Carassiti, V.: Photochemistry of Coordination Compounds, New York: Academic Press, 1970, p. 324.
15. Malatesta, L., Cariello, C.: J. Chem. Soc., 2323 (1958).
16. Ugo, R., Cariati, F., La Monica, G.: Chem. Commun., 868 (1966).
17. Cotton, F. A., Wilkinson, G.: Advanced Inorganic Chemistry, New York: Interscience, 1966, p. 719.
18. Coates, G. E., Green, M. L. H., Powell, P., Wade, K.: Principles of Organometallic Chemistry, London: Methuen, 1968, Chapt. 7.
19. Trost, B. M., Chen, F.: Tetrahedron Lett., 2603 (1971).
20. Ito, T., Kitazume, S., Yamamoto, A., Ikeda, S.: J. Am. Chem. Soc. **92**, 3011 (1970).
21. Gosser, L. W., Knoth, W. H., Parshall, G. W.: J. Am. Chem. Soc. **95**, 3436 (1973).
22. Vaska, L.: Accounts Chem. Res. **1**, 335 (1968).
23. Shunn, R. A.: Transition Metal Hydrides, Edited by Muetterties, E. L., New York: Marcel Dekker, 1971, p. 203.
24. Kaesz, H. D., Saillant, R. B.: Chem. Rev. **72**, 231 (1972).
25. Vaska, L., DiLuzio, J. W.: J. Am. Chem. Soc. **84**, 679 (1962).
26. James, B. R.: Homogeneous hydrogenation, New York: Wiley Interscience, 1973.
27. Discussion Faraday Society, No. **46**, 1968.
28. Harmon, R. E., Gupta, S. K.: Chem. Rev. **73**, 21 (1973).
29. Kokes, R. J.: Catal. Rev. **6**, 1 (1972).
30. Sloan, M. F., Matlack, A. S., Breslow, D. S.: J. Am. Chem. Soc. **85**, 4014 (1963).
31. Meakin, P., Jesson, J. P., Tolman, C. A.: J. Am. Chem. Soc. **94**, 3240 (1972).
32. Halpern, J., Wong, C. S.: Chem. Commun., 629 (1973).
33. Robinson, S. D., Shaw, B. L.: Tetrahedron Lett., 1301 (1964).
34. Nakamura, A., Otsuka, S.: J. Am. Chem. Soc. **94**, 1886 (1972), **95**, 5091, 7263 (1973).
35. Trocha-Grimshaw, J., Henbest, H. B.: Chem. Commun., 757 (1968).
36. Montelatici, S., van der Ent, A., Osborn, J. A., Wilkinson, G.: J. Chem. Soc. A, 1054 (1968).
37. Hussey, A. S., Takeuchi, Y.: J. Am. Chem. Soc. **91**, 672 (1969).
38. Hussey, A. S., Takeuchi, Y.: J. Org. Chem. **35**, 643 (1970).
39. Candlin, J. P., Oldham, A. R.: Discuss. Faraday Soc. **46**, 60 (1968).
40. Brown, M., Piszkiewicz, L. W.: J. Org. Chem. **32**, 2013 (1967).

41. Birch, A.J., Walker, K.A.M.: J. Chem. Soc. C, 1894 (1966).
42. Birch, A.J., Walker, K.A.M.: Tetrahedron Lett., 4939 (1966).
43. Djerassi, C., Gutzwiller, J.: J. Am. Chem. Soc. **88**, 4537 (1966).
44. Voelter, W., Djerassi, C.: Chem. Ber. **101**, 58 (1968).
45. Biellman, J.F., Liesenfelt, H.: Compt. Red. **263**, 251 (1966).
46. Biellman, J.F., Liesenflet, H.: Bull. Soc. Chim. France, 4029 (1966).
47. Simes, J.J., Honwad, V.K., Selman, L.H.: Tetrahedron Lett., 87 (1969).
48. Harmon, R.E., Parsons, J.L., Cooke, D.W., Gupta, S.K., Schoolenberg, J.: J. Org. Chem. **34**, 3684 (1969).
49. Hornfeldt, A.B., Gronowitz, J.S., Gronowitz, S.: Acta Chim. Scan. **22**, 2725 (1968).
50. Emery, A., Oehlschlager, A.C., Unrau, A.M.: Tetrahedron Lett., 4401 (1970).
51. Bennett, M.A., Longstaff, P.A.: Chem. Ind., 846 (1965).
52. Jardine, F.H., Osborn, J.A., Wilkinson, G., Young, J.F.: Chem. Ind., 560 (1965).
53. Cope, A.C., Ganellin, C.R., Johnson, Jr., H.W., Van Auken, T.V., Winkler, H.J.S.: J. Am. Chem. Soc. **85**, 3276 (1963).
54. Cope, A.C., Ganellin, C.R., Johnson, Jr. H.W.: J. Am. Chem. Soc. **84**, 3191 (1962).
55. Cope, A.C., Hecht, J.K., Johnson, Jr. H.W., Keller, H., Winkler, H.J.S.: J. Am. Chem. Soc. **88**, 761 (1966).
56. Paiaro, G.: Organometal. Chem. Rev. A, **6**, 319 (1970).
57. Paiaro, G., Corradini, P., Palumbo, P., Panunzi, A.: Makromol. Chem. **71**, 184 (1964).
58. Paiaro, G., Panunzi, A.: J. Am. Chem. Soc. **86**, 5148 (1964).
59. Corradini, P., Paiaro, G., Panunzi, A., Mason, S.F., Searle, G.H.: J. Am. Chem. Soc. **88**, 2863 (1966).
60. Panunzi, A., Paiaro, G.: J. Am. Chem. Soc. **88**, 4841 (1966).
61a. Horner, L., Winkler, H., Rapp, A., Mentrup, A., Hoffmann, H., Beck, P., Tetrahedron Lett., 161 (1961).
61b. Naumann, K., Zon, G., Mislow, K.: J. Am. Chem. Soc. **91**, 7012 (1969).
62a. Horner, L., Büthe, H., Suegel, H., Tetrahedron Lett., 4023 (1968).
62b. Knowles, W.S., Sabacky, M.J., Vineyard, B.D.: Chem. Tech., 590 (1972).
63. Knowles, W.S., Sabacky, M.J.: Chem. Commun., 1445 (1968).
64. Knowles, W.S., Sabacky, M.J., Vineyard, B.D.: Chem. Commun., 10 (1972).
65. Morrison, J.D., Burnett, R.E., Aguiar, A.M., Morrow, C.J., Phillips, C.: J. Am. Chem. Soc. **93**, 1301 (1971).
66. Dang, T.P., Kagan, H.B.: Chem. Commun., 481 (1971).
67. Hallman, P.S., McGarvey, B.R., Wilkinson, G.: J. Chem. Soc. A, 3143 (1968).
68. Hallman, P.S., Evans, D., Osborn, J.A., Wilkinson, G.: Chem. Commun., 305 (1967).
69. Evans, D., Osborn, J.A., Jardine, F.H., Wilkinson, G.: Nature **208**, 1203 (1965).
70. Fahey, D.R.: J. Org. Chem. **35**, 80 (1973).
71. Fahey, D.R.: J. Org. Chem. **35**, 3343 (1973).
72. Nakamura, A., Otsuka, S.: Tetrahedron Lett., 4529 (1973).
73. Baird, M.C., Nyman, C.J., Wilkinson, G.: J. Chem. Soc. A, 348 (1968).
74. Tsuji, J., Ohno, K.: J. Am. Chem. Soc. **90**, 94 (1968).
75. Ohno, K., Tsuji, J.: J. Am. Chem. Soc. **90**, 99 (1968).
76. Harvie, I., Kemmitt, R.D.W.: Chem. Commun., 198 (1970).
77. Tripathy, P.B., Roundhill, D.M.: J. Am. Chem. Soc. **92**, 3825 (1970).
78. Collman, J.P., Kang, J.W.: J. Am. Chem. Soc. **89**, 844 (1967).
79. Roundhill, D.M., Jonassen, H.B.: Chem. Commun., 1233 (1968).
80. Nelson, J.H., Jonassen, H.B., Roundhill, D.M.: Inorg. Chem. **8**, 2591 (1969).
81. Chatt, J., Davidson, J.M.: J. Chem. Soc., 843 (1965).
82. Parshall, G.W.: Accounts Chem. Res. **3**, 139 (1970).
83. Bennett, M.A., Milner, D.L.: Chem. Commun., 581 (1967).
84. Bennett, M.A., Milner, D.L.: J. Am. Chem. Soc. **91**, 6983 (1969).
85. Hata, G., Kondo, H., Miyake, A.: J. Am. Chem. Soc. **90**, 2278 (1968).
86. Fenton, D.M.: J. Org. Chem. **38**, 3192 (1973).
87. Levison, J.J., Robinson, S.D.: J. Chem. Soc. A, 639 (1970).
88. Parshall, G.W., Knoth, W.H., Schunn, R.A.: J. Am. Chem. Soc. **91**, 4990 (1969).
89. Keim, W.: J. Organometal. Chem. **14**, 179 (1968), **19**, 161 (1969).
90. Bradford, C.W., Nyholm, R.S.: J. Chem. Soc. Dalton, 529 (1973).

91. Cope, A.C., Friedrich, E.C.: J. Am. Chem. Soc. **90**, 909 (1968).
92. Bruce, M.I., Goodall, B.L., Stone, F.G.A.: Chem. Commun., 529 (1973).
93. Alper, H., Chan, A.S.K.: J. Am. Chem. Soc. **95**, 4905 (1973).
94. Saito, T., Uchida, Y., Misono, A., Yamamoto, A., Morifuji, K., Ikeda, S.: J. Organometal. Chem. **6**, 572 (1966).
95. Vaska, L., DiLuzio, J.W.: J. Am. Chem. Soc. **83**, 2784 (1961).
96. Deeming, A.J., Shaw, B.L.: J. Chem. Soc. A, 1887 (1968).
97. Sacco, A., Ugo, R., Moles, A.: J. Chem. Soc. A, 1670 (1966).
98. Jonas, K., Wilke, G.: Angew. Chem. **81**, 534 (1969).
99. Deeming, A.J., Shaw, B.L.: J. Chem. Soc. A, 1802 (1969).
100. Cariati, F., Ugo, R., Bonati, F.: Inorg. Chem. **5**, 1128 (1966).
101. Singer, H., Wilkinson, G.: J. Chem. Soc. A, 2516 (1968).
102. Roundhill, D.M.: Inorg. Chem. **9**, 254 (1970).
103. Roundhill, D.M.: Chem. Commun., 567 (1969).
104. Chatt, J., Eaborn, C., Kapoor, P.N.: J. Chem. Soc. A, 881 (1970).
105. Haszeldine, R.N., Parish, R.V., Parry, J.D.: J. Chem. Soc. A, 683 (1969).
106. Bennett, M.A., Robertson, G.R., Whimp, P.O., Yoshida, T.: J. Am. Chem. Soc. **95**, 3028 (1973).
107. Appleton, T.G., Bennett, M.A.: J. Organometal. Chem. **55**, C88 (1973).
108. Hüttel, R., Christ, H.: Chem. Ber. **96**, 3101 (1963).
109. Morelli, D., Ugo, R., Conti, F., Donat, M.: Chem. Commun., 801 (1967).
110. Volger, H.C.: Rec. Trav. Chim. **88**, 225 (1969).
111. Ketty, A.D., Braatz, J.: Chem. Commun., 169 (1969).
112. Tsuji, J., Imamura, S., Kiji, J.: J. Am. Chem. Soc. **86**, 4491 (1964).
113. Tsuji, J., Imamura, S.: Bull. Chem. Soc. Japan **40**, 197 (1967).
114. Trost, B.M., Fullerton, T.J.: J. Am. Chem. Soc. **95**, 292 (1973).
115. Schott, H., Wilke, G.: Angew. Chem. **81**, 896 (1969).
116. Schott, H., Schott, A., Wilke, G., Brandt, J., Hoberg, H., Hoffmann, G.: Ann. Chem., 508 (1973).
117. Kolomnikov, I.S., Svoboda, P., Vol'pin, M.E.: Izv. Akad. Nauk, ser. Khim., 1972, 2818. Chem. Abstr. **78**, 97789k (1973).
118. Evans, J.A., Everitt, G.F., Kemmitt, R.D.W., Russell, D.R.: Chem. Commun., 158 (1973).
119. Argento, B.J., Fitton, P., McKeon, J.E., Rick, E.A.: Chem. Commun., 1427 (1969).
120. Gerlach, D.H., Kane, A.R., Parshall, G.W., Jesson, J.P., Muetterties, E.L.: J. Am. Chem. Soc. **93**, 3543 (1971).
121. Burmeister, J.L., Edwards, L.M.: J. Chem. Soc. A, 1663 (1971).
122. Beck, W., Schorpp, K., Oetker, C., Schlodder, R., Smedal, H.S.: Chem. Ber. **106**, 2144 (1973).
123. Traverso, O., Carassiti, V., Graziani, M., Belluco, U.: J. Organometal. Chem. **57**, C22 (1973).
124. Guvogny, T., Larcherveque, M., Normant, H.: Bull. soc. chim. France, 1174 (1973).
125. Van Tamelen, E.E., Rudler, H., Bjorklund, C.: J. Am. Chem. Soc. **93**, 7113 (1971).
126. Cassar, L., Eaton, P.E., Halpern, J.: J. Am. Chem. Soc. **92**, 3515 (1970).
127. Wiberg, K.B., Bishop III, K.C.: Tetrahedron Lett., 2727 (1973).
128. Tsuji, J., Ohno, K.: J. Am. Chem. Soc. **88**, 3452 (1966).
129. Baird, M.C., Mague, J.T., Osborn, J.A., Wilkinson, G.: J. Chem. Soc. A, 1347 (1967).
130. Otsuka, S., Nakamura, A., Yoshida, T.: J. Am. Chem. Soc. **91**, 7196 (1969).
131. Angelici, R.J.: Accounts Chem. Res. **5**, 335 (1972).
132. Deeming, A.J., Shaw, B.L.: J. Chem. Soc. A, 443 (1969).
133. Fitton, P., Johnson, M.P., McKeon, J.E.: Chem. Commun., 6 (1968).
134. Otsuka, S., Nakamura, A., Yoshida, T., Naruto, M., Ataka, K.: J. Am. Chem. Soc. **95**, 3180 (1973).
135. Commereuc, D., Douek, I., Wilkinson, G.: J. Chem. Soc. A, 1771 (1970).
136. Dent, S.P., Eaborn, C., Pidcock, A.: Chem. Commun., 1703 (1970).
137. Citron, J.D.: J. Org. Chem. **34**, 1977 (1969).
138. Tsuji, J., Ohno, K., Kajimoto, T.: Tetrahedron Lett., 4565 (1965).
139. Chock, P.B., Halpern, J.: J. Am. Chem. Soc. **88**, 3511 (1966).
140. Kudo, K., Sato, M., Hidai, M., Uchida, Y.: Bull. Chem. Soc. Japan **46**, 2820 (1973).
141. Collman, J.P., Murphey, D.W., Dolcetti, G.: J. Am. Chem. Soc. **95**, 2687 (1973).
142. Bland, W.J., Kemmitt, R.D.: J. Chem. Soc. A, 1278 (1968).
143. Fahey, D.R.: J. Am. Chem. Soc. **92**, 402 (1970).

144. Fahey, D.R.: Organometal. Chem. Rev. **7**, 245 (1972).
145. Semmelhack, M.F., Helquist, P.M., Gorzynski, J.D.: J. Am. Chem. Soc. **94**, 9234 (1972).
146. Semmelhack, M.F., Helquist, P.M., Jones, L.D.: J. Am. Chem. Soc. **93**, 5908 (1971).
147. Semmelhack, M.P., Stauffer, R.D., Rogerson, T.D.: Tetrahedron Lett., 4519 (1973).
148. Ryang, M.: Organometal. Chem. Rev. A, **5**, 67 (1970).
149. Ryang, M., Tsutsumi, S.: Synthesis, 55 (1971).
150. Baker, R.: Chem. Rev. **73**, 487 (1973).
151. Heimbach, P., Jolly, P.W., Wilke, G.: Adv. Organometal. Chem. **8**, 29 (1970).
152. Buchholz, H., Heimbach, P., Hey, H.J., Selbeck, H., Wiese, W.: Coordination Chem. Rev. **8**, 129 (1972).
153. Semmelhack, M.F.: Org. Reactions **19**, 117 (1972).
154. Corey, E.J., Hegedus, L.F., Semmelhack, M.F.: J. Am. Chem. Soc. **90**, 2417 (1968).
155. Chiusoli, G.P.: Angew. Chem., Intern. Ed. **6**, 124 (1967).
156. Tsuji, J., Imamura, S., Kiji, J.: J. Am. Chem. Soc. **86**, 4350 (1964).
157. Dent, W.T., Long, R., Whitfield, G.H.: J. Chem. Soc. 1588 (1964).
158. Corey, E.J., Wat, E.K.W.: J. Am. Chem. Soc. **89**, 2757 (1967).
159. Corey, E.J., Hamanaka, E.: J. Am. Chem. Soc. **89**, 2758 (1967).
160. Corey, E.J., Broger, E.A.: Tetrahedron Lett., 1779 (1969).
161. Corey, E.J., Semmelhack, M.F., Hegedus, L.S.: J. Am. Chem. Soc. **90**, 2416 (1968).
162. Chiusoli, G.P.: 23rd Intern. Cong. Pure Appl. Chem. **6**, 169 (1971).
163. Sato, K., Inoue, S., Ota, S., Fujita, Y.: J. Org. Chem. **37**, 462 (1972).
164. Corey, E.J., Semmelhack, M.F.: J. Am. Chem. Soc. **89**, 2755 (1967).
165. Hodgson, G.L., MacSweeney, D.F., Mills, R.W., Money, T.: Chem. Commun; 235 (1973).
166. Sato, K., Inoue, S., Saito, K.: Chem. Commun., 953 (1972) and J. Chem. Soc. Perkin I. 2289 (1973).
167. Dubini, M., Montino, F.: J. Organometal. Chem. **6**, 188 (1966).
168. Hegedus, L.S., Waterman, E.L., Catlin, J.: J. Am. Chem. Soc. **94**, 7155 (1972).
169. Fischer, E.O., Burger, G.: Z. Naturforsch. **16 b**, 702 (1961).
170. Agnes, G., Chiusoli, G.P., Marraccini, A.: J. Organometal. Chem. **49**, 239 (1973).
171. Bauld, N.C.: Tetrahedron, Lett., 1841 (1963).
172. Rhee, I., Mizuta, N., Ryang, M., Tsutsumi, S.: Bull. Chem. Soc. Japan **41**, 1417 (1968).
173. Yoshisato, E., Tsutsumi, S.: J. Am. Chem. Soc. **90**, 4488 (1968).
174. Yoshisato, E., Tsutsumi, S.: Chem. Commun., 33 (1968).
175. Miller, A., Durand, M.H., Dubois, J.E.: Tetrahedron Lett., 2831 (1965).
176. Dubois, J.E., Itzkowitch, J.: Tetrahedron Lett., 2839 (1965).
177. Spencer, T.A., Britton, R.W., Watt, D.S.: J. Am. Chem. Soc. **89**, 5727 (1967).
178. Alper, H., Keung, E.C.H.: J. Org. Chem. **37**, 2566 (1972).
179. Noyori, R., Makino, S., Takaya, H.: J. Am. Chem. Soc. **93**, 1272 (1971).
180. Noyori, R., Hayakawa, Y., Makino, S., Takaya, H.: Chem. Lett., 3 (1973).
181. Noyori, R., Baba, Y., Makino, S., Takaya, H.: Tetrahedron Lett., 1741 (1973).
182. Noyori, R., Makino, S., Takaya, H.: Tetrahedron Lett., 1745 (1973).
183. Noyori, R., Yokoyama, K., Makino, S., Hayakawa, Y.: J. Am. Chem. Soc. **94**, 1772 (1972).
184. Noyori, R., Yokoyama, K., Hayakawa, Y.: J. Am. Chem. Soc. **95**, 2722 (1973).
185. Noyori, R., Hayakawa, Y., Makino, S., Takaya, H.: J. Am. Chem. Soc. **95**, 4103 (1973).
186. Noyori, R., Baba, Y., Hayakawa, Y.: J. Am. Chem. Soc. **96**, 3336 (1974).
187. Noyori, R., Hayakawa, Y., Funakura, M., Takaya, H., Murai, S., Kobayashi, R., Tsutsumi, S.: J. Am. Chem. Soc. **94**, 7202 (1972).
188. Zimmerman, H.E., Crumrine, D.S., Dopp, D., Huyffer, P.S.: J. Am. Chem. Soc. **91**, 434 (1969).
189. Hoffman, H.M.R., Clemens, K.E., Schmidt, E.A., Smithers, R.H.: J. Am. Chem. Soc. **94**, 3940 (1972).
190. Ghera, E., Perry, D.H., Shoua, S.: Chem. Commun., 858 (1973).
191. Coffey, C.E.: J. Am. Chem. Soc. **83**, 1623 (1961).
192. Emerson, G.F., Watts, L., Pettit, R.: J. Am. Chem. Soc. **87**, 131 (1965).
193. Fitzpatrick, J.D., Watts, L., Emerson, G.F., Pettit, R.: J. Am. Chem. Soc. **87**, 3255 (1965).
194. Emerson, G.F., Pettit, R.: J. Am. Chem. Soc. **84**, 4591 (1962).
195. Fitzpatrick, J.D., Watts, L., Pettit, R.: Tetrahedron Lett., 1299 (1966).
196. Reeves, P., Henery, J., Pettit, R.: J. Am. Chem. Soc. **91**, 5888 (1969).

197. Grubbs, R. H., Grey, R. A.: J. Am. Chem. Soc. **95**, 5765 (1973).
198. Schmidt, E. K. G.: Angew. Chem., Intern. Ed. **12**, 777 (1973).
199. Watts, L., Fritzpatrick, J. D., Pettit, R.: J. Am. Chem. Soc. **87**, 3253 (1965).
200. Burt, G. D., Pettit, R.: Chem. Commun., 517 (1965).
201. McKenneis, J. C., Brenner, L., Ward, J. S., Pettit, R.: J. Am. Chem. Soc. **93**, 4957 (1971).
202. Barborak, J. C., Watts, L., Pettit, R.: J. Am. Chem. Soc. **88**, 1328 (1966).
203. Ward, J. S., Pettit, R.: J. Am. Chem. Soc. **93**, 262 (1971).
204. Paquette, L. A., Epstein, M. J.: J. Am. Chem. Soc. **95**, 6717 (1973).
205. Peterson, J. L., Nappier, Jr. T. E., Meek, D. W.: J. Am. Chem. Soc. **95**, 8195 (1973).
206. Strope, D., Shriver, D. F.: J. Am. Chem. Soc. **95**, 8197 (1973).
207. Deane, M., Lalor, F. J.: J. Organometal. Chem. **57**, C61 (1973).
208. Vaska, L.: Science **140**, 809 (1963).
209. Sohn, Y. S., Balch, A. L.: J. Am. Chem. Soc. **93**, 1290 (1971).
210. Cenini, S., Ugo, R., La Monica, G.: J. Chem. Soc. A, 416 (1971).
211. Valentine, J. S., Valentine, D.: J. Am. Chem. Soc. **92**, 5795 (1970).
212. Lam, C. T., Senoff, C. V.: J. Organometal. Chem. **57**, 207 (1973).
213. Otsuka, S., Nakamura, A., Tatsuno, T., Miki, M.: J. Am. Chem. Soc. **94**, 3761 (1972).
214. Castro, C. E., Kray, Jr. W. C.: J. Am. Chem. Soc. **85**, 2768 (1963).
215. Kochi, J. K., Davis, D. D.: J. Am. Chem. Soc. **86**, 5264 (1964).
216. Bradley, J. S., Conner, D. E., Dolphin, D., Labinger, J. A., Osborn, J. A.: J. Am. Chem. Soc. **94**, 4044 (1972).
217. Labinger, J. A., Kramer, A. V., Osborn, J. A.: J. Am. Chem. Soc. **95**, 7908 (1973).
218. Lappert, M. F., Lednor, P. W.: Chem. Commun., 948 (1973).
219. Hargreaves, N. G., Puddephatt, R. J., Sutcliffe, L. H., Thompson, R. J.: Chem. Commun., 861 (1973).
220. Nelson, S. J., Detre, G., Tanabe, M.: Tetrahedron Lett., 447 (1973).
221. Hopgood, D., Jenkins, R. A.: J. Am. Chem. Soc. **95**, 4462 (1973).
222. Nesmeyanov, A. N., Chukovskaya, E. T., Kamyshova, A. A., Freidlina, R. Kh.: Tetrahedron **17**, 61 (1962).
223. Chukovskaya, E. C., Kamyshova, A. A., Freidlina, R. Kh.: Dokl. Akad. Nauk SSSR **164**, 602 (1965) and the references cited therein.
224. Susuki, T., Tsuji, J.: J. Org. Chem. **35**, 2982 (1970).
225. Susuki, T., Tsuji, J.: Tetrahedron Lett., 913 (1968).
226. Asahara, T., Seno, M., Wu, C. C.: Bull. Chem. Soc. Japan **43**, 1127 (1970).
227. Asscher, M., Vofsi, D.: Chem. Ind., 209 (1962).
228. Asscher, M., Vofsi, D.: J. Chem. Soc., 1887 and 3921 (1963).
229. Freidlina, R. Kh., Belyavski, A. B.: Dokl. Akad. Nauk SSSR **127**, 1027 (1959).
230. Freidlina, R. Kh., Chukovskaya, Englin, B. A.: Dokl. Akad. Nauk SSSR **159**, 1346 (1964).
231. Mori, Y., Tsuji, J.: Tetrahedron **28**, 29 (1972).
232. Freidlina, R. Kh., Chukovskaya, E. T., Terentev, A. B.: Izv. Akad. Nauk SSSR, 2474 (1967).
233. Bamford, C. H., Eastmond, G. C., Whittle, D.: J. Organometal. Chem. **17**, P33 (1969), and the references cited therein.
234. Mori, Y., Tsuji, J.: Tetrahedron **27**, 3811 (1971).
235. Mori, Y., Tsuji, J.: Tetrahedron **27**, 4039 (1971).
236. Cross, R. J.: Organometal. Chem. Rev. **2**, 97 (1967).
237. Yamamoto, A., Morifuji, K., Ikeda, S., Saito, T., Uchida, Y., Misono, A.: J. Am. Chem. Soc. **87**, 4652 (1965).
238. Saito, T., Uchida, Y., Misono, A., Yamamoto, A., Morifuji, K., Ikeda, S.: J. Am. Chem. Soc. **88**, 5198 (1966).
239. Yamamoto, A., Morifuji, K., Ikeda, S., Saito, T., Uchida, Y., Misono, A.: J. Am. Chem. Soc. **90**, 1878 (1968).
240. Yagupsky, G., Mowat, W., Shortland, A., Wilkinson, G.: Chem. Commun., 1369 (1970).
241. Cross, R. J., Wardle, R.: J. Chem. Soc. A, 840 (1970).
242. Tamura, M., Kochi, J.: J. Am. Chem. Soc. **93**, 1483 (1971).
243. Tamura, M., Kochi, J.: Synthesis, 303 (1971).
244. Chuit, C., Felkin, H., Frajerman, C., Roussi, G., Swierczewski, G.: Chem. Commun., 1604 (1968).
245. Felkin, H., Swierczewski, G.: Tetrahedron Lett., 1433 (1972).

180 References

246. Corriu, R. J., Masse, J. P.: Chem. Commun., 144 (1972).
247. Tamao, K., Sumitani, K., Kumada, M.: J. Am. Chem. Soc. **94**, 4374 (1972).
248. Tamao, K., Kiso, Y., Sumitani, K., Kumada, M.: J. Am. Chem. Soc. **94**, 9268 (1972).
249. Uchino, M., Yamamoto, A., Ikeda, S.: J. Organometal. Chem. **24**, C63 (1970).
250. Chatt, J., Shaw, B. L.: J. Chem. Soc., 1718 (1960).
251. Yamazaki, H., Nishido, T., Matsumoto, Y., Sumida, S., Hagihara, N.: J. Organometal. Chem. **6**, 86 (1966).
252. Tamao, K., Zembayashi, M., Kiso, Y., Kumada, M.: J. Organometal. Chem. **55**, C91 (1973).
253. Hegedus, L. S., Lo, S. M., Bloss, D. E.: J. Am. Chem. Soc. **95**, 3040 (1973).
254. Semmelhack, M. F., Ryono, L.: Tetrahedron Lett., 2967 (1973).
255. Bönnemann, H.: Angew. Chem. Intern. Ed. **9**, 736 (1970).
256. Heck, R. F.: J. Am. Chem. Soc. **90**, 5518 (1968).
257. Heck, R. F.: J. Am. Chem. Soc. **90**, 5526 (1968).
258. Heck, R. F.: J. Am. Chem. Soc. **90**, 5531 (1968).
259. Heck, R. F.: J. Am. Chem. Soc. **90**, 5535 (1968).
260. Heck, R. F.: J. Am. Chem. Soc. **90**, 5538 (1968).
261. Heck, R. F.: J. Am. Chem. Soc. **90**, 5542 (1968).
262. Heck, R. F.: J. Am. Chem. Soc. **90**, 5546 (1968).
263. Heck, R. F.: J. Am. Chem. Soc. **91**, 6707 (1969).
264. Henry, P. M.: Tetrahedron Lett., 2285 (1968).
265. Seyferth, D., Spohn, R. J.: J. Am. Chem. Soc. **90**, 540 (1968).
266. Seyferth, D., Spohn, R. J.: J. Am. Chem. Soc. **91**, 3037, 6192 (1969).
267. Espenson, J. H., Shveima, J. S.: J. Am. Chem. Soc. **95**, 4468 (1973), and references cited therein.
268. Seebach, D.: Angew. Chem. Intern. Ed. **8**, 639 (1969).
269. Trzupek, L. S., Newirth, T. L., Kelly, E. G., Sbarbati, N. E., Whitesides, G. M.: J. Am. Chem. Soc. **95**, 8118 (1973).
270. Myeong, S. K., Sawa, Y., Ryang, M., Tsutsumi, S.: Bull. Chem. Soc. Japan **38**, 330 (1965).
271. Fischer, E. O., Maasböl, A.: Chem. Ber. **100**, 2445 (1967).
272. Ryang, M., Song, K. M., Sawa, Y., Tsutsumi, S.: J. Organometal. Chem. **5**, 305 (1966).
273. Sawa, Y., Ryang, M., Tsutsumi, S.: Tetrahedron Lett., 5189 (1969).
274. Corey, E. J., Hegedus, L. S.: J. Am. Chem. Soc. **91**, 4926 (1969).
275. Sawa, Y., Hashimoto, I., Ryang, M., Tsutsumi, S.: J. Org. Chem. **33**, 2159 (1968).
276. Fukuoka, S., Ryang, M., Tsutsumi, S.: J. Org. Chem. **33**, 2973 (1968).
277. Fukuoka, S., Ryang, M., Tsutsumi, S.: J. Org. Chem. **36**, 2721 (1971).
278. Fischer, E. O., Beck, H. J., Kreiter, C. G., Lynch, J., Muller, J., Winkler, E.: Chem. Ber. **105**, 162 (1972).
279. Fischer, E. O., Kiener, V.: J. Organometal. Chem. **23**, 215 (1970).
280. King, R. B.: Accounts Chem. Res. **3**, 417 (1970).
281. King, R. B.: Advan. Organometal. Chem. **2**, 157 (1964),
282. Pankowsky, M., Bigorgne, M.: J. Organometal. Chem. **30**, 227 (1971).
283. Corey, E. J., Hegedus, L. S.: J. Am. Chem. Soc. **91**, 1233 (1969).
284. Corey, E. J., Kirst, H. A., Katzenellenbogen, J. A.: J. Am. Chem. Soc. **92**, 6314 (1970).
285. Crandall, J. K., Michaely, W. J.: J. Organometal. Chem. **51**, 375 (1973).
286. Blanchard, A. W., Coleman, G. W.: Inorg. Synthesis **2**, 243 (1946). McGraw-Hill.
287. Takegami, Y., Watanabe, Y., Masada, H., Kanaya, I.: Bull. Chem. Soc. Japan **40**, 1456 (1967).
288. Cooke, M. P.: J. Am. Chem. Soc. **92**, 6080 (1970).
289. Collman, J. P., Cawse, J. N., Brauman, J. I.: J. Am. Chem. Soc. **94**, 5905 (1972).
290. Watanabe, Y., Mitsudo, T., Tanaka, M., Yamamoto, K., Okajima, T., Takegami, Y.: Bull. Chem. Soc. Japan **44**, 2569 (1971).
291. Watanabe, Y., Yamashita, M., Mitsudo, T., Tanaka, M., Takegami, Y.: Tetrahedron Lett., 3535 (1973).
292. Wakamatsu, H., Furukawa, J., Yamakami, N.: Bull. Chem. Soc. Japan **44**, 288 (1971).
293. Masada, H., Mitsudo, M., Suga, S., Watanabe, Y., Takegami, Y.: Bull. Chem. Soc. Japan **43**, 3824 (1970).
294. Watanabe, Y., Mitsudo, T., Yamashita, M., Tanaka, M., Takegami, Y.: Chem. Lett., 475 (1973).
295. Mitsudo, T., Watanabe, Y., Tanaka, M., Yamamoto, K., Takegami, Y.: Bull. Chem. Soc. Japan, **45**, 305 (1972).

296. Collman, J. P., Winter, S. R., Komoto, R. G.: J. Am. Chem. Soc. **95**, 249 (1973).
297. Collman, J. P., Winter, S. R., Clark, D. R.: J. Am. Chem. Soc. **94**, 1788 (1972).
298. Collman, J. P., Hoffman, N. W.: J. Am. Chem. Soc. **95**, 2689 (1973).
299. Takegami, Y., Watanabe, Y., Mitsudo, T., Masada, H.: Bull. Chem. Soc. Japan **42**, 202 (1969).
300. Masada, H., Mizuno, M., Suga, S., Watanabe, Y., Takegami, Y.: Bull. Chem. Soc. Japan **43**, 3824 (1970).
301. Noyori, R., Umeda, I., Ishigami, T.: J. Org. Chem. **37**, 1542 (1972).
302. Merour, Y. Y., Roustan, J. L., Charrier, C., Collin, J.: J. Organometal. Chem. **51**, C24 (1973).
303. Collman, J. P., Komoto, R. G., Siegel, W. O.: J. Am. Chem. Soc. **96**, 2390 (1973).
304. Burgess, W. H., Eastes, J. W.: Inorg. Synthesis **5**, 197 (1957).
305. Jarchow, O., Schultz, H., Nast, R.: Angew. Chem. **82**, 43 (1970).
306. Hashimoto, I., Ryang, M., Tsutsumi, S.: Tetrahedron Lett., 3291 (1969).
307. Hashimoto, I., Tsuruta, N., Ryang, M., Tsutsumi, S.: J. Org. Chem. **35**, 3748 (1970).
308. Hashimoto, I.,Ryang, M., Tsutsumi, S.: Tetrahedron Lett., 4567 (1970).
309. Fischer, E. O.: Pure Appl. Chem. **30**, 353 (1972).
310. Fischer, E. O.: Pure Appl. Chem. **24**, 407 (1970).
311. Cardin, D. J., Cetinkaya, B., Lappert, M. F.: Chem. Rev. **72**, 545 (1972).
312. Cardin, D. J., Cetinkaya, B., Lappert, M. F., Doyle, M. J.: Chem. Soc. Rev. **2**, 99 (1973).
313. Cotton, F. A., Lukehart, C. M.: Progr. Inorg. Chem. **16**, 487 (1972).
314. Fischer, E. O., Maasböl, A.: Angew. Chem. **76**, 645 (1964).
315. Fischer, E. O., Kreis, G., Kreissl, F. R.: J. Organometal. Chem. **56**, C37 (1973).
316. Fischer, E. O., Dötz, K. H.: J. Organometal. Chem. **36**, C4 (1972).
317. Fischer, E. O., Heckl, B., Dötz, K. H., Müller, J., Werner, H.: J. Organometal. Chem. **16**, P29 (1969).
318. Fischer, E. O., Dötz, K. H.: Chem. Ber. **103**, 1273 (1970).
319. Dötz, K. H., Fischer, E. O.: Chem. Ber. **105**, 1356 (1972).
320. Cooke, M. D., Fischer, E. O.: J. Organometal. Chem. **56**, 279 (1973).
321. Fischer, E. O., Weiss, K., Burger, K.: Chem. Ber. **106**, 1581 (1973).
322. Aumann, R., Fischer, E. O.: Chem. Ber. **101**, 954 (1968).
323. Fischer, E. O., Aumann, R.: Chem. Ber. **101**, 963 (1968).
324. Fischer, E. O., Maasböl, A.: J. Organometal. Chem. **12**, P15 (1968).
325. Weiss, K., Fischer, E. O.: Chem. Ber. **106**, 1277 (1973).
326. Casey, C. P., Burkhardt, T. J.: J. Am. Chem. Soc. **95**, 5833 (1973).
327. Casey, C. P., Burkhardt, T. J.: J. Am. Chem. Soc. **94**, 6543 (1972).
328. Casey, C. P., Berts, S. H., Burkhardt, T. J.: Tetrahedron Lett., 1421 (1973).
329. Casey, C. P., Boggs, R. A., Anderson, R. L.: J. Am. Chem. Soc. **94**, 8947 (1972).
330. Werner, H., Fischer, E. O., Heckl, B., Kreiter, C. G.: J. Organometal. Chem. **28**, 367 (1971).
331. Rees, C. W., von Angerer, E.: Chem. Commun., 420 (1972).
332. Kreissl, F. R., Fischer, E. O., Kreiter, C. G., Weiss, K.: Angew. Chem., Intern. Ed. **12**, 563 (1973).
333. Fischer, E. O., Kreis, G., Kreiter, C. G., Müller, J., Huttner, G., Lorenz, H.: Angew. Chem. Intern. Ed. **12**, 564 (1973).
334. Fischer, E. O., Kreis, G., Kreissl, F. R., Kolbfus, W., Winkler, E.: J. Organometal. Chem. **65**, C53 (1974).
335. Chiusoli, G. P.: Pure Appl. Chem., 23rd Intern. Congr. **6**, 169 (1971).
336. Tsuji, J., Morikawa, M., Takahashi, H.: Tetrahedron Lett., 4387 (1965).
337. Takahashi, H., Morikawa, M., Tsuji, J.: Kogyo Kagaku Zasshi **69**, 920 (1966).
338. Dubini, M., Montino, F., Chiusoli, G. P.: Chim. Ind. (Milan) **49**, 839 (1965).
339. Corey, E. J., Semmelhack, M. F.: J. Am. Chem. Soc. **89**, 2755 (1967).
340. Heimbach, P., Jolly, P. W., Wilke, G.: Adv. Organometal. Chem. **8**, 29 (1970).
341. Baker, R., Blackett, B. N., Cookson, R. C., Cross, R. C., Madden, D. P.: Chem. Commun., 343 (1972).
342. Tsuji, J.: Accounts Chem. Res. **6**, 8 (1973).
343. Takahashi, S., Shibano, T., Hagihara, N.: Tetrahedron Lett., 2451 (1967).
344. Smutny, E. J.: J. Am. Chem. Soc. **89**, 6793 (1967).
345. Takahashi, S., Yamazaki, H., Hagihara, N.: Bull. Chem. Soc. Japan **41**, 254 (1968).
346. Shields, T. C., Walker, W. E.: Chem. Commun., 193 (1971).
347. Heimbach, P.: Angew. Chem., Intern. Ed. **7**, 882 (1968).

348. Baker, R., Halliday, D.E., Smith, T.N.: Chem. Commun., 1583 (1971).
349. Baker, R., Halliday, D.E., Smith, T.N.: J. Organometal. Chem. **35**, C61 (1972).
350. Atkins, K.E., Walker, W.E., Manyik, R.M.: Chem. Commun., 330 (1971).
351. Walker, W.E., Manyik, R.M., Atkins, K.E., Farmer, M.L.: Tetrahedron Lett., 3817 (1970).
352. Mitsuyasu, T., Hara, M., Tsuji, J.: Chem. Commun., 345 (1971).
353. Ohno, K., Mitsuyasu, T., Tsuji, J.: Tetrahedron Lett., 67 (1971).
354. Ohno, K., Mitsuyasu, T., Tsuji, J.: Tetrahedron **28**, 3705 (1972).
355. Haynes, P.: Tetrahedron Lett., 3687 (1970).
356. Manyik, R.M., Walker, W.E., Atkins, K.E., Hammack, E.S.: Tetrahedron Lett., 3813 (1970).
357. Shier, G.D.: J. Organometal. Chem. **10**, P15 (1967).
358. Coulson, D.R.: J. Org. Chem. **38**, 1483 (1973).
359. Baker, R., Cook, A.H.: Chem. Commun., 472 (1973).
360. Pettit, R., Emerson, G.F.: Adv. Organometal. Chem. **1**, 1 (1964).
361. Whitesides, T.H., Arhart, R.W., Slaven, R.W.: J. Am. Chem. Soc. **95**, 5792 (1973).
362. Birch, A.J., Chamberlain, K.B., Haas, M.A., Thompson, D.J.: J. Chem. Soc., Perkin I, 1882 (1973).
363. Birch, A.J., Chamberlain, K.B., Thompson, D.J.: J. Chem. Soc. Perkin I, 1900 (1973).
364. Noyori, R., Suzuki, T., Kumagai, Y., Takaya, H.: J. Am. Chem. Soc. **93**, 5894 (1971).
365. Gassman, P.G., Williams, F.J.: J. Am. Chem. Soc. **94**, 7733 (1972).
366. Gassman, P.G., Williams, F.J.: J. Am. Chem. Soc. **92**, 7631 (1970).
367. Gassman, P.G., Meyer, G., Williams, F.J.: J. Am. Chem. Soc. **94**, 7741 (1972).
368. Gassman, P.G., Atkins, T.J., Lumb, J.T.: J. Am. Chem. Soc. **94**, 7757 (1972).
369. Gassman, P.G., Atkins, T.J.: J. Am. Chem. Soc. **94**, 7748 (1972).
370. Noyori, R.: Tetrahedron Lett., 1691 (1973).
371. Gassman, P.G., Nakai, T.: J. Am. Chem. Soc. **93**, 5897 (1971).
372. Gassman, P.G., Nakai, T.: J. Am. Chem. Soc. **94**, 2877 (1972).
373. Hong, K., Sonogashira, K., Hagihara, N., Nippon Kagaku Zasshi **89**, 74 (1968).
374. Ulrich, H., Trucker, B., Sayigh, A.A.: Tetrahedron Lett., 1731 (1967).
375. Campbell, T.W., Monagle, J.J., Foldi, V.S.: J. Am. Chem. Soc. **84**, 3673 (1962).
376. Monagle, J.J., Campbell, T.W., McShane, Jr. H.F.: J. Am. Chem. Soc. **84**, 4288 (1962).
377. Lobeeva, T.S., Vol'pin, M.E., Kolomnikov, I.S., Koreshkov, Yu.D.: Chem. Commun., 1432 (1970).
378. Drapier, J., Hubert, A.J., Teyssie, Ph.: Chem. Commun., 484 (1972).
379. Semmelhack, M.F., Stauffer, R.D.: Tetrahedron Lett., 2667 (1973).
380. Daub, T., Trautz, V., Erhardt, U.: Tetrahedron Lett., 4435 (1972).
381. Corey, E.J., Winter, R.A.E.: J. Am. Chem. Soc. **85**, 2677 (1963).
382. Wilkinson, G.: Pure Appl. Chem. **30**, 627 (1972).
383. Braterman, P.S., Cross, R.J.: J. Chem. Soc. Dalton 657 (1972).
384. Schrauzer, G.N.: Accounts Chem. Res. **1**, 97 (1968).
385. Penfold, B.R., Robinson, B.H.: Accounts Chem. Res. **6**, 73 (1973).
386. Seyferth, D., Hallgren, J.E., Spohn, R.J., Wehman, A.T., Williams, G.H.: Pure Appl. Chem., 23rd Intern. Congress **6**, 133 (1971).
387. Sakamoto, N., Kitamura, T., Joh, T.: Chem. Lett., 583 (1973).
388. Seyferth, D., Hallgren, J.E., Spohn, R.J.: J. Organometal. Chem. **23**, C55 (1970).
389. Seyferth, D., Wehman, A.T.: J. Am. Chem. Soc. **92**, 5520 (1970).
390. Hallgren, J.E., Eschbach, C.S., Seyferth, D.: J. Am. Chem. Soc. **94**, 2547 (1972).
391. Hieber, W., Wagner, C.: Z. Naturforsch. **12b**, 478 (1957).
392. Hieber, W., Wagner, C.: Z. Naturforsch. **13b**, 339 (1958).
393. Hieber, W., Lindner, E.: Chem. Ber. **94**, 1417 (1961).
394. Hieber, W., Duchatsh, H.: Chem. Ber. **98**, 2933 (1965).
395. Otsuka, S., Rossi, M.: J. Chem. Soc. A, 497 (1969).
396. Slaugh, L.H., Mullineaux, R.D.: J. Organometal. Chem. **13**, 469 (1968).
397. Kniese, W., Nienberg, H.J., Fischer, R.: J. Organometal. Chem. **17**, 133 (1969).
398. Tucci, E.R.: Ind. Eng. Chem. **7**, 32 (1968).
399. Tucci, E.R.: Ind. Eng. Chem. **8**, 286 (1969).
400. Imjanitov, N.S., Rudkovski, D.M.: J. Prak. Chem. **311**, 712 (1969).
401. Joh, T., Hagihara, N.: Nippon Kagaku Zasshi **91**, 378, 383 (1970).

402. Hieber, W., Wagner, G.: Ann. **618**, 24 (1958).
403. Kudo, K., Hidai, M., Uchida, Y.: J. Organometal. Chem. **56**, 413 (1973).
404. Lappert, M.F., Prokai, B.: Adv. Organometal. Chem. **5**, 225 (1967).
405. Wojcicki, A.: Adv. Organometal. Chem. **11**, 88 (1973).
406. Tsutsui, M., Hancock, M., Ariyoshi, J., Levy, M.N.: Angew. Chem. Intern. Ed. **8**, 410 (1969).
407. Iwamoto, M., Yuguchi, S.: Bull. Chem. Soc. Japan **41**, 150 (1968).
408. Iwamoto, M., Yuguchi, S.: J. Org. Chem. **31**, 4290 (1966).
409. Hata, G., Miyake, A.: Bull. Chem. Soc. Japan **41**, 2762 (1968).
410. Bogdanovic, B., Henc, B., Meister, B., Pauling, H., Wilke, G.: Angew. Chem. Intern. Ed. **11**, 1023 (1972).
411. Tsuji, J.: Accounts Chem. Res. **2**, 144 (1969).
412. Tsuji, J.: Adv. Org. Chem. **6**, 109. New York: Interscience Pub. 1969.
413. Hüttel, R.: Synthesis, 225 (1970).
414. Aguilo, A.: Adv. Organometal. Chem. **5**, 321 (1967).
415. Stern, E.W.: Cata. Rev. **1**, 73 (1967).
416. Jira, R., Freiesleben, W.: Organometallic Reactions **3**, 5 (1972).
417. Clement, W.H., Selwitz, C.M.: J. Org. Chem. **29**, 241 (1964).
418. Jira, R., Sedlmeier, J., Smidt, J.: Ann. **693**, 99 (1966).
419. Ariyaratne, J.K.P., Green, M.L.H.: J. Chem. Soc., 1 (1964).
420. Wakatsuki, Y., Nozakura, S., Murahashi, S.: Bull. Chem. Soc. Japan **42**, 273 (1969).
421. Tsutsui, M., Ori, M., Francis, J.: J. Am. Chem. Soc. **94**, 1414 (1972).
422. Thyret, H.: Angew. Chem., Intern. Ed. **11**, 520 (1972).
423. Hillis, J., Tsutsui, M.: J. Am. Chem. Soc. **95**, 7907 (1973).
424. Takahashi, H., Tsuji, J.: J. Am. Chem. Soc. **90**, 2387 (1968).
425. Moritani, I., Fujiwara, Y.: Synthesis, 524 (1973).
426. Chalk, A.J., Harrod, J.F.: J. Am. Chem. Soc. **87**, 16 (1965).
427. Yamamoto, K., Hayashi, T., Kumada, M.: J. Am. Chem. Soc. **93**, 5301 (1971).
428. Yamamoto, K., Uramoto, Y., Kumada, M.: J. Organometal. Chem. **31**, C9 (1971).
429. Kiso, Y., Yamamoto, K., Tamao, K., Kumada, M.: J. Am. Chem. Soc. **94**, 4373 (1972).
430. Takahashi, S., Shibano, T., Hagihara, N.: Chem. Commun., 161 (1969).
431. Hara, M., Ohno, K., Tsuji, J.: Chem. Commun., 247 (1971).
432. Cunico, R.F., Dexheimer, E.M.: Organometal. Syn. **1**, 253 (1971).
433. Falbe, J.: Synthesis with carbon monoxide, Berlin-Heidelberg-New York: Springer-Verlag, 1967.
434. Bird, C.W.: Chem. Rev. **62**, 283 (1962).
435. Thompson, D.T., Whyman, R.: in: Transition metals in homogeneous catalysis. New York: Marcel Dekker, 1971, 147.
436. Cassar, L., Chiusoli, G.P., Guerrieri, F.: Synthesis, 509 (1973).
437. Day, J.P., Basolo, F., Pearson, R.G.: J. Am. Chem. Soc. **90**, 6927 (1968).
438. Heck, R.F.: J. Am. Chem. Soc. **85**, 2013 (1963).
439. Bittler, K., Kutepow, N.V., Neubauer, D., Reis, H.: Angew. Chem. **80**, 352 (1968).
440. Heck, R.F.: J. Am. Chem. Soc. **93**, 6896 (1971).
441. Tsuji, J., Hosaka, S., Susuki, T., Kiji, J.: Bull. Chem. Soc., Japan **39**, 141 (1966).
442. Brewis, S., Hughes, P.R.: Chem. Commun., 6 (1966).
443. Iwashita, Y., Sakuraba, M.: J. Org. Chem. **36**, 3927 (1971).
444. Orchin, M., Rupilius, P.R.: Catalysis Rev. **6**, 85 (1972).
445. Noack, K., Calderazzo, F.: J. Organometal. Chem. **10**, 101 (1967), and references cited therein.
446. Wakamatsu, H., Uda, J., Yamakami, N.: Chem. Commun., 1540 (1971).
447. Medema, D., van Helden, R., Kohll, C.F.: Inorg. Chim. Acta **3**, 255 (1969).
448. Chiusoli, G.P., Bottaccio, G.: Chim. Ind. (Milan) **47**, 165 (1965).
449. Cassar, L., Chiusoli, G.P.: Chim. Ind. (Milan) **48**, 323 (1966).
450. Cassar, L., Foa, M., Chiusoli, G.P.: Organometal. Chem. Syn. **1**, 302 (1972).
451. Chiusoli, G.P., Merzoni, S.: Chem. Commun., 522 (1971).
452. Chiusoli, G.P., Commetti, G., Merzoni, S.: Organometal. Chem. Syn. **1**, 439 (1972).
453. Corey, E.J., Jautelat, M.: J. Am. Chem. Soc. **89**, 3912 (1967).
454. Cassar, L., Chiusoli, G.P.: Tetrahedron Lett., 2805 (1966).
455. Hosaka, S., Tsuji, J.: Tetrahedron **27**, 3821 (1971).
456. Tsuji, J., Hara, M., Mori, Y.: Tetrahedron **28**, 3721 (1972).

184 References

457. Billups, W. E., Walker, W. E., Shields, T. C.: Chem. Commun., 1067 (1971).
458. Roth, J. F., Craddock, J. H., Hershman, A., Paulik, F. E.: Chem. Tech., 347 (1971).
459. Tsuji, J., Morikawa, M., Iwamoto, N.: J. Am. Chem. Soc. **86**, 2095 (1964).
460. Chiusoli, G. P., Venturello, C., Merzoni, S.: Chem. Ind., 977 (1968).
461. Chiusoli, G. P.: Chim. Ind. (Milan) **41**, 513 (1959).
462. Rosenthal, R. W., Schwartzman, L. H.: J. Org. Chem. **24**, 836 (1959).
463. Rosenthal, R. W., Schwartzman, L. H., Greco, N. P., Proper, R.: J. Org. Chem. **28**, 2835 (1963).
464. Jones, E. R. H., Whitham, G. H., Whiting, M. G.: J. Chem. Soc., 4628 (1957).
465. Tsuji, J., Nogi, T.: Tetrahedron Lett., 1801 (1966).
466. Heck, R. F.: Adv. Organometal. Chem. **4**, 243 (1966).
467. Heck, R. F.: J. Am. Chem. Soc. **87**, 4727 (1965).
468. Heck, R. F.: J. Am. Chem. Soc. **85**, 3383 (1963).
469. Heck, R. F.: J. Am. Chem. Soc. **86**, 2819 (1964).
470. Takahashi, H., Tsuji, J.: J. Organometal. Chem. **10**, 511 (1967).
471. Murahashi, S., Horiie, S.: J. Am. Chem. Soc. **78**, 4816 (1956).
472. Horiie, S., Murahashi, S.: Bull. Chem. Soc. Japan **33**, 88 (1960).
473. Bennett, M. A., Yoshida, T.: J. Am. Chem. Soc. **95**, 3030 (1973).
474. Heck, R. F.: J. Am. Chem. Soc. **85**, 1460 (1963).
475. Yamamoto, Y., Yamazaki, H.: Coord. Chem. Rev. **8**, 225 (1972).
476. Yamamoto, Y., Yamazaki, H., Hagihara, N.: J. Organometal. Chem. **18**, 189 (1969).
477. Kajimoto, T., Tsuji, J.: J. Organometal. Chem. **23**, 275 (1970).
478. Boschi, T., Crociani, B.: Inorg. Chim. Acta **5**, 477 (1971).
479. Yamamoto, Y., Yamazaki, H.: Inorg. Chem. **11**, 211 (1972).
480. Wojcicki, A.: Accounts Chem. Res. **4**, 344 (1971).
481. Kitching, W., Fong, C. W.: Organometal. Chem. Rev. **5**, 281 (1970).
482. Jacobsen, S. T., Reich-Rohrwig, P., Wojcicki, A.: Inorg. Chem. **12**, 717 (1973).
483. Flood, T. C., Miles, D. L.: J. Am. Chem. Soc. **95**, 6460 (1973).
484. Klein, H. S.: Chem. Commun., 377 (1968).
485. O'Brien, S.: J. Chem. Soc. A, 9 (1970).
486. Calas, R.: Pure Appl. Chem. **13**, 61 (1966).
487. Frainnet, E.: Pure Appl. Chem. **19**, 489 (1969).
488. Ojima, I., Nihonyanagi, M., Nagai, Y.: Chem. Commun., 938 (1972).
489. Ojima, I., Kogure, T., Nihonyanagi, M., Nagai, Y.: Bull. Chem. Soc., Japan **45**, 3506 (1972).
490. Corriu, R. J. P., Moreau, J. J. E.: Chem. Commun., 38 (1973).
491. Ojima, I., Kogure, T., Nagai, Y.: Tetrahedron Lett., 5035 (1972).
492. Ojima, I., Kogure, T., Nagai, Y.: Chem. Lett., 541 (1973).
493. Yamamoto, K., Hayashi, T., Kumada, M., J. Organometal. Chem. **54**, C45 (1973).
494. Poulin, J. C., Dumont, W., Dang, T. P., Kagan, H. B.: Compt. Rend. C **277**, 41 (1973).
495. Corriu, R. J. P., Moreau, J. J. E.: J. Organometal. Chem. **64**, C51 (1974).
496. Yamamoto, K., Hayashi, T., Kumada, M.: J. Organometal. Chem. **42**, C65 (1972).
497. Ojima, I., Kogure, T., Nagai, Y.: Tetrahedron Lett., 2475 (1973).
498. Langlois, N., Dang, T. P., Kagan, H. B.: Tetrahedron Lett., 4865 (1973).
499. Nagai, Y., Uetake, K., Yoshikawa, T., Matsumoto, H.: J. Syn. Org. Chem. Japan **31**, 759 (1973).
500. Vol'pin, M. E.: Pure Appl. Chem. **30**, 607 (1972).
501. Vol'pin, M. E., Kolomnikov, I. S., Lobejeva, T. S.: Izv. Akad. Nauk SSSR, Ser., Khim., 2084 (1969).
502. Iwashita, Y., Hayata, A.: J. Am. Chem. Soc. **91**, 2525 (1969).
503. Pu, L. S., Yamamoto, A., Ikeda, S.: J. Am. Chem. Soc. **90**, 3896 (1968).
504. Misono, A., Uchida, Y., Hidai, M., Kuse, T.: Chem. Commun., 981 (1968).
505. Haynes, P., Slaugh, L. H., Kohnle, J. F.: Tetrahedron Lett., 365 (1970).
506. Kolomnikov, I. S., Lobejeva, T. S., Vol'pin, M. E.: Izv. Akad. Nauk, SSSR, 2650 (1970).
507. Schrauzer, G. N., Sibert, J. W.: J. Am. Chem. Soc. **92**, 3509 (1970).
508. De Pasquale, R. J.: Chem. Commun., 157 (1973).
509. Ugo, R., Conti, F., Cenini, S., Mason, R., Robertson, G. B.: Chem. Commun., 1498 (1968).
510. Flynn, B. R., Vaska, L.: J. Am. Chem. Soc. **95**, 5081 (1973).
511. Riepe, M. E., Wang, J. H.: J. Am. Chem. Soc. **89**, 4229 (1967).
512. Commerenc, D., Douek, I., Wilkinson, G.: J. Chem. Soc. A, 1771 (1970).

513. Lindner, E., Grimmer, R., Weber, H.: Angew. Chem. **82**, 639 (1970).
514. Jetz, W., Angelici, R.J.: J. Am. Chem. Soc. **94**, 3799 (1972).
515. Cooke, J., Cullen, W.R., Green, M., Stone, F.G.A.: Chem. Commun., 170 (1968).
516. Moritani, I., Yamamoto, Y., Konishi, H.: Chem. Commun., 1457 (1969).
517. Chen, K.S., Kleinberg, J., Landgrebe, J.A.: Chem. Commun., 295 (1972).
518. Yoshimura, N., Murahashi, S., Moritani, I.: J. Organometal. Chem. **52**, C59 (1973).
519. Miller, R.G., Kuhlman, D.P.: J. Organometal. Chem. **26**, 401 (1971).
520. Sneeden, R.P.A., Zeiss, H.H.: J. Organometal. Chem. **22**, 713 (1970).
521. Light, J.R.C., Zeiss, H.H.: J. Organometal. Chem. **21**, 517 (1970).
522. Whitesides, G.M., Gaasch, J.F., Stedronsky, E.S.: J. Am. Chem. Soc. **94**, 5258 (1972).
523. Ruddick, J.D., Shaw, B.L.: J. Chem. Soc. A, 2969 (1969).
524. Belluco, U., Giustiniani, M., Graziani, M.: J. Am. Chem. Soc. **89**, 6494 (1967).
525. Takahashi, Y., Sakai, S., Ishii, Y.: Chem. Commun., 1092 (1967).
526. Yamazaki, H., Nishido, T., Matsumoto, Y., Sumida, S., Hagihara, N.: J. Organometal. Chem. **6**, 86 (1966).
527. Yamazaki, H., Hagihara, N.: Bull. Chem. Soc. Japan **37**, 907 (1964).
528. Ustynyuk, Yu. A., Barinov, I.V.: J. Organometal. Chem. **23**, 551 (1970).
529. Wilke, G., Kroner, M., Bogdanovic, M.: Angew. Chem. **73**, 755 (1961).
530. Breil, H., Wilke, G.: Angew. Chem. **82**, 355 (1970).
531. Yamamoto, T., Yamamoto, A., Ikeda, S.: J. Am. Chem. Soc. **93**, 3350 (1971).
532. Yamamoto, T., Yamamoto, A., Ikeda, S.: J. Am. Chem. Soc. **89**, 5989 (1967).
533. Schwartz, J., Hart, D.W., Holden, J.L.: J. Am. Chem. Soc. **94**, 9269 (1972).
534. Traunmuller, R., Polansky, O.E., Heimbach, P., Wilke, G.: Chem. Phys. Lett. **3**, 300 (1969).
535. Tsuji, J., Ohno, K.: Synthesis, 157 (1969).
536. Dawson, D.J., Ireland, R.E.: Tetrahedron Lett., 1899 (1968).
537. Ireland, R.E., Pfister, G.: Tetrahedron Lett., 2145 (1969).
538. Walborsky, H.M., Allen, L.B.: Tetrahedron Lett., 823 (1970).
539. Fries, R.W., Stille, J.K.: Syn. Inorg. Metal. Org. Chem. **1**, 295 (1971).
539a. Blum, J., Oppenheimer, E., Bergmann, E.D.: J. Am. Chem. Soc. **89**, 2338 (1967).
540. Cramer, R.: Accounts Chem. Res. **1**, 186 (1968).
541. Yamamoto, A., Yamamoto, T., Saruyama, T., Nakamura, Y.: J. Am. Chem. Soc. **95**, 4073 (1973).
542. Anderson, S.N., Fong, C.W., Johnson, M.D.: J. Am. Chem. Soc. **94**, 659 (1972).
543. Nicholas, K.M., Rosenblum, M.: J. Am. Chem. Soc. **95**, 4449 (1973).
544. Corey, E.J., Moinet, G.: J. Am. Chem. Soc. **95**, 7185 (1973).
545. Nicholas, K.M., Pettit, R.: Tetrahedron Lett., 3475 (1971).
546. Bogdanovic, B., Heimbach, P., Lroner, M.G., Wilke, G.: Ann. **727**, 143 (1969).
547. Brenner, W., Heimbach, P., Hey, H.J., Müller, E.W., Wilke, G.: Ann. **727**, 161 (1969).
548. Kiji, J., Masui, K., Furukawa, J.: Tetrahedron Lett., 2561 (1970).
549. Kiji, J., Masui, K., Furukawa, J.: Bull. Chem. Soc. Japan **44**, 1956 (1971).
550. Kiji, J., Yamamoto, K., Mitani, S., Yoshikawa, S., Furukawa, J.: *ibid.* **46**, 1791 (1973).
551. Miyake, A., Kondo, H., Nishino, M.: Angew. Chem. Intern. Ed. **10**, 802 (1971).
552. Billups, W.E., Cross, J.H., Smith, C.V.: J. Am. Chem. Soc. **95**, 3438 (1973).
553. Heimbach, P., Wilke, G.: Ann. **727**, 183 (1969).
554. Cannell, L.G.: J. Am. Chem. Soc. **94**, 6867 (1972).
555. Brenner, W., Heimbach, P., Wilke, G.: Angew. Chem. **78**, 983 (1966).
556. Brenner, W., Heimbach, P.: Ann. **727**, 194 (1969).
557. Heimbach, P., Ploner, K.J., Thomel, F.: Angew. Chem. Intern. Ed. **10**, 276 (1971).
558. Fahey, D.R.: J. Org. Chem. **37**, 4471 (1972).
559. Carbonaro, A., Greco, A., Dall'Asta, G.: J. Org. Chem. **33**, 3948 (1968).
560. Carbonaro, A., Greco, A., Dall'Asta, G.: Tetrahedron Lett., 5129 (1968).
561. Carbonaro, A., Greco, A., Dall'Asta, G.: J. Organometal. Chem. **20**, 177 (1969).
562. Otsuka, S., Nakamura, A., Yamagata, T., Tani, K.: J. Am. Chem. Soc. **94**, 1037 (1972).
563. Otsuka, S., Nakamura, A., Tani, K., Ueda, S.: Tetrahedron Lett., 297 (1969).
564. Otsuka, S., Nakamura, A., Ueda, S., Tani, K.: Chem. Commun., 863 (1971).
565. Jones, F.N., Lindsey, Jr., R.V.: J. Org. Chem. **33**, 3838 (1968).
566. Baker, R., Blackett, B.N., Cookson, R.C.: Chem. Commun., 802 (1972).
567. Reppe, W., Kutepow, N.V., Magin, A.: Angew. Chem. **81**, 717 (1969).

568. Hoogzand, C., Hubel, W.: Organic Synthesis via metal carbonyls, Vol. 1, p. 343 (1968), New York: Interscience.
569. Meriwether, L. S., Leto, M. F., Colthup, E. C., Kennerly, G. W.: J. Org. Chem. **27**, 3930 (1962).
570. Maitlis, P. M.: Pure Appl. Chem. **33**, 489 (1973).
571. Moseley, K., Maitlis, P. M.: Chem. Commun., 1604 (1971).
572. Wakatsuki, Y., Yamazaki, H.: Chem. Commun., 280 (1973).
573. Wakatsuki, Y., Yamazaki, H.: Tetrahedron Lett., 3383 (1973).
574. Chalk, A. J.: J. Am. Chem. Soc. **94**, 5928 (1972).
575. Cairns, T. L., Engelhardt, V. A., Jackson, H. L., Kalb, G. H., Sauer, J. C.: J. Am. Chem. Soc. **74**, 5636 (1952).
576. Singleton, D. M.: Tetrahedron Lett., 1245 (1973).
577. Ohshiro, Y., Kinugasa, K., Minami, T., Agawa, T.: J. Org. Chem. **35**, 2136 (1970).
578. Kinugasa, K., Agawa, T.: Organometal. Chem. Syn. **1**, 427 (1972).
579. Kinugasa, K., Agawa, T.: J. Organometal. Chem. **51**, 329 (1973).
580. Noyori, R., Ishigami, T., Hayashi, N., Takaya, H.: J. Am. Chem. Soc. **95**, 167 (1973).
581. Noyori, R., Odagi, T., Takaya, H.: J. Am. Chem. Soc. **92**, 5780 (1970).
582. Noyori, R., Kumagai, Y., Umeda, I., Takaya, H.: J. Am. Chem. Soc. **94**, 4018 (1972).
583. Binger, P.: Angew. Chem. **84**, 352 (1972).
584. Binger, P.: Synthesis, 427 (1973).
585. Takaya, H., Hayashi, N., Ishigami, T., Noyori, R.: Chem. Lett., 813 (1973).
586. Noyori, R., Kumagai, Y., Takaya, H.: J. Am. Chem. Soc. **96**, 634 (1974).
587. Calderon, N.: Accounts Chem. Res. **5**, 127 (1972).
588. Calderon, N., Chen, H. Y., Scott, K. W.: Tetrahedron Lett. 3327 (1967).
589. Calderon, N., Ofstead, E. A., Ward, J. P., Judy, W. A., Scott, K. W.: J. Am. Chem. Soc. **90**, 4133 (1968).
590. Banks, R. L., Bailey, G. C.: Ind. Eng. Chem., Prod. Res. Develop. **3**, 170 (1964).
591. Dall'Asta, G., Mazzanti, G., Natta, G., Porii, L.: Makrom. Chem. **56**, 224 (1962).
592. Natta, G., Dall'Asta, G., Mazzanti, G., Motroni, G.: Makrom. Chem. **69**, 163 (1963).
593. Calderon, N., Ofstead, E. A., Judy, W. A.: J. Polym. Sci. A 5, 2209 (1967).
594. Wasserman, W., BenEfraim, D. A., Wolovsky, R.: J. Am. Chem. Soc. **90**, 3286 (1968).
595. Wolovsky, R., Nir, Z.: Synthesis, 134 (1972).
596. Wideman, L. G.: J. Org. Chem. **33**, 4541 (1968).
597. Wolovsky, R.: J. Am. Chem. Soc. **92**, 2132 (1970).
598. van Dam, P. B., Mittelmeijer, M. C., Boelhouwer, C.: Chem. Commun., 1221 (1972).
599. Mango, F. D.: Adv. Catalysis **20**, 291 (1969).
600. Mango, F. D., Schachtschneider, J. H.: Transition metal in homogeneous catalysis, 1971, p. 223. New York: Marcel Dekker.
601. Mango, F. D.: Chem. Tech., 758 (1971).
602. Lewanda, G. S., Pettit, R.: J. Am. Chem. Soc. **93**, 7087 (1971).
603. O'Neill, P. P. O., Rooney, J. J.: Chem. Commun., 104 (1972).
604. Grubbs, R. H., Brunck, T. K.: J. Am. Chem. Soc. **94**, 2538 (1972).
605. Ohno, K., Tsuji, J.: Chem. Commun., 247 (1971).
606. J. Falbe: Angew. Chem. **78**, 532 (1966).
607. Falbe, J.: Carbon monoxide in organic synthesis, p. 147, 1970, Berlin-Heidelberg-New York: Springer.
608. Rosenthal, R., Wender, I.: Organic synthesis via metal carbonyls. Vol. 1, 405 (1968). New York: Interscience.
609. Falbe, J., Schulze-Steinen, H. J., Korte, F.: Chem. Ber. **98**, 886 (1965).
610. Falbe, J., Korte, F.: Chem. Ber. **98**, 1928 (1965).
611. Falbe, J., Weitkamp, H., Korte, F.: Tetrahedron Lett., 2677 (1965).
612. Falbe, J., Korte, F.: Chem. Ber. **95**, 2680 (1962).
613. Murahashi, S., Horiie, S., Joh, T.: Bull. Chem. Soc. Japan **33**, 81 (1960).
614. Murahashi, S., Horiie, S.: Bull. Chem. Soc. Japan **33**, 247 (1960).
615. Candlin, J. P., Taylor, K. A., Thompson, D. T.: Reactions of transition metal complexes. Amsterdam: Elsevier 1968.
616. Maitlis, P. M.: Organic chemistry of palladium, New York: Academic Press, 1971.
617. Bird, C. W.: Transition metal intermediates in organic synthesis. London: Logos Press, 1967.
618. Rylander, P. N.: Organic syntheses with noble metal catalysts, New York: Academic Press, 1973.

Author Index

Subject Index

A. Gossauer

Die Chemie der Pyrrole

17 Abbildungen. XX, 433 Seiten. 1974
(Organische Chemie in Einzeldarstellungen,
Band 15) Gebunden DM 158,—; US $68.00
ISBN 3-540-06603-9

Preisänderungen vorbehalten

Das Pyrrol und seine Derivate haben als technische Grundstoffe wie auch als Naturprodukte wachsendes Interesse gewonnen.
Diese Monographie ist eine umfassende Übersicht über die seit 1934 erschienene Literatur (ausgenommen Porphyrine). Bedingt durch die seitdem ständig wachsende Anzahl der Veröffentlichungen, die sich mit den physikalischen Eigenschaften dieser Verbindungsklasse befassen, weichen Konzeption und Gliederung dieses Buches von denjenigen des klassischen Werkes von H. Fischer und H. Orth grundsätzlich ab. Die Anwendung quantenmechanischer Rechenverfahren zur Deutung der Eigenschaften des Pyrrol-Moleküls wird im ersten Kapitel ausführlich erörtert. Die entscheidende Bedeutung der physikalischen Methoden zur Untersuchung der Konstitution und Reaktivität des Pyrrols und seiner Derivate ist durch zahlreiche tabellarisch geordnete Datenangaben, deren Interpretation im Text diskutiert wird, hervorgehoben. Dem präparativ arbeitenden Chemiker soll die Systematisierung der synthetischen Methoden bei der Suche nach der einschlägigen Literatur helfen: Ringsynthesen sind nach dem Aufbaumodus des Heterocyclus, die Einführung von Substituenten nach funktionellen Gruppen klassifiziert und anhand von Schemata übersichtlich zusammengefaßt worden. Bei der Zusammenstellung der Abbildungen wurden neben den trivialen Beispielen, die zum besseren Verständnis des Textes dienen, beonders jene Reaktionen ausgewählt, bei denen Pyrrole Ausgangsverbindungen zur Darstellung anderer Heterocyclen (Indole, Pyrrolizine, Azepine, u.a.) sind. Besondere Sorgfalt gilt der Beschreibung von Reaktionsmechanismen. (2621 Literaturzitate.)

Springer-Verlag
Berlin
Heidelberg
New York

J. Falbe

Carbon Monoxide in Organic Synthesis

Translator: Ch. R. Adams

21 figures. IX, 219 pages. 1970
(Deutsche Ausgabe s. Organische Chemie in
Einzeldarstellungen, Band 10)
Cloth DM 72,—; US $31.00   ISBN 3-540-04814-6

Prices are subject to change without notice

Contents: The Hydroformylation Reaction (Oxo
Reaction/Roelen Reaction). — Metal Carbonyl
Catalyzed Carbonylation (Reppe Reactions). —
Carbonylation with Acid Catalysts (Koch Reac-
tion). — Ring Closures with Carbon Monoxide

The importance of carbon monoxide chemistry has
grown apace in the last few years, both as regards
scientific research and chemical processing. This
has made it necessary to revise the book, Synthesen
mit Kohlenmonoxyd, published in German in 1967.
This covered Roelen's discovery of hydroformylation or the oxo reaction, Reppe's carbonylation
process, Koch's carbonic acid synthesis and ring
closure with carbon monoxide. The author has
included in the new edition the latest research findings in these fields and has in particular given more
space to the discussion of reaction mechanisms.
The latest developments in industry are also mentioned and there have been numerous additions to
the list of references.
This book will be an important reference source,
both for the established expert and for those aspir-
ing to enter the fields of petrochemicals, organic
chemicals and chemical engineering, particularly as
results tend to be published in patents and thus
remain outside the ken of a wide range of readers.

Springer-Verlag
Berlin
Heidelberg
New York